THE BOOK
OF
THE ARTICLES ON FOURIER SERIES

THE NICE TEXTBOOK ON SOME USEFUL RESEARCH

BY
Aitzaz Imtiaz
Beaconhouse School System, Senior Boys Branch BSTRB

FIRST EDITION
ON CALCULUS

RAWALPINDI
NEUROSTOL PAKISTAN & AMAZON INTERNATIONAL
FRANCE: NEUROSTOL FRANCE & AMAZON FRANCE
2023

Copyright, 2023
BY
AITZAZ IMTIAZ
PROJECT STARTED IN 2023

PRESENTED TO YOU BY NEUROSTOL PUBLISHING

In Dedication for

Ma'am Sadia Khan

The original picture shot courtesy of Saad Muneer (thanks Saad!) was given to GIA (my personal AI) to generate above and depicts her in Universe 3, at MIT Doom as a student in Bachelor of Science in Mathematics with Computer Science, as an alternative to Universe 4, where she currently taught me Pakistan Studies at BSTRB in 2020-21.

In acknowledgment for

Sophie Germain

for inspiring me to write this work, much of this edition though dealing with Fourier Analysis inspired by the book of William Elwood Byerly, but I was still motivated by Sophie to write this work.

Contents

1 INTRODUCTION 1
2 DEVELOPMENT IN TRIGONOMETRIC SERIES 25
3 CONVERGENCE OF FOURIER SERIES 46
4 PHYSICS APPLICATIONS BY FOURIER INTEGRAL 57
5 ZONAL HARMONICS 116
6 SPHERICAL HARMONICS 159
7 CYLINDRICAL HARMONICS 179
8 LAPLACE'S EQUATION 196
9 HISTORICAL SUMMARY 219

Chapter 1

INTRODUCTION

1. In numerous significant issues in numerical physical science we are obliged to manage *partial differential equations* of a similarly straightforward structure.

For instance, in the Scientific Hypothesis of Intensity we have for the difference in temperature of any strong because of the progression of intensity inside the strong, the condition

$$D_t u = a^2(D_x^2 u + D_y^2 u + D_z^2 u),\,^1 \qquad [\text{I}]$$

where u addresses the temperature anytime of the strong and t the time.

In the most straightforward case, that of a section of boundless degree with equal plane faces, where the temperature can be viewed as an element of one coÃ¶rdinate, [I] diminishes to

$$D_t u = a^2 D_x^2 u, \qquad [\text{II}]$$

a type of impressive significance in the thought of the issue of the cooling of the world's outside layer.

In the issue of the extremely durable condition of temperatures in a slim rectangular plate, the equation [I] becomes

$$D_x^2 u + D_y^2 u = 0. \qquad [\text{III}]$$

In *polar* or *spherical* coÃ¶rdinates [I] is less basic, it is

$$D_t u = \frac{a^2}{r^2}\left[D_r(r^2 D_r u) + \frac{1}{\sin\theta}D_\theta(\sin\theta D_\theta u) + \frac{1}{\sin^2\theta}D_\phi^2 u\right]. \qquad [\text{IV}]$$

For the situation where the strong being referred to is a circle and the temperature anytime relies just upon the distance of the point from the middle [IV] lessens to

$$D_t(ru) = a^2 D_r^2(ru). \qquad [\text{V}]$$

In *cylindrical* coÃ¶rdinates [I] becomes

$$D_t u = a^2[D_r^2 u + \frac{1}{r}D_r u + \frac{1}{r^2}D_\phi^2 u + D_z^2 u]. \qquad [\text{VI}]$$

In taking into account the progression of intensity in a chamber when the temperature at any point relies only upon the distance r of the point from the hub [VI] becomes

$$D_t u = a^2(D_r^2 u + \frac{1}{r}D_r u). \qquad [\text{VII}]$$

---

[1] For quickness we will frequently utilize the image ∇^2 for the activity $D_x^2 + D_y^2 + D_z^2$; also, with this documentation equation [I] would be composed $D_t u = a^2 \nabla^2 u$.

INTRODUCTION.

In Acoustics in a few issues we have the condition

$$D_t^2 y = a^2 D_x^2 y; \qquad [\text{VIII}]$$

for example, in taking into account the cross over or the longitudinal vibrations of a extended flexible string, or the transmission of plane sound waves through the air.

In the event that in considering the cross over vibrations of an extended string we take record of the obstruction of the air [VIII] is supplanted by

$$D_t^2 y + 2k D_t y = a^2 D_x^2 y. \qquad [\text{IX}]$$

In managing the vibrations of an extended versatile layer, we have the condition

$$D_t^2 z = c^2 (D_x^2 z + D_y^2 z), \qquad [\text{X}]$$

or on the other hand in *cylindrical coördinates*

$$D_t^2 z = c^2 \left(D_r^2 z + \frac{1}{r} D_r z + \frac{1}{r^2} D_\phi^2 z \right). \qquad [\text{XI}]$$

In the hypothesis of *Potential* we continually meet Laplace's Condition

and $D_x^2 V +$
$$D_y^2 V + D_z^2 V = 0 \qquad [\text{XII}]$$
or and
$$\nabla^2 V = 0$$

which in *spherical coördinates* becomes

$$\frac{1}{r^2} \left[r D_r^2 (rV) + \frac{1}{\sin\theta} D_\theta (\sin\theta D_\theta V) + \frac{1}{\sin^2\theta} D_\phi^2 V \right] = 0, \qquad [\text{XIII}]$$

also, in *cylindrical coördinates*

$$D_r^2 V + \frac{1}{r} D_r V + \frac{1}{r^2} D_\phi^2 V + D_z^2 V = 0. \qquad [\text{XIV}]$$

In *curvilinear coördinates* it is

$$h_1 h_2 h_3 \left[D_{\rho_1} \left(\frac{h_1}{h_2 h_3} D_{\rho_1} V \right) + D_{\rho_2} \left(\frac{h_2}{h_3 h_1} D_{\rho_2} V \right) + D_{\rho_3} \left(\frac{h_3}{h_1 h_2} D_{\rho_3} V \right) \right] = 0; \qquad [\text{XV}]$$

where
$$f_1(x,y,z) = \rho_1, \; f_2(x,y,z) = \rho_2, \; f_3(x,y,z) = \rho_3$$

address a bunch of surfaces what cut each other at right points, regardless what values are given to ρ_1, ρ_2, and ρ_3; and where

$$h_1^2 = (D_x \rho_1)^2 + (D_y \rho_1)^2 + (D_z \rho_1)^2$$
$$h_2^2 = (D_x \rho_2)^2 + (D_y \rho_2)^2 + (D_z \rho_2)^2$$
$$h_3^2 = (D_x \rho_3)^2 + (D_y \rho_3)^2 + (D_z \rho_3)^2,$$

also, obviously, should be communicated as far as ρ_1, ρ_2, and ρ_3.

In the event that it happens that $\nabla^2 \rho_1 = 0$, $\nabla^2 \rho_2 = 0$, and $\nabla^2 \rho_3 = 0$, Laplace's Equation [XV] expects the exceptionally straightforward structure

$$h_1^2 D_{\rho_1}^2 V + h_2^2 D_{\rho_2}^2 V + h_3^2 D_{\rho_3}^2 V = 0. \qquad [\text{XVI}]$$

INTRODUCTION.

2. A *differential equation* is a condition containing subordinates or differentials regardless of the crude factors from which they are determined.

The *general solution* of a differential condition is the condition communicating the most broad connection between the crude factors which is reliable with the given differential condition and which doesn't include differentials or subordinates. An overall arrangement will constantly contain erratic (*i. e.*, unsure) *constants* or *arbitrary functions*.

A *particular solution* of a differential condition is a connection between the crude factors which is steady with the given differential condition, in any case, which is less broad than the overall arrangement, albeit remembered for it.

Hypothetically, every specific arrangement can be acquired from the general arrangement by subbing in the overall arrangement specific qualities for the erratic constants or specific capabilities for the erratic capabilities; however in practice it is frequently simple to acquire specific arrangements straightforwardly from the differential condition when it would be troublesome or difficult to get the general arrangement.

3. On the off chance that an issue expecting for its answer the tackling of a differential condition is *determinate*, there must continuously be given notwithstanding the differential condition an adequate number of outside conditions for the assurance of all the inconsistent constants or erratic capabilities that go into the overall arrangement of the condition; and in managing such an issue, in the event that the differential condition can be promptly tackled the regular technique for methodology is to acquire its general arrangement, and afterward to decide the constants or capabilities by the guide of the given conditions.

It frequently works out, nonetheless, that the overall arrangement of the differential condition being referred to can't be acquired, and afterward, since the issue *if determinate* will be settled on the off chance that using any and all means an answer of the situation can be tracked down which will likewise fulfill the given external circumstances, it is worth while to attempt to get *particular solutions* thus to consolidate them as to shape an outcome which will fulfill the given circumstances consistently to fulfill the differential condition.

4. A differential condition is *linear* when it would be of the principal degree if the reliant variable and every one of its subsidiaries were viewed as logarithmic obscure amounts. On the off chance that it is straight and contains no term which doesn't include the reliant variable or one of its subordinates, it is supposed to be direct also, *homogeneous*.

Every one of the differential conditions gathered in Art. 1 are direct and homogeneous.

5. *If a worth of the reliant variable has been found which fulfills a given homogeneous, straight, differential condition, the item framed by increasing this worth by any consistent will likewise be a worth of the reliant variable which will fulfill the equation.*

For assuming every one of the conditions of the given condition are rendered to the principal part, the replacement of the first-named esteem should lessen that part to zero; subbing the subsequent worth is identical to duplicating each term of the consequence of the principal replacement by a similar consistent component, which in this way might be taken out as a variable of the entire first part. The leftover factor being zero, the item is zero and the condition is fulfilled.

If a few upsides of the reliant variable have been tracked down every one of which fulfills the given differential condition, their aggregate will fulfill the equation; for on the off chance that the amount of the qualities being referred to is subbed in the situation each term of the aggregate will lead to a bunch of terms which should be equivalent to nothing, and in this way the amount of these sets should be zero.

6. It is by and large conceivable to scrape by some basic gadget *particular solutions* of such differential conditions as those we have gathered in Art. 1. The object of the part of science with which we are going to bargain is to track down techniques for so consolidating these specific arrangements as to fulfill any given conditions which are reliable with the idea of the issue being referred to.

INTRODUCTION.

This frequently expects us to have the option to foster any given capability of the factors which go into the declaration of these circumstances as far as *normal forms* fit to the issue with which we end up being managing, and proposed by the type of specific arrangement that we can acquire for the differential condition.

These ordinary structures are every now and again sines and cosines, yet they are frequently significantly more convoluted capabilities known as *Legendre's Coefficients,* or *Zonal Music; Laplace's Coefficients,* or *Spherical Sounds: Bessel's Functions,* or then again *Cylindrical Music; LamÃ©'s Functions,* or *Ellipsoidal Harmonics,* &c.

7. As a delineation, let us take Fourier's concern of the long-lasting state of temperatures in a slight rectangular plate of broadness π and of boundless length whose countenances are impenetrable to warm. We will guess that the two long edges of the plate are kept at the consistent temperature zero, that one of the short edges, which we will call the foundation of the plate, is kept at the temperature solidarity, and that the temperatures of focuses in the plate decline endlessly as we subside from the base; we will endeavor to track down the temperature anytime of the plate.

Allow us to accept the base as the pivot of X and one finish of the base as the beginning. Then, at that point, to take care of the issue we are to find the temperature u of any point from the condition

$$D_x^2 u + D_y^2 u = 0 \qquad \text{[III] Art. 1}$$

dependent upon the circumstances

$$u = 0 \quad \text{when} \quad and\, x = 0 \qquad (1)$$
$$u = 0 \quad \text{"} and\, x = \pi \qquad (2)$$
$$u = 0 \quad \text{"} and\, y = \infty \qquad (3)$$
$$u = 1 \quad \text{"} and\, y = 0. \qquad (4)$$

We will start by getting a specific arrangement of [III], and we will utilize a gadget which generally succeeds when the condition is *linear* and *homogeneous* what's more, has *constant coefficients.*

Assume[2] $u = e^{\alpha y + \beta x}$, where α and β are constants, substitute in [III] and partition by $e^{\alpha y + \beta x}$, and we have $\alpha^2 + \beta^2 = 0$. On the off chance that, this condition is fulfilled $u = e^{\alpha y + \beta x}$ is an answer.

Subsequently $u = e^{\alpha y \pm \alpha x i}$ [3] is an answer of [III], regardless of what worth might be given to α.

This structure is shocking, since it includes a fanciful. We can, notwithstanding, promptly further develop it.

Take $u = e^{\alpha y} e^{\alpha x i}$, an answer of [III], and $u = e^{\alpha y} e^{-\alpha x i}$, another arrangement of [III]; add these upsides of u and partition the total by 2 and we have $e^{\alpha y} \cos \alpha x$. (v. Int. Cal. Art. 35, [1].) Hence by Art. 5

$$u = e^{\alpha y} \cos \alpha x \qquad (5)$$

is an answer of [III]. Take $u = e^{\alpha y} e^{\alpha x i}$ and $u = e^{\alpha y} e^{-\alpha x i}$, deduct the second worth of u from the first and separation by $2i$ and we have $e^{\alpha y} \sin \alpha x$. (v. Int. Cal. Art. 35, [2]). In this manner by Art. 5

$$u = e^{\alpha y} \sin \alpha x \qquad (6)$$

is an answer of [III].

Allow us now to check whether out of these specific arrangements we can develop an answer which will fulfill the circumstances (1), (2), (3), and (4).

Consider $$u = e^{\alpha y} \sin \alpha x. \qquad (6)$$

It is zero when $x = 0$ for all upsides of α. It is zero when $x = \pi$ on the off chance that α is a entire number. It is zero when $y = \infty$ assuming that α is negative. On the off chance that, we compose

---

[2] This presumption should be viewed as absolutely conditional. It should tried by substitute in the situation, and is legitimate on the off chance that it prompts a solution.

[3] We will routinely utilize the image i for $\sqrt{-1}$.

INTRODUCTION.

u equivalent to an amount of terms of the structure $Ae^{-my}\sin mx$, where m is a positive whole number, we will have an answer of [III] which fulfills conditions (1), (2) what's more (3). Leave this arrangement alone

$$u = A_1 e^{-y}\sin x + A_2 e^{-2y}\sin 2x + A_3 e^{-3y}\sin 3x + A_4 e^{-4y}\sin 4x + \cdots \qquad (7)$$

A_1, A_2, A_3, A_4, &c., being unsure constants.

When $y = 0$ (7) decreases to

$$u = A_1\sin x + A_2\sin 2x + A_3\sin 3x + A_4\sin 4x + \cdots. \qquad (8)$$

On the off chance that now it is feasible to form solidarity into a progression of the structure (8), our issue is tackled; we have just to substitute the coefficients of that series for A_1, A_2, A_3, &c. in (7).

It will be demonstrated later that

$$1 = \frac{4}{\pi}\left(\sin x + \frac{1}{3}\sin 3x + \frac{1}{5}\sin 5x + \frac{1}{7}\sin 7x + \cdots\right)$$

for all upsides of x among 0 and π; thus our expected arrangement is

$$u = \frac{4}{\pi}\left[e^{-y}\sin x + \frac{1}{3}e^{-3y}\sin 3x + \frac{1}{5}e^{-5y}\sin 5x + \frac{1}{7}e^{-7y}\sin 7x + \cdots\right] \qquad (9)$$

for this fulfills the differential condition and every one of the given circumstances.

In the event that the given temperature of the foundation of the plate as opposed to being solidarity is a component of x, we can tackle the issue as in the past on the off chance that we can communicate the given capability of x as an amount of terms of the structure $A\sin mx$, where m is a entire number.

The issue of tracking down the worth of the *potential function* anytime of a long, slender, rectangular leading sheet, of expansiveness π, through which an electric flow is streaming, when the two long edges are kept at likely zero, what's more, one short edge at expected solidarity, is numerically indistinguishable with the issue we have recently tackled.

EXAMPLE.

Taking the temperature of the foundation of the plate depicted above as 100Â° centigrade, and that of the sides of the plate as 0Â°, figure the temperatures of the places

$$(a)\ \left(\frac{\pi}{6}, 1\right);\ (b)\ \left(\frac{\pi}{3}, 2\right);\ (c)\ \left(\frac{\pi}{2}, 3\right),$$

right to the closest degree. *Ans.* (a) 26Â°; (b) 15Â°; (c) 6Â°.

8. As another outline, we will take the issue of the cross over vibrations of an extended string secured at the finishes, at first contorted into some given bend and afterward permitted to swing.

Let the length of the string be l. Take the place of balance of the string as the hub of X, and one of the closures as the beginning, and assume the string at first mutilated into a bend whose condition $y = f(x)$ is given.

We have then to track down an articulation for y which will be an answer of the condition

$$D_t^2 y = a^2 D_x^2 y \qquad \text{[VIII] Art. 1,}$$

while fulfilling the circumstances

$$y = 0 \quad \text{when} \quad x = 0 \qquad (1)$$
$$y = 0 \quad \text{"and} \quad x = l \qquad (2)$$
$$y = f(x) \quad \text{"and} \quad t = 0 \qquad (3)$$
$$D_t y = 0 \quad \text{"and} \quad t = 0, \qquad (4)$$

INTRODUCTION.

the last condition meaning just that the string begins from rest.

As in the keep going issue let[4] $y = e^{\alpha x + \beta t}$ and substitute in [VIII]. Partition by $e^{\alpha x + \beta t}$ and we have $\beta^2 = a^2\alpha^2$ as the condition that our expected worth of y will fulfill the condition.

$$y = e^{\alpha x \pm a\alpha t} \tag{5}$$

is, then, at that point, an answer of (VIII) whatever the worth of α.

It is more helpful to have a geometrical than a dramatic structure to manage, and we can promptly acquire one by involving a fanciful incentive for α in (5). Supplant α by αi and (5) becomes $y = e^{(x \pm at)\alpha i}$, an answer of [VIII]. Supplant α by $-\alpha i$ and (5) becomes $y = e^{-(x \pm at)\alpha i}$, one more arrangement of [VIII]. Add these upsides of y and partition by 2 and we have $\cos\alpha(x \pm at)$. Deduct the second worth of y from the first and gap by $2i$ and we have $\sin\alpha(x \pm at)$.

$$y = \cos\alpha(x + at)$$
$$y = \cos\alpha(x - at)$$
$$y = \sin\alpha(x + at)$$
$$y = \sin\alpha(x - at)$$

are, then, at that point, arrangements of [VIII]. Composing y progressively equivalent to around 50of the principal sets of values, a portion of their distinction, a portion of the amount of the last sets of values, and around 50arrangements of [VIII].

$$y = \cos\alpha x \cos\alpha at$$
$$y = \sin\alpha x \sin\alpha at$$
$$y = \sin\alpha x \cos\alpha at$$
$$y = \cos\alpha x \sin\alpha at.$$

Assuming that we take the third structure

$$y = \sin\alpha x \cos\alpha at$$

it will fulfill conditions (1) and (4), regardless of what worth might be given to α, and it will fulfill (2) if $\alpha = \frac{m\pi}{l}$ where m is a whole number.

In the event that, we take

$$y = A_1 \sin\frac{\pi x}{l} \cos\frac{\pi at}{l} + A_2 \sin\frac{2\pi x}{l} \cos\frac{2\pi at}{l} + A_3 \sin\frac{3\pi x}{l} \cos\frac{3\pi at}{l} + \cdots \tag{6}$$

where $A_1, A_2, A_3 \cdots$ are unsure constants, we will have an answer of [VIII] which fulfills (1), (2), and (4). When $t = 0$ it diminishes to

$$y = A_1 \sin\frac{\pi x}{l} + A_2 \sin\frac{2\pi x}{l} + A_3 \sin\frac{3\pi x}{l} + \cdots \tag{7}$$

In the event that now it is feasible to form $f(x)$ into a progression of the structure (7), we can take care of our concern totally. We have just to take the coefficients of this series as upsides of $A_1, A_2, A_3 \cdots$ in (6), and we will have an answer of [VIII] which fulfills all our given circumstances. In every one of the previous issues the *normal function*, as far as which a given capability must be communicated, is the sine of a basic numerous of the variable. It would be not difficult to adjust the issue so that the *normal form* ought to be a cosine.

We will presently take two or three issues which are considerably more confounded furthermore, where the ordinary capability is a new one.

---

[4]See note on page 4.

INTRODUCTION.

9. Let it be expected to track down the likely capability because of a round wire ring of little cross segment and of given range c, assuming the issue of the ring to draw in as per the law of nature.

We can promptly find, by direct incorporation, the worth of the expected capability anytime of the hub of the ring. We get for it

$$V = \frac{M}{\sqrt{c^2 + x^2}} \qquad (1)$$

where M is the mass of the ring, and x the distance of the point from the focal point of the ring.

Allow us to utilize round coÃ¶rdinates, taking the focal point of the ring as beginning and the pivot of the ring as the polar hub.

To acquire the worth of the expected capability anytime in space, we should fulfill the condition

$$rD_r^2(rV) + \frac{1}{\sin\theta}D_\theta(\sin\theta D_\theta V) + \frac{1}{\sin^2\theta}D_\phi^2 V = 0, \qquad \text{[XIII] Art. 1,}$$

dependent upon the condition

$$V = \frac{M}{(c^2 + r^2)^{\frac{1}{2}}} \quad \text{when} \quad \theta = 0. \qquad (1)$$

From the balance of the ring, obviously the worth of the potential capability should be autonomous of ϕ, so [XIII] will diminish to

$$rD_r^2(rV) + \frac{1}{\sin\theta}D_\theta(\sin\theta D_\theta V) = 0. \qquad (2)$$

We should now attempt to get specific arrangements of (2), and as the coefficients are not steady, we are headed to another gadget.

Let[5] $V = r^m P$, where P is an element of θ just, and m is a positive whole number, furthermore, substitute in (2), which becomes

$$m(m+1)r^m P + \frac{r^m}{\sin\theta}D_\theta(\sin\theta D_\theta P) = 0.$$

Partition by r^m and utilize the documentation of common subordinates since P depends upon θ just, and we have the condition

$$m(m+1)P + \frac{1}{\sin\theta}\frac{d\left(\sin\theta \frac{dP}{d\theta}\right)}{d\theta} = 0, \qquad (3)$$

from which to acquire P.

Condition (3) can be worked on by changing the autonomous variable. Let $x = \cos\theta$ and (3) becomes

$$\frac{d}{dx}\left[(1-x^2)\frac{dP}{dx}\right] + m(m+1)P = 0. \qquad (4)$$

Assume[6] now that P can be communicated as a total or as a progression of terms including entire powers of x increased by consistent coefficients.

Let $P = \sum a_n x^n$ and substitute this worth of P in (4). We get

$$\sum[n(n-1)a_n x^{n-2} - n(n+1)a_n x^n + m(m+1)a_n x^n] = 0, \qquad (5)$$

where the image $\sum$ demonstrates that we are to frame every one of the terms we can by taking progressive entire numbers for n.

---

[5] See note on page 4.
[6] See note on page 4.

INTRODUCTION.

As (5) should be valid regardless of what the worth of x, the coefficient of any given force of x, concerning occurrence x^k, should evaporate. Thus

$$(k+2)(k+1)a_{k+2} - k(k+1)a_k + m(m+1)a_k = 0 \quad \text{and} \tag{6}$$

and

$$a_{k+2} = -\frac{m(m+1) - k(k+1)}{(k+1)(k+2)}a_k. \quad \text{and} \tag{7}$$

Assuming that currently any arrangement of coefficients fulfilling the connection (7) be taken, $P = \sum a_k x^k$ will be an answer of (4).

If $\quad$ and $k = m$, $\quad a_{k+2} = 0$, $\quad a_{k+4} = 0$, &c.

Since it will answer our motivation on the off chance that we select the least complex arrangement of coefficients that will submit to the condition (7), we can take a set including a_m.

Allow us to change (7) in the structure

$$a_k = -\frac{(k+2)(k+1)}{(m-k)(m+k+1)}a_{k+2}. \tag{8}$$

We get from (8), starting with $k = m - 2$,

$$a_{m-2} = -\frac{m(m-1)}{2.(2m-1)}a_m$$

$$a_{m-4} = \frac{m(m-1)(m-2)(m-3)}{2.4.(2m-1)(2m-3)}a_m$$

$$a_{m-6} = -\frac{m(m-1)(m-2)(m-3)(m-4)(m-5)}{2.4.6.(2m-1)(2m-3)(2m-5)}a_m, \quad \text{&c.}$$

Assuming m is even we see that the set will end with a_0, on the off chance that m is odd, with a_1.

$$P = a_m \left[x^m - \frac{m(m-1)}{2.(2m-1)}x^{m-2} + \frac{m(m-1)(m-2)(m-3)}{2.4.(2m-1)(2m-3)}x^{m-4} - \cdots \right]$$

where a_m is totally erratic, is, then, an answer of (4). It is seen as advantageous to take a_m equivalent to

$$\frac{(2m-1)(2m-3)\cdots 1}{m!}$$

also, it very well may be shown that with this worth of a_m $P = 1$ when $x = 1$.

P is a component of x and contains no higher powers of x than x^m. It is normal to compose it as $P_m(x)$.

We continue to register a couple of upsides of $P_m(x)$ from the equation

$$P_m(x) = \frac{(2m-1)(2m-3)\cdots 1}{m!} \left[x^m - \frac{m(m-1)}{2.(2m-1)}x^{m-2} \right.$$

$$\left. + \frac{m(m-1)(m-2)(m-3)}{2.4.(2m-1)(2m-3)}x^{m-4} - \cdots \right]. \tag{9}$$

INTRODUCTION.

We have:

$$\left.\begin{array}{ll} P_0(x) = 1 & \text{or} \quad P_0(\cos\theta) = 1 \\ P_1(x) = x & \text{''} \quad P_1(\cos\theta) = \cos\theta \\ P_2(x) = \tfrac{1}{2}(3x^2 - 1) & \text{''} \quad P_2(\cos\theta) = \tfrac{1}{2}(3\cos^2\theta - 1) \\ P_3(x) = \tfrac{1}{2}(5x^3 - 3x) & \text{''} \quad P_3(\cos\theta) = \tfrac{1}{2}(5\cos^3\theta - 3\cos\theta) \\ P_4(x) = \tfrac{1}{8}(35x^4 - 30x^2 + 3) \text{ or} \\ \qquad P_4(\cos\theta) = \tfrac{1}{8}(35\cos^4\theta - 30\cos^2\theta + 3) \\ P_5(x) = \tfrac{1}{8}(63x^5 - 70x^3 + 15x) \text{ or} \\ \qquad P_5(\cos\theta) = \tfrac{1}{8}(63\cos^5\theta \, and \, 70\cos^3\theta + 15\cos\theta). \end{array}\right\} \quad (10)$$

We have gotten $P = P_m(x)$ as a specific arrangement of (4) and $P = P_m(\cos\theta)$ as a specific arrangement of (3). $P_m(x)$ or $P_m(\cos\theta)$ is a new capability, known as a *Legendre's Coefficient*, or as a *Surface Zonal Harmonic*, what's more, happens as an ordinary structure in numerous significant issues.

$V = r^m P_m(\cos\theta)$ is a specific arrangement of (2) and $r^m P_m(\cos\theta)$ is in some cases called a *Solid Zonal Harmonic*.

We can now continue to the arrangement of our unique issue.

$$V = A_0 r^0 P_0(\cos\theta) + A_1 r P_1(\cos\theta) + A_2 r^2 P_2(\cos\theta) + A_3 r^3 P_3(\cos\theta) + \cdots \quad (11)$$

where A_0, A_1, A_2, &c., are completely erratic, is an answer of (2) (v. Art. 5). When $\theta = 0$ (11) diminishes to

$$V = A_0 + A_1 r + A_2 r^2 + A_3 r^3 + \cdots,$$

since, as we have said, $P_m(x) = 1$ when $x = 1$, or $P_m(\cos\theta) = 1$ when $\theta = 0$.

By our condition (1)

$$V = \frac{M}{(c^2 + r^2)^{\frac{1}{2}}}$$

when $\theta = 0$.

By the Binomial Hypothesis

$$\frac{M}{(c^2 + r^2)^{\frac{1}{2}}} = \frac{M}{c}\left[1 - \frac{1}{2}\frac{r^2}{c^2} + \frac{1.3}{2.4}\frac{r^4}{c^4} - \frac{1.3.5}{2.4.6}\frac{r^6}{c^6} + \cdots\right]$$

given $r < c$. Subsequently

$$V = \frac{M}{c}\left[P_0(\cos\theta) - \frac{1}{2}\frac{r^2}{c^2}P_2(\cos\theta) + \frac{1.3}{2.4}\frac{r^4}{c^4}P_4(\cos\theta) \right. \\ \left. - \frac{1.3.5}{2.4.6}\frac{r^6}{c^6}P_6(\cos\theta) + \cdots\right] \quad (12)$$

is our necessary arrangement if $r < c$; for it is an answer of equation (2) and fulfills condition (1).

Example.

Taking the mass of the ring as one pound and the span of the ring as one foot, figure to two decimal places the worth of the possible capability due to the ring at the places

(a) $(r = .2, \theta = 0)$; and

(b) $\left(r = .2, \theta = \dfrac{\pi}{4}\right)$; and

(c) $\left(r = .2, \theta = \dfrac{\pi}{2}\right)$;

(d) $(r = .6, \theta = 0)$; and (f) $\left(r = .6, \theta = \dfrac{\pi}{3}\right)$;

(e) $\left(r = .6, \theta = \dfrac{\pi}{6}\right)$; and (g) $\left(r = .6, \theta = \dfrac{\pi}{2}\right)$;

INTRODUCTION.

Ans. (a) .98; (b) .99; (c) 1.01; (d) .86; (e) .90; (f) 1.00; (g) 1.10.

The unit utilized is the potential because of a pound of mass gathered at a point what's more, drawing in a moment pound of mass gathered at a point, the two focuses being a foot separated.

10. A marginally unique issue calling for improvement as far as Zonal Sounds is the accompanying:

Required the extremely durable temperatures inside a strong circle of radius 1, one portion of the surface being kept at the consistent temperature zero, and the other half at the steady temperature solidarity.

Allow us to take the width opposite to the plane isolating the inconsistent warmed surfaces as our hub and let us utilize round coördinates. As in the last issue, we should address the condition

$$rD_r^2(ru) + \frac{1}{\sin\theta}D_\theta(\sin\theta D_\theta u) + \frac{1}{\sin^2\theta}D_\phi^2 u = 0 \qquad \text{[XIII] Art. 1}$$

which as before lessens to

$$rD_r^2(ru) + \frac{1}{\sin\theta}D_\theta(\sin\theta D_\theta u) = 0 \qquad (1)$$

from the thought that the temperatures should be autonomous of ϕ.

Our condition of condition is

$$u = 1 \text{ from } \theta = 0 \text{ to } \theta = \frac{\pi}{2} \text{ and } u = 0 \text{ from } \theta = \frac{\pi}{2} \text{ to } \theta = \pi, \qquad (2)$$

when $r = 1$.

As we have seen $u = r^m P_m(\cos\theta)$ is a specific arrangement of (1), m being any sure entire number, and

$$u = A_0 r^0 P_0(\cos\theta) + A_1 r P_1(\cos\theta) + A_2 r^2 P_2(\cos\theta) + A_3 r^3 P_3(\cos\theta) + \cdots \qquad (3)$$

where $A_0, A_1, A_2, A_3 \cdots$ are unsure constants, is an answer of (1).

When $r = 1$ (3) decreases to

$$u = A_0 P_0(\cos\theta) + A_1 P_1(\cos\theta) + A_2 P_2(\cos\theta) + A_3 P_3(\cos\theta) + \cdots \qquad (4)$$

In the event that then we can foster our capability of θ which goes into equation (2) in a progression of the structure (4), we have just to take the coefficients of that series as the upsides of A_0, A_1, A_2, &c., in (3) and we will have our expected arrangement.

11. As a last model we will take the issue of the vibration of an extended roundabout film secured at the circuit, or at least, of a conventional drumhead. We will guess the layer at first mutilated into some random structure which has roundabout symmetry[7] about a pivot through the middle opposite to the plane of the limit, and afterward permitted to vibrate.

Here we need to tackle

$$D_t^2 z = c^2 \left(D_r^2 z + \frac{1}{r}D_r z + \frac{1}{r^2}D_\phi^2 z \right) \qquad \text{[XI] Art. 1}$$

dependent upon the circumstances

$$z = f(r) \quad \text{when} \quad and\, t = 0 \qquad (1)$$
$$D_t z = 0 \quad \quad\text{"} and\, t \quad = 0 \qquad (2)$$
$$z = 0 \quad \quad\text{"} and\, r \quad = a. \qquad (3)$$

---

[7] An element of the coördinates of a point has *circular symmetry* about a pivot when its esteem isn't impacted by pivoting the point through any point about the hub. A surface has round evenness about a hub when it is a surface of unrest about the axis.

INTRODUCTION.

From the balance of the alleged starting twisting z should be autonomous of ϕ, along these lines [XI] decreases to

$$D_t^2 z = c^2 \left(D_r^2 z + \frac{1}{r} D_r z \right) \qquad (4)$$

furthermore, this is the condition for which we wish to track down a specific arrangement.

We will utilize a gadget much the same as that utilized in Art. 9.

Assume[8] $z = R.T$ where R is an element of r alone and T is a component of t alone. Substitute this worth of z in (4) and we get

$$R D_t^2 T = c^2 T \left(D_r^2 R + \frac{1}{r} D_r R \right)$$

or

$$\frac{1}{c^2 T} \frac{d^2 T}{dt^2} = \frac{1}{R} \left(\frac{d^2 R}{dr^2} + \frac{1}{r} \frac{dR}{dr} \right). \qquad (5)$$

The second individual from (5) doesn't include t, hence its equivalent the first part should be free of t. The main individual from (5) doesn't include r, and subsequently since it contains neither t nor r, it should be steady. Let it equivalent $-\mu^2$, where μ obviously is a dubious consistent.

Then (5) separates into the two differential conditions

$$\frac{d^2 T}{dt^2} + \mu^2 c^2 T = 0 \qquad (6)$$

$$\frac{d^2 R}{dr^2} + \frac{1}{r} \frac{dR}{dr} + \mu^2 R = 0. \qquad (7)$$

(6) can be addressed by natural techniques, and we get $T = \cos \mu c t$ and $T = \sin \mu c t$ as basic specific arrangements (v. Int. Cal. p. 319, § 21).

To settle (7) is really difficult. We will initially improve on it by a difference in free variable. Let $r = \frac{x}{\mu}$. (7) becomes

$$\frac{d^2 R}{dx^2} + \frac{1}{x} \frac{dR}{dx} + R = 0. \qquad (8)$$

Expect, as in Art. 9, that R can be communicated as far as entire powers of x. Let $R = \sum a_n x^n$ and substitute in (8). We get

$$\sum [n(n-1) a_n x^{n-2} + n a_n x^{n-2} + a_n x^n] = 0,$$

a condition which should be valid regardless of what the worth of x. The coefficient of some random force of x, as x^{k-2}, must, then, at that point, disappear, and

$$k(k-1) a_k + k a_k + a_{k-2} = 0$$

or

$$k^2 a_k + a_{k-2} = 0$$

whence we get

$$a_{k-2} = -k^2 a_k \qquad (9)$$

as the main connection that need be fulfilled by the coefficients all together that $R = \sum a_k x^k$ will be an answer of (8).

If $\qquad k = 0, \quad a_{k-2} = 0, \quad a_{k-4} = 0, \quad$ &c.

---
[8]See note on page 4.

INTRODUCTION.

We can then start with $k=0$ as our most reduced addendum.

From (9) $$a_k = -\frac{a_{k-2}}{k^2}.$$

Then $$a_2 = -\frac{a_0}{2^2}$$
$$a_4 = \frac{a_0}{2^2 . 4^2}$$
$$a_6 = -\frac{a_0}{2^2 . 4^2 . 6^2}, \&c.$$

Hence $$R = a_0 \left[1 - \frac{x^2}{2^2} + \frac{x^4}{2^2 . 4^2} - \frac{x^6}{2^2 . 4^2 . 6^2} + \cdots \right]$$

where a_0 might be taken at joy, is an answer of (8), gave the series is focalized.

Take $a_0 = 1$, and afterward $R = J_0(x)$ where

$$J_0(x) = 1 - \frac{x^2}{2^2} + \frac{x^4}{2^2 . 4^2} - \frac{x^6}{2^2 . 4^2 . 6^2} + \frac{x^8}{2^2 . 4^2 . 6^2 . 8^2} - \cdots \tag{10}$$

is an answer of (8).

$J_0(x)$ is handily demonstrated to be united for all values genuine or nonexistent of x, since the series comprised of the moduli of the terms of $J_0(x)$ (v. Int. Cal. Art. 30)

$$1 + \frac{r^2}{2^2} + \frac{r^4}{2^2 . 4^2} + \frac{r^6}{2^2 . 4^2 . 6^2} + \cdots,$$

where r is the modulus of x, is joined for all upsides of r. For the proportion of the $n + 1$st term of this series to the nth term is $\frac{r^2}{4n^2}$ and approaches zero as its cutoff as n is endlessly expanded, regardless of what the worth of r. In this manner $J_0(x)$ is *absolutely convergent*.

$J_0(x)$ is a new and significant structure. It is known as a *Bessel's Function* of the zeroth request, or a *Cylindrical Harmonic*.

Condition (8) was acquired from (7) by the replacement of $x = \mu r$, hence

$$R = J_0(\mu r) = 1 - \frac{(\mu r)^2}{2^2} + \frac{(\mu r)^4}{2^2 . 4^2} - \frac{(\mu r)^6}{2^2 . 4^2 . 6^2} + \cdots$$

is an answer of (7), regardless of what the worth of μ, and $z = J_0(\mu r) \cos \mu c t$ or on the other hand $z = J_0(\mu r) \sin \mu c t$ is an answer of (4).

$z = J_0(\mu r) \cos \mu c t$ fulfills condition (2) whatever the worth of μ. In request that it ought to likewise fulfill condition (3) μ should be taken to the point that

$$J_0(\mu a) = 0; \tag{11}$$

that is, μ should be a foundation of (11) viewed as a situation in μ.

It very well may be shown that $J_0(x) = 0$ has an endless number of genuine positive roots, any of which can be acquired to any expected level of estimation easily. Let $x_1, x_2, x_3, \cdots$ be these roots. Then, at that point, if

$$\frac{x_1}{a} = \mu_1, \quad \frac{x_2}{a} = \mu_2, \quad \frac{x_3}{a} = \mu_3, \quad \&c.$$

$$z = A_1 J_0(\mu_1 r) \cos \mu_1 c t + A_2 J_0(\mu_2 r) \cos \mu_2 c t + A_3 J_0(\mu_3 r) \cos \mu_3 c t + \cdots, \tag{12}$$

where A_1, A_2, A_3, &c., are any constants, is an answer of (4) which fulfills conditions (2) and (3).

When $t = 0$ (12) decreases to

$$z = A_1 J_0(\mu_1 r) + A_2 J_0(\mu_2 r) + A_3 J_0(\mu_3 r) + \cdots. \tag{13}$$

In the event that, $f(r)$ can be communicated as a progression of the structure recently given, the arrangement of our concern can be acquired by subbing the coefficients of that series for A_1, A_2, A_3, &c., in (12).

INTRODUCTION. 13

EXAMPLE.

The temperature of a long chamber is at first solidarity all through. The raised surface is then kept at the steady temperature zero. Show that the temperature of any point in the chamber at the termination of the time t is

$$u = A_1 e^{-a^2 \mu_1^2 t} J_0(\mu_1 r) + A_2 e^{-a^2 \mu_2^2 t} J_0(\mu_2 r) + A_3 e^{-a^2 \mu_3^2 t} J_0(\mu_3 r) + \cdots$$

where μ_1, μ_2, &c., are the underlying foundations of $J_0(\mu c) = 0$, and where

$$1 = A_1 J_0(\mu_1 r) + A_2 J_0(\mu_2 r) + A_3 J_0(\mu_3 r) + \cdots,$$

c being the range of the chamber.

12. Every one of the five issues which we have taken up powers upon us the thought of the improvement of a given capability as far as some *normal form*, and in two of them the ordinary structure proposed is an new capability. It is clear, then, that a total treatment of our subject will require the examination of the properties and relations of specific new and significant capabilities, as well as the thought of strategies for creating with regards to them.

13. In every one of the issues just taken up we need to manage a homogeneous straight incomplete differential condition including two autonomous factors, and we are content on the off chance that we can acquire specific arrangements. For each situation the suspicion made in the last issue, that there exists an answer of the situation in which the subordinate variable is the result of two factors every one of which includes yet one of the free factors, will lessen the question to addressing two conventional differential conditions which can be treated independently.

In the event that these conditions are recognizable ones their answers can be composed down on the double; if new, the gadget utilized in issues 3 and 5 is frequently useful, specifically, that of expecting to be that the ward variable can be communicated as a total or series of terms including entire powers of the free factor, and afterward deciding the coefficients.

Allow us to rethink the conditions utilized in the first, second and third issues.

(a) $$D_x^2 u + D_y^2 u = 0 \tag{1}$$

Expect $u = X.Y$ where X includes x yet not y, and Y includes y however not x.

Substitute in (1), $$Y D_x^2 X + X D_y^2 Y = 0,$$

or on the other hand, since we are currently managing elements of a solitary variable,

$$\frac{1}{X}\frac{d^2 X}{dx^2} + \frac{1}{Y}\frac{d^2 Y}{dy^2} = 0,$$

or $$\frac{1}{Y}\frac{d^2 Y}{dy^2} = -\frac{1}{X}\frac{d^2 X}{dx^2}. \tag{2}$$

Starting from the main individual from (2) doesn't contain x, and the subsequent part doesn't contain y, and the two individuals should be indistinguishably equivalent, neither of them can contain either x or y, and each should be equivalent to a steady, say α^2.

Then $$\frac{d^2 Y}{dy^2} - \alpha^2 Y = 0 \tag{3}$$

and $$\frac{d^2 X}{dx^2} + \alpha^2 X = 0; \tag{4}$$

what's more, if (3) and (4) can be settled, we can address (1). They have for their total arrangements

$$Y = A e^{\alpha y} + B e^{-\alpha y}$$

and $$X = C \sin \alpha x + D \cos \alpha x. \quad \text{(v. Int. Cal. p. 319, § 21.)}$$

INTRODUCTION.

Thus $Y = e^{\alpha y}$ and $Y = e^{-\alpha y}$ are specific arrangements of (3), $X = \sin \alpha x$ what's more, $X = \cos \alpha x$ are specific arrangements of (1), and subsequently

$$u = e^{\alpha y} \sin \alpha x, \quad u = e^{\alpha y} \cos \alpha x, \quad u = e^{-\alpha y} \sin \alpha x, \text{ and } u = e^{-\alpha y} \cos \alpha x$$

are specific arrangements of (1). These concur with the consequences of Art. 7.

(b) and $D_t^2 y = a^2 D_x^2 y$ \hfill (1)

Expect $y = T.X$ where T is a component of t just and X an element of x just; substitute in (1) and gap by $a^2 T X$. We get

$$\frac{1}{a^2 T}\frac{d^2 T}{dt^2} = \frac{1}{X}\frac{d^2 X}{dx^2}; \tag{2}$$

hence as in the last case $\frac{1}{X}\frac{d^2 X}{dx^2}$ is a steady; call it $-\alpha^2$, and (2) separates into

$$\frac{d^2 X}{dx^2} + \alpha^2 X = 0 \tag{3}$$

$$\frac{d^2 T}{dt^2} + \alpha^2 a^2 T = 0. \tag{4}$$

The total arrangements of (3) and (4) are

$$X = A \sin \alpha x + B \cos \alpha x$$

and $\qquad T = C \sin \alpha a t + D \cos \alpha a t,$ \quad (v. Int. Cal. p. 319, § 21).

$$y = \sin \alpha x \cos \alpha a t, \; y = \sin \alpha x \sin \alpha a t, \; y = \cos \alpha x \cos \alpha a t, \; y = \cos \alpha x \sin \alpha a t$$

are specific arrangements of (1), and concur with the aftereffects of Art. 8.

(c) $\qquad r D_r^2 (rV) + \dfrac{1}{\sin \theta} D_\theta (\sin \theta D_\theta V) = 0.$ \hfill (1)

Accept $V = R.\Theta$ where R includes r alone, and Θ includes θ alone; substitute in (1), partition by $R.\Theta$, and render; we get

$$\frac{r}{R}\frac{d^2(rR)}{dr^2} = -\frac{1}{\Theta \sin \theta}\frac{d\left(\sin \theta \frac{d\Theta}{d\theta}\right)}{d\theta}. \tag{2}$$

Since by the thinking utilized in (a) and (b) every individual from (2) should be a consistent, say α^2, we have

$$r\frac{d^2(rR)}{dr^2} = \alpha^2 R \tag{3}$$

and $\qquad \dfrac{1}{\sin \theta}\dfrac{d\left(\sin \theta \frac{d\Theta}{d\theta}\right)}{d\theta} + \alpha^2 \Theta = 0.$ \hfill (4)

(3) can be ventured into

$$r^2 \frac{d^2 R}{dr^2} + 2r\frac{dR}{dr} - \alpha^2 R = 0. \tag{5}$$

(5) can be addressed (v. Int. Cal. p. 321, § 23), and has for its finished arrangement

$$R = Ar^m + Br^n,$$

where $\qquad m = -\tfrac{1}{2} + \sqrt{\alpha^2 + \tfrac{1}{4}} \qquad$ and $\qquad n = -\tfrac{1}{2} - \sqrt{\alpha^2 + \tfrac{1}{4}}.$

INTRODUCTION.

Subsequently $n = -m-1$, and α^2 might be composed $m(m+1)$, m being completely inconsistent; and

$$R = Ar^m + Br^{-m-1}.$$

$$R = r^m, \quad \text{and} \quad R = \frac{1}{r^{m+1}}$$

are, then, specific arrangements of

$$r^2 \frac{d^2 R}{dr^2} + 2r \frac{dR}{dr} - m(m+1)R = 0. \tag{6}$$

With the new worth of α^2 (4) becomes

$$\frac{1}{\sin\theta} \frac{d\left(\sin\theta \frac{d\Theta}{d\theta}\right)}{d\theta} + m(m+1)\Theta = 0. \tag{7}$$

which has been treated in Art. 9 for the situation where m is a positive number, furthermore, the specific arrangement $\Theta = P_m(\cos\theta)$ has been acquired.

Hence
$$V = r^m P_m(\cos\theta)$$
and
$$V = \frac{1}{r^{m+1}} P_m(\cos\theta),$$

m being a positive whole number, are specific arrangements of (1). The first of these was gotten in Art. 9, however the second is new and extremely significant.

14. Tgy for getting a specific arrangement of a conventional straight differential condition, which we have utilized in Articles 9 and 11, is of very broad application, and frequently prompts the overall arrangement of the situation being referred to.

As an exceptionally straightforward model, let us take the condition Art. 13 (a) (4), which we will compose

$$\frac{d^2 z}{dx^2} + \alpha^2 z = 0. \tag{1}$$

Expect that there is an answer which can be communicated regarding powers of x; that is, let $z = \sum a_n x^n$, where the coefficients are not set in stone. Substitute this incentive for z in (1) and we get

$$\sum [n(n-1)a_n x^{n-2} + \alpha^2 a_n x^n] = 0.$$

Since this situation should be valid from its structure, without reference to the worth of x, or at least, since it should be an indistinguishable condition, the coefficient of each force of x should approach zero, and we have

$$(n+1)(n+2)a_{n+2} + \alpha^2 a_n = 0; \text{ and}$$

whence
$$a_n = -\frac{(n+1)(n+2)}{\alpha^2} a_{n+2}$$

is the main connection that need hold between the coefficients all together that $z = \sum a_n x^n$ ought to be an answer of (1).

On the off chance that $n+2 = 0$ or $n+1 = 0$, a_n will be zero and a_{n-2}, a_{n-4}, &c., will be zero. In the main case the series will start with a_0, in the second with a_1.

$$a_{n+2} = -\frac{\alpha^2}{(n+1)(n+2)} a_n.$$

Assuming we start with a_0 we have

$$a_2 = -\frac{\alpha^2}{2!} a_0, \text{ and } a_4 = \frac{\alpha^4}{4!} a_0, \text{ and } a_6 = -\frac{\alpha^6}{6!} a_0, \&c., \cdots$$

INTRODUCTION.

and
$$z = a_0 \left(1 - \frac{\alpha^2 x^2}{2!} + \frac{\alpha^4 x^4}{4!} - \frac{\alpha^6 x^6}{6!} + \cdots \right) \tag{2}$$

or
$$z = a_0 \cos \alpha x \tag{3}$$

is a specific arrangement of (1).

Assuming that we start with a_1 we have

$$a_3 = -\frac{\alpha^2}{3!} a_1, \text{ and } a_5 = \frac{\alpha^4}{5!} a_1, \text{ and } a_7 = -\frac{\alpha^6}{7!} a_1, \&c., \cdots$$

and
$$z = a_1 \left(x - \frac{\alpha^2 x^3}{3!} + \frac{\alpha^4 x^5}{5!} - \frac{\alpha^6 x^7}{7!} + \cdots \right) \tag{4}$$

is an answer of (1); a_1 can be taken at joy. Let $a_1 = \alpha$, (4) becomes

$$z = \alpha x - \frac{\alpha^3 x^3}{3!} + \frac{\alpha^5 x^5}{5!} - \frac{\alpha^7 x^7}{7!} + \cdots$$

or and $z = \sin \alpha x$

which, then, at that point, is a specific arrangement of (1).

$$z = A \sin \alpha x + B \cos \alpha x \tag{5}$$

is, then, at that point, an answer of (1), and since it contains two erratic constants it is the overall arrangement.

15. As another model we will take the condition

$$x^2 \frac{d^2 z}{dx^2} + 2x \frac{dz}{dx} - m(m+1)z = 0, \tag{1}$$

which is active condition (6), Art. 13 (c), and let m be a positive number.

Expect $z = \sum a_n x^n$ and substitute in (1). We get

$$\sum [n(n+1) - m(m+1)] a_n x^n = 0.$$

This is an indistinguishable condition, consequently

$$[n(n+1) - m(m+1)] a_n = 0.$$

Consequently $a_n = 0$ for all upsides of n with the exception of those which make

$$n(n+1) - m(m+1) = 0,$$

that is, for all upsides of n aside from $n = m$ and $n = -m - 1$. Then

$$z = Ax^m + Bx^{-m-1} \tag{2}$$

is the overall arrangement of (1) and

$$z = x^m \quad \text{and} \quad z = \frac{1}{x^{m+1}}$$

are specific arrangements. In the event that m is definitely not a positive number this technique will not lead to an outcome, and we are driven back to that utilized in Art. 13 (c).

INTRODUCTION.

16. Let us presently take the condition

$$\frac{d}{dx}\left[(1-x^2)\frac{dz}{dx}\right] + m(m+1)z = 0 \tag{1}$$

which is active condition (4), Art. 9, and is known as *Legendre's Equation*. (1) might be composed

$$(1-x^2)\frac{d^2z}{dx^2} - 2x\frac{dz}{dx} + m(m+1)z = 0. \tag{2}$$

Expect $z = \sum a_n x^n$ and substitute in (2). We get

$$\sum\{n(n-1)a_n x^{n-2} + [m(m+1) - n(n+1)]a_n x^n\} = 0.$$

Hence and $(n+1)(n+2)a_{n+2} + [m(m+1) - n(n+1)]a_n = 0,$

or and $a_n = -\dfrac{(n+1)(n+2)}{m(m+1)-n(n+1)}a_{n+2}. \tag{3}$

If $a_n = 0$, then $a_{n-2} = 0$, $a_{n-4} = 0$, &c.; however $a_n = 0$ if $n = -2$ or $n = -1$. For the main case we have the succession of coefficients

$$a_2 = -\frac{m(m+1)}{2!}a_0$$
$$a_4 = \frac{m(m-2)(m+1)(m+3)}{4!}a_0$$
$$a_6 = -\frac{m(m-2)(m-4)(m+1)(m+3)(m+5)}{6!}a_0, \quad \&c.$$

Allow us to take a_0, which is erratic, as 1. Then $z = p_m(x)$ where

$$p_m(x) = \left[1 - \frac{m(m+1)}{2!}x^2 + \frac{m(m-2)(m+1)(m+3)}{4!}x^4 - \cdots\right] \tag{4}$$

is an answer of Legendre's Situation on the off chance that $p_m(x)$ is a limited total or a joined series.

For the second case we have the arrangement of coefficients

$$a_3 = -\frac{(m-1)(m+2)}{3!}a_1$$
$$a_5 = \frac{(m-1)(m-3)(m+2)(m+4)}{5!}a_1$$
$$a_7 = -\frac{(m-1)(m-3)(m-5)(m+2)(m+4)(m+6)}{7!}a_1, \quad \&c.$$

Allow us to take a_1, which is erratic, as 1. Then, at that point, $z = q_m(x)$ where

$$q_m(x) = \left[x - \frac{(m-1)(m+2)}{3!}x^3 + \frac{(m-1)(m-3)(m+2)(m+4)}{5!}x^5 - \cdots\right] \tag{5}$$

is an answer of Legendre's Situation in the event that $q_m(x)$ is a limited total or a joined series.

On the off chance that m is a positive even entire number, $p_m(x)$ will end with the term containing x^m, and is handily seen to be indistinguishable with

$$(-1)^{\frac{m}{2}} \frac{2^m \left[\Gamma\left(\frac{m}{2}+1\right)\right]^2}{\Gamma(m+1)} P_m(x). \qquad [\text{v. Art. 9 (9)}]$$

INTRODUCTION.

For any remaining upsides of m, $p_m(x)$ is a series.

The proportion of the $(n+1)$st term of $p_m(x)$ to the nth, when m is certainly not a positive indeed, even number, is

$$\frac{(2n-2-m)(2n-1+m)}{(2n-1)(2n)}x^2.$$

Its restricting worth, as n is expanded, is x^2, and the series is hence focalized on the off chance that $-1 < x < 1$. It is unique for any remaining upsides of x.

In the event that m is a positive odd entire number $q_m(x)$ will end with the term containing x^m, and is effortlessly seen to be indistinguishable with

$$(-1)^{\frac{m-1}{2}}\frac{2^{m-1}\left[\Gamma\left(\frac{m+1}{2}\right)\right]^2}{\Gamma(m+1)}P_m(x).$$

For any remaining upsides of m, $q_m(x)$ is a series, and can be demonstrated to be focalized on the off chance that $-1 < x < 1$, and disparate for any remaining upsides of x.

$$z = Ap_m(x) + Bq_m(x) \qquad (6)$$

is the overall arrangement of Legendre's Situation if $-1 < x < 1$, regardless what the worth of m. From Art. 13 (c) that's what it follows

$$\left. \begin{aligned} V &= r^m p_m(\cos\theta) \\ V &= \frac{1}{r^{m+1}} p_m(\cos\theta) \\ V &= r^m q_m(\cos\theta) \\ V &= \frac{1}{r^{m+1}} q_m(\cos\theta) \end{aligned} \right\} \qquad (7)$$

are specific arrangements of

$$rD_r^2(rV) + \frac{1}{\sin\theta}D_\theta(\sin\theta D_\theta V) = 0,$$

regardless of what the worth of m, if $\cos\theta$ is neither one nor short one.

In the work we will have to do with Laplace's and Legendre's Conditions, it is for the most part conceivable to confine m to being a positive number, and from now on we will normally keep our regard for that case.

With this understanding let us return to (3), which might be revised

$$a_{n+2} = -\frac{(m-n)(m+n+1)}{(n+1)(n+2)}a_n.$$

If $\qquad a_{n+2} = 0$, then, at that point, $a_{n+4} = 0$, $a_{n+6} = 0$, &c.;

but $\qquad a_{n+2} = 0$ if $n = m$, or $n = -m-1$.

In the event that in (3) we start with $n = m - 2$, we get the succession of coefficients as of now acquired in Art. 9, and we have $z = P_m(x)$, where

$$P_m(x) = \frac{(2m-1)(2m-3)\cdots 1}{m!}\left[x^m - \frac{m(m-1)}{2(2m-1)}x^{m-2}\right.$$

$$+ \frac{m(m-1)(m-2)(m-3)}{2.4.(2m-1)(2m-3)}x^{m-4}$$

$$\left. - \frac{m(m-1)(m-2)(m-3)(m-4)(m-5)}{2.4.6.(2m-1)(2m-3)(2m-5)}x^{m-6} + \cdots \right], \qquad (8)$$

INTRODUCTION.

as a specific arrangement of Legendre's Situation.

If, nonetheless, we start with $n = -m - 3$, we have

$$a_{-m-3} = \frac{(m+1)(m+2)}{2(2m+3)} a_{-m-1}$$

$$a_{-m-5} = \frac{(m+1)(m+2)(m+3)(m+4)}{2.4.(2m+3)(2m+5)} a_{-m-1}$$

$$a_{-m-7} = \frac{(m+1)(m+2)(m+3)(m+4)(m+5)(m+6)}{2.4.6.(2m+3)(2m+5)(2m+7)} a_{-m-1}, \quad \&c.$$

a_{-m-1} might be taken at joy, and is typically taken as $\frac{m!}{1.3.5.\cdots(2m+1)}$, what's more, $z = Q_m(x)$ where

$$Q_m(x) = \frac{m!}{(2m+1)(2m-1)\cdots 1} \left[\frac{1}{x^{m+1}} + \frac{(m+1)(m+2)}{2.(2m+3)} \frac{1}{x^{m+3}} \right.$$
$$\left. + \frac{(m+1)(m+2)(m+3)(m+4)}{2.4.(2m+3)(2m+5)} \frac{1}{x^{m+5}} + \cdots \right] \quad (9)$$

is a second specific arrangement of Legendre's Situation, gave the series is focalized. $Q_m(x)$ is known as a *Surface Zonal Harmonic* of the *second kind*. It is handily seen to be merged if $x < -1$ or $x > 1$, and unique if $-1 < x < 1$.

Consequently in the event that m is a positive number,

$$z = AP_m(x) + BQ_m(x) \quad (10)$$

is the overall arrangement of Legendre's Situation if $x < -1$ or $x > 1$.

We have seen that for $-1 < x < 1$

$$P_m(x) = (-1)^{\frac{m}{2}} \frac{\Gamma(m+1)}{2^m \left[\Gamma\left(\frac{m}{2}+1\right)\right]^2} p_m(x) \quad (11)$$

if m is an even number, and

$$P_m(x) = (-1)^{\frac{m-1}{2}} \frac{\Gamma(m+1)}{2^{m-1} \left[\Gamma\left(\frac{m+1}{2}\right)\right]^2} q_m(x) \quad (12)$$

in the event that m is an odd number.

In the event that now we characterize $Q_m(x)$ as follows when $-1 < x < 1$

$$Q_m(x) = (-1)^{\frac{m+1}{2}} \frac{2^{m-1} \left[\Gamma\left(\frac{m+1}{2}\right)\right]^2}{\Gamma(m+1)} p_m(x) \quad (13)$$

if m is an odd whole number, and

$$Q_m(x) = (-1)^{\frac{m}{2}} \frac{2^m \left[\Gamma\left(\frac{m}{2}+1\right)\right]^2}{\Gamma(m+1)} q_m(x) \quad (14)$$

on the off chance that m is an even whole number, (10) will be the overall arrangement of Legendre's Condition in the event that m is a positive number when $-1 < x < 1$, as well as when $x < -1$ or on the other hand $x > 1$.

INTRODUCTION.

17. Let us last think about the situation

$$\frac{d^2z}{dx^2} + \frac{1}{x}\frac{dz}{dx} + \left(1 - \frac{m^2}{x^2}\right)z = 0 \tag{1}$$

which is known as Bessel's Situation, and which diminishes to (8) Art. 11, that is, to

$$\frac{d^2z}{dx^2} + \frac{1}{x}\frac{dz}{dx} + z = 0$$

when $m = 0$;[9] (1) can be improved on by a difference in the reliant variable. Let $z = x^m v$ and we get

$$\frac{d^2v}{dx^2} + \frac{2m+1}{x}\frac{dv}{dx} + v = 0 \tag{2}$$

to decide v.

Expect $v = \sum a_n x^n$, and substitute in (2). We get

$$\sum [n(2m+n)a_n x^{n-2} + a_n x^n] = 0;$$

whence
$$a_{n-2} = -n(2m+n)a_n.$$

On the off chance that we start with $n = 0$, $a_{-2} = 0$, $a_{-4} = 0$, &c., and we have the set of values

$$a_2 = -\frac{a_0}{2(2m+2)} = -\frac{a_0}{2^2(m+1)}$$

$$a_4 = \frac{a_0}{2.4(2m+2)(2m+4)} = \frac{a_0}{2^4.2!(m+1)(m+2)}$$

$$a_6 = -\frac{a_0}{2.4.6(2m+2)(2m+4)(2m+6)} = -\frac{a_0}{2^6.3!(m+1)(m+2)(m+3)};$$

whence
$$z = a_0 x^m \left[1 - \frac{x^2}{2^2(m+1)} + \frac{x^4}{2^4.2!(m+1)(m+2)} \right.$$
$$\left. - \frac{x^6}{2^6.3!(m+1)(m+2)(m+3)} + \cdots \right] \tag{3}$$

is an answer of Bessel's Situation. a_0 is generally taken as $\frac{1}{2^m m!}$ in the event that m is a positive whole number, or as $\frac{1}{2^m \Gamma(m+1)}$ on the off chance that m is unlimited in esteem, and the second individual from (3) is addressed by $J_m(x)$ and is known as a *Bessel's Function* of the mth request, or a *Cylindrical Harmonic* of the mth request.

If $m = 0$, $J_m(x)$ becomes $J_0(x)$ and is the worth of z got in Art. 11 as the arrangement of equation (8) of that article.

In the event that in condition (1) we substitute $x^{-m}v$ instead of $x^m v$ for z, we get set up of (2) the condition

$$\frac{d^2v}{dx^2} + \frac{1-2m}{x}\frac{dv}{dx} + v = 0$$

furthermore, instead of (3)

$$z = a_0 x^{-m} \left[1 - \frac{x^2}{2^2(1-m)} + \frac{x^4}{2^4.2!(1-m)(2-m)} \right.$$
$$\left. - \frac{x^6}{2^6.3!(1-m)(2-m)(3-m)} + \cdots \right] \tag{4}$$

---

[9]This condition was first concentrated on by Fourier in thinking about the cooling of a chamber. We will assign it as "Fourier's Equation."

INTRODUCTION. 21

In the event that a_0 is taken equivalent to $\frac{1}{2^{-m}\Gamma(1-m)}$ the second individual from (4) is something similar capability of $-m$ and x that $J_m(x)$ is of $+m$ and x and might be composed $J_{-m}(x)$.

In this manner
$$z = AJ_m(x) + BJ_{-m}(x) \qquad (5)$$
is the overall arrangement of (1) except if $J_m(x)$ and $J_{-m}(x)$ ought to demonstrate not to be free.

It is handily seen that when $m = 0$, $J_{-m}(x)$ and $J_m(x)$ become indistinguishable also (5) lessens to
$$z = (A+B)J_0(x)$$
what's more, contains however a solitary erratic consistent and isn't the overall arrangement of Fourier's Equation (8) Art. (11).

It tends to be shown that $J_{-m}(x) = (-1)^m J_m(x)$ at whatever point m is a number, also, thus that the solution (5) is general just when m assuming genuine is partial or then again incommensurable.

The overall answer for the significant situation where $m = 0$ is, notwithstanding, without any problem acquired. Allow $F(m, x)$ to be the worth which the second individual from (3) expects when $a_0 = 1$; then, at that point, the worth which the second individual from (4) accepts when $a_0 = 1$ will be $F(-m, x)$, and it has been shown that $z = F(m, x)$ and $z = F(-m, x)$ are arrangements of Bessel's Situation; $z = F(m, x) - F(-m, x)$ is, then, at that point, an answer, as is too
$$z = \frac{F(m, x) - F(-m, x)}{2m}, \qquad (6)$$
however, the restricting worth which $\frac{F(m,x)-F(-m,x)}{2m}$ methodologies as m approaches zero is $[D_m F(m, x)]_{m=0}$ and thus
$$z = [D_m F(m, x)]_{m=0} \qquad (7)$$
is an answer of the situation
$$\frac{d^2 z}{dx^2} + \frac{1}{x}\frac{dz}{dx} + z = 0, \qquad (8)$$
also, the overall arrangement of (8) is
$$z = AJ_0(x) + B[D_m F(m, x)]_{m=0}.$$

$$F(m, x) = x^m \left[1 - \frac{x^2}{2^2(m+1)} + \frac{x^4}{2^4 \cdot 2!(m+1)(m+2)} - \frac{x^6}{2^6 \cdot 3!(m+1)(m+2)(m+3)} + \cdots\right]$$

$$D_m F(m, x) = x^m \log x \left[1 - \frac{x^2}{2^2(m+1)} + \frac{x^4}{2^4 \cdot 2!(m+1)(m+2)} - \cdots\right] \text{ and}$$

$$+x^m D_m \left[1 \text{ and } \frac{x^2}{2^2(m+1)} + \frac{x^4}{2^4 \cdot 2!(m+1)(m+2)} + \cdots\right].$$

The general term of the last bracket can be composed
$$(-1)^k \frac{x^{2k}}{2^{2k} \cdot k!(m+1)(m+2)\cdots(m+k)},$$
also, its incomplete subsidiary concerning m is
$$(-1)^k \frac{x^{2k}}{2^{2k} \cdot k!} D_m \frac{1}{(m+1)(m+2)\cdots(m+k)}.$$

$$\log \frac{1}{(m+1)(m+2)\cdots(m+k)} = -[\log(m+1) + \log(m+2) + \cdots + \log(m+k)].$$

INTRODUCTION.

Take the D_m of the two individuals and we have

$$D_m \frac{1}{(m+1)(m+2)\cdots(m+k)}$$
$$= -\frac{1}{(m+1)(m+2)\cdots(m+k)}\left[\frac{1}{m+1} + \frac{1}{m+2} + \cdots \frac{1}{m+k}\right].$$

$$D_m\left[1 - \frac{x^2}{2^2(m+1)} + \frac{x^4}{2^4.2!(m+1)(m+2)} - \frac{x^6}{2^6.3!(m+1)(m+2)(m+3)}\right.$$
$$\left.+\cdots\right] = \frac{x^2}{2^2}\frac{1}{(m+1)^2} - \frac{x^4}{2^4.2!}\frac{1}{(m+1)(m+2)}\left[\frac{1}{m+1} + \frac{1}{m+2}\right]$$
$$+ \frac{x^6}{2^6.3!}\frac{1}{(m+1)(m+2)(m+3)}\left[\frac{1}{m+1} + \frac{1}{m+2} + \frac{1}{m+3}\right] + \cdots$$

furthermore, we have

$$[D_m F(m,x)]_{m=0} = J_0(x)\log x + \frac{x^2}{2^2(1!)^2}\frac{1}{1} - \frac{x^4}{2^4(2!)^2}\left(\frac{1}{1} + \frac{1}{2}\right)$$
$$+ \frac{x^6}{2^6(3!)^2}\left(\frac{1}{1} + \frac{1}{2} + \frac{1}{3}\right) - \frac{x^8}{2^8(4!)^2}\left(\frac{1}{1} + \frac{1}{2} + \frac{1}{3} + \frac{1}{4}\right) + \cdots ;$$

and

$$z = AJ_0(x) + BK_0(x), \qquad (9)$$

where

$$K_0(x) = J_0(x)\log x + \frac{x^2}{2^2} - \frac{x^4}{2^4(2!)^2}\left(\frac{1}{1} + \frac{1}{2}\right)$$
$$+ \frac{x^6}{2^6(3!)^2}\left(\frac{1}{1} + \frac{1}{2} + \frac{1}{3}\right) - \frac{x^8}{2^8(4!)^2}\left(\frac{1}{1} + \frac{1}{2} + \frac{1}{3} + \frac{1}{4}\right) + \cdots \quad (10)$$

is the overall arrangement of Fourier's Situation (8).

$K_0(x)$ is known as a *Bessel's Function* of the *Second Kind*.

18. It is worth while to affirm the consequences of the last couple of articles by getting the overall arrangements of the situations being referred to by an alternate and recognizable strategy.

The overall arrangement of any common straight differential condition of the second request can be gotten when a specific arrangement of the situation has been found [v. Int. Cal. p. 321, §24 (a)].

The most broad type of a homogeneous customary straight differential condition of the subsequent request is

$$\frac{d^2y}{dx^2} + P\frac{dy}{dx} + Qy = 0 \qquad (1)$$

where P and Q are elements of x. Assume that

$$y = v \qquad (2)$$

is a specific arrangement of (1). Substitute $y = vz$ in (1) and we get

$$v\frac{d^2z}{dx^2} + \left(2\frac{dv}{dx} + Pv\right)\frac{dz}{dx} = 0. \qquad (3)$$

Call $\frac{dz}{dx} = z'$. Then, at that point (3) becomes

$$v\frac{dz'}{dx} + \left(2\frac{dv}{dx} + Pv\right)z' = 0, \qquad (4)$$

INTRODUCTION.

a differential condition of the primary request wherein the factors can be isolated. Increase by dx and partition by vz' and (4) lessens to

$$\frac{dz'}{z'} + 2\frac{dv}{v} + Pdx = 0.$$

Coordinate and we have

$$\log z' + \log v^2 + \int Pdx = C$$

or *and* $z'v^2 = e^{C-\int Pdx} = Be^{-\int Pdx},$

$$z' = \frac{dz}{dx} = B\frac{e^{-\int Pdx}}{v^2},$$

$$z = A + B\int \frac{e^{-\int Pdx}}{v^2}dx;$$

and *and* $y =$

$$v\left(A + B\int \frac{e^{-\int Pdx}}{v^2}dx\right) \qquad (5)$$

is the overall arrangement of (1), the main erratic constants in the subsequent part of (5) being those unequivocally composed, in particular, A and B.

(a) Apply this recipe to (1) Art. 14,

$$\frac{d^2z}{dx^2} + \alpha^2 z = 0; \qquad (1)$$

given: $z = \cos \alpha x$, as a specific arrangement. Subbing in (5) we have since $P = 0$

$$z = \cos \alpha x \left(A + B\int \frac{dx}{\cos^2 \alpha x}\right)$$
$$= \cos \alpha x \left(A + \frac{B}{\alpha} \tan \alpha x\right)$$
$$= A\cos \alpha x + B_1 \sin \alpha x, \qquad (2)$$

as the overall arrangement of (1), and this concurs impeccably with (5) Art. 14.

(b) Take condition (1) Art. 15.

$$x^2 \frac{d^2z}{dx^2} + 2x\frac{dz}{dx} - m(m+1)z = 0; \qquad (1)$$

given: $z = x^m$, as a specific arrangement.
Here $P = \frac{2}{x}$, $\int Pdx = 2\log x = \log x^2$, and $e^{-\int Pdx} = \frac{1}{x^2}$. Subsequently by (5)

$$z = x^m \left(A + B\int \frac{dx}{x^{2m+2}}\right) = x^m\left(A + \frac{B}{-2m-1}\frac{1}{x^{2m+1}}\right),$$

that is *and* $z =$

$$Ax^m + \frac{B_1}{x^{m+1}} \qquad (2)$$

is the overall arrangement of (1), and concurs with (2) Art. 15.

(c) Take Legendre's Condition, (2) Art. 16.

$$(1-x^2)\frac{d^2z}{dx^2} - 2x\frac{dz}{dx} + m(m+1)z = 0; \qquad (1)$$

INTRODUCTION.

given: $z = P_m(x)$, as a specific arrangement.

Here $P = \frac{-2x}{1-x^2}$, $\int P dx = \log(1-x^2)$, and $e^{-\int P dx} = \frac{1}{1-x^2}$.

Hence by (5)
$$z = P_m(x)\left(A + B\int \frac{dx}{(1-x^2)[P_m(x)]^2}\right) \qquad (2)$$

is the overall arrangement of (1) and should concur with (10) Art. 16, assuming m is an number, and along these lines
$$Q_m(x) = CP_m(x)\int \frac{dx}{(1-x^2)[P_m(x)]^2} \qquad (3)$$

where C is at this point dubious, and no consistent term is to be perceived with the essential in the second member.

(d) Take Bessel's Condition, (1) Art. 17.
$$\frac{d^2z}{dx^2} + \frac{1}{x}\frac{dz}{dx} + \left(1 - \frac{m^2}{x^2}\right)z = 0; \qquad (1)$$

given: $z = J_m(x)$, as a specific arrangement.

Here $P = \frac{1}{x}$, $\int P dx = \log x$, and $e^{-\int P dx} = \frac{1}{x}$. Thus by (5)
$$z = J_m(x)\left(A + B\int \frac{dx}{x[J_m(x)]^2}\right) \qquad (2)$$

is the overall arrangement of Bessel's Situation.

In the event that $m = 0$ (2) becomes
$$z = J_0(x)\left(A + B\int \frac{dx}{x[J_0(x)]^2}\right) \qquad (3)$$

furthermore, should concur with (9) Art. 17. In this manner
$$K_0(x) = CJ_0(x)\int \frac{dx}{x[J_0(x)]^2}, \qquad (4)$$

where C is at present unsure, and no consistent term is to be taken with the essential.

The principal significant subject proposed by the issues which we have taken up in this early on part is that of advancement in Mathematical Series (v. Arts. 7 and 8).

Chapter 2

DEVELOPMENT IN TRIGONOMETRIC SERIES

19. We have found in Section I. that it is at times vital to be capable to communicate a given capability of a variable x, concerning the sines or of the cosines of products of x. The issue in its general structure was first settled by Fourier in his "Logical Hypothesis of Intensity" (1822), and its answer plays a vital part in many parts of current Physical science. Series including just sines and cosines of entire products of x, that is series of the structure

$$b_0 + b_1 \cos x + b_2 \cos 2x + \cdots + a_1 \sin x + a_2 \sin 2x + \cdots$$

are for the most part known as Fourier's series.

Allow us to try to foster a given capability of x as far as $\sin x$, $\sin 2x$, $\sin 3x$, &c., so that the capability and the series will be equivalent for all upsides of x among $x = 0$ and $x = \pi$.

To fix our thoughts let us guess that we have a bend,

$$y = f(x),$$

given, and that we wish to frame the condition,

$$y = a_1 \sin x + a_2 \sin 2x + a_3 \sin 3x + \cdots,$$

of a bend which will concur with such a great deal the given bend as lies between the focuses comparing to $x = 0$ and $x = \pi$. Obviously in the situation

$$y = a_1 \sin x \tag{1}$$

a_1 might be resolved with the goal that the bend addressed will go through any given point. For in the event that we substitute in (1) the coördinates of the point being referred to we will have a condition of the main degree where a_1 is the as it were obscure amount and which will in this manner give us one and only one worth for a_1.

Likewise the bend

$$y = a_1 \sin x + a_2 \sin 2x$$

might be made to go through any two randomly picked focuses whose abscissas lie among 0 and π given that the abscissas are not equivalent; and

$$y = a_1 \sin x + a_2 \sin 2x + a_3 \sin 3x + \cdots + a_n \sin nx$$

might be made to go through any n randomly picked focuses whose abscissas lie among 0 and π gave as before that their abscissas are unique.

DEVELOPMENT IN TRIGONOMETRICAL SERIES.

On the off chance that, the given capability $f(x)$ is of such a person that for each worth of x among $x = 0$ and $x = \pi$ it has one and only one worth, and if between $x = 0$ and $x = \pi$ it is limited and persistent, or then again if irregular has as it were *finite discontinuities* (v. Int. Cal. Art. 83, p. 78), the coefficients in

$$y = a_1 \sin x + a_2 \sin 2x + a_3 \sin 3x + \cdots + a_n \sin nx \tag{2}$$

can be resolved so the bend addressed by (2) will go through any n for arbitrary reasons picked points of the bend

$$y = f(x) \tag{3}$$

whose abscissas lie among 0 and π and are unique, and these coefficients will have however one bunch of values.

For effortlessness guess that the n focuses are decided to the point that their projections on the pivot of X are equidistant.

Call $\frac{\pi}{n+1} = \Delta x$; then, at that point, the coÃ¶rdinates of the n focuses will be $[\Delta x, f(\Delta x)]$, $[2\Delta x, f(2\Delta x)]$, $[3\Delta x, f(3\Delta x)]$, $\cdots$ $[n\Delta x, f(n\Delta x)]$. Substitute them in (2) and we have

$$\left.\begin{aligned}
f(\Delta x) &= a_1 \sin \Delta x + a_2 \sin 2\Delta x + a_3 \sin 3\Delta x + \cdots + a_n \sin n\Delta x \text{ and}\\
f(2\Delta x) &= a_1 \sin 2\Delta x + a_2 \sin 4\Delta x + a_3 \sin 6\Delta x + \cdots + a_n \sin 2n\Delta x \text{ and}\\
f(3\Delta x) &= a_1 \sin 3\Delta x + a_2 \sin 6\Delta x + a_3 \sin 9\Delta x + \cdots + a_n \sin 3n\Delta x \text{ and}\\
&\vdots \qquad \vdots \qquad \vdots \quad \text{and:} \qquad \vdots \quad \text{and}\\
f(n\Delta x) &= a_1 \sin n\Delta x + a_2 \sin 2n\Delta x + a_3 \sin 3n\Delta x + \cdots + a_n \sin n^2 \Delta x, \text{ and}
\end{aligned}\right\} \tag{4}$$

n conditions of the principal degree to decide the n coefficients $a_1, a_2, a_3, \cdots a_n$.

Not exclusively can conditions (4) be tackled in principle, however they can be in fact settled in some random case by an exceptionally basic and clever technique due to Lagrange.

Allow us to take as an illustration the straightforward issue to decide the coefficients a_1, a_2, a_3, a_4, and a_5, so that

$$y = a_1 \sin x + a_2 \sin 2x + a_3 \sin 3x + a_4 \sin 4x + a_5 \sin 5x \tag{5}$$

will go through the five places of the line

$$y = x$$

which have the abscissas $\frac{\pi}{6}$, $\frac{2\pi}{6}$, $\frac{3\pi}{6}$, $\frac{4\pi}{6}$, and $\frac{5\pi}{6}$, $\frac{\pi}{6}$ here being Δx.

We should now settle the conditions

$$\left.\begin{aligned}
\frac{\pi}{6} &= a_1 \sin \frac{\pi}{6} + a_2 \sin \frac{2\pi}{6} + a_3 \sin \frac{3\pi}{6} + a_4 \sin \frac{4\pi}{6} + a_5 \sin \frac{5\pi}{6}\\
\frac{2\pi}{6} &= and\, a_1 \sin \frac{2\pi}{6} + a_2 \sin \frac{4\pi}{6} + a_3 \sin \frac{6\pi}{6} + a_4 \sin \frac{8\pi}{6} + a_5 \sin \frac{10\pi}{6}\\
\frac{3\pi}{6} &= and\, a_1 \sin \frac{3\pi}{6} + a_2 \sin \frac{6\pi}{6} + a_3 \sin \frac{9\pi}{6} + a_4 \sin \frac{12\pi}{6} + a_5 \sin \frac{15\pi}{6}\\
\frac{4\pi}{6} &= a_1 \sin \frac{4\pi}{6} + a_2 \sin \frac{8\pi}{6} + a_3 \sin \frac{12\pi}{6} + a_4 \sin \frac{16\pi}{6} + a_5 \sin \frac{20\pi}{6}\\
\frac{5\pi}{6} &= a_1 \sin \frac{5\pi}{6} + a_2 \sin \frac{10\pi}{6} + a_3 \sin \frac{15\pi}{6} + a_4 \sin \frac{20\pi}{6} + a_5 \sin \frac{25\pi}{6}.
\end{aligned}\right\} \tag{6}$$

Duplicate the principal condition by $2 \sin \frac{\pi}{6}$, the second by $2 \sin \frac{2\pi}{6}$, the third by $2 \sin \frac{3\pi}{6}$, the fourth by $2 \sin \frac{4\pi}{6}$, the fifth by $2 \sin \frac{5\pi}{6}$ and add the conditions.

The coefficient of a_2 is

$$2 \sin \frac{\pi}{6} \sin \frac{2\pi}{6} + 2 \sin \frac{2\pi}{6} \sin \frac{4\pi}{6} + 2 \sin \frac{3\pi}{6} \sin \frac{6\pi}{6} + 2 \sin \frac{4\pi}{6} \sin \frac{8\pi}{6} \text{ and}$$
$$+ 2 \sin \frac{5\pi}{6} \sin \frac{10\pi}{6};$$

but

$$2 \sin \frac{\pi}{6} \sin \frac{2\pi}{6} = \cos \frac{\pi}{6} - \cos \frac{3\pi}{6}, \&c.$$

DEVELOPMENT IN TRIGONOMETRICAL SERIES.

Consequently the coefficient of a_2 becomes

$$\left.\begin{array}{l}\cos\dfrac{\pi}{6}+\cos\dfrac{2\pi}{6}+\cos\dfrac{3\pi}{6}+\cos\dfrac{4\pi}{6}+\cos\dfrac{5\pi}{6}\\-\cos\dfrac{3\pi}{6}\text{ and}\cos\dfrac{6\pi}{6}\text{ and }\cos\dfrac{9\pi}{6}\text{ and}\cos\dfrac{12\pi}{6}\text{ and }\cos\dfrac{15\pi}{6}\end{array}\right\} \quad (7)$$

also, this might be diminished by the guide of a significant Mathematical recipe which we continue to lay out.

20. Lemma.

$$\cos\theta + \cos 2\theta + \cos 3\theta + \cdots + \cos n\theta = -\frac{1}{2} + \frac{1}{2}\frac{\sin(2n+1)\frac{\theta}{2}}{\sin\frac{\theta}{2}}. \quad (1)$$

For let $S = \cos\theta + \cos 2\theta + \cos 3\theta + \cdots + \cos n\theta$ and increase by $2\cos\theta$.

$$2S\cos\theta = 2\cos^2\theta + 2\cos\theta\cos 2\theta + 2\cos\theta\cos 3\theta + \cdots + 2\cos\theta\cos n\theta$$
$$= 1 + \cos\theta + \cos 2\theta + \cdots + \cos(n-1)\theta$$
$$\qquad + \cos 2\theta + \cos 3\theta + \cos 4\theta + \cdots + \cos(n+1)\theta$$
$$= 2S + 1 + \cos(n+1)\theta - \cos\theta - \cos n\theta. \quad \text{Hence}$$

$$S = -\frac{1}{2} + \frac{\cos n\theta - \cos(n+1)\theta}{2(1-\cos\theta)}$$

or
$$S = -\frac{1}{2} + \frac{1}{2}\frac{\sin(2n+1)\frac{\theta}{2}}{\sin\frac{\theta}{2}}. \qquad\text{Q.E.D.}$$

21. Applying (1) Art. 20 to (7) Art. 19 the coefficient of a_2 diminishes to

$$\frac{\sin\frac{11\pi}{12}}{2\sin\frac{\pi}{12}} - \frac{\sin\frac{33\pi}{12}}{2\sin\frac{3\pi}{12}};$$

but $\qquad \dfrac{11\pi}{12} = \pi - \dfrac{\pi}{12}, \text{ and } \dfrac{33\pi}{12} = 3\pi - \dfrac{3\pi}{12};$

therefore
$$\frac{\sin\left(\pi - \frac{\pi}{12}\right)}{2\sin\frac{\pi}{12}} - \frac{\sin\left(3\pi - \frac{3\pi}{12}\right)}{2\sin\frac{3\pi}{12}} = \frac{1}{2} - \frac{1}{2} = 0,$$

furthermore, a_2 disappears.

In this way it very well might be shown that the coefficients of a_3, a_4, and a_5 disappear.

The coefficient of a_1 is

$$2\sin^2\frac{\pi}{6} + 2\sin^2\frac{2\pi}{6} + 2\sin^2\frac{3\pi}{6} + 2\sin^2\frac{4\pi}{6} + 2\sin^2\frac{5\pi}{6}$$
$$= 1\ +\ 1\ +\ 1\ +\ 1\ +\ 1$$
$$-\cos\frac{2\pi}{6} - \cos\frac{4\pi}{6} - \cos\frac{6\pi}{6} - \cos\frac{8\pi}{6} - \cos\frac{10\pi}{6}$$
$$= 5 + \frac{1}{2} - \frac{\sin\frac{11\pi}{6}}{2\sin\frac{\pi}{6}} = 5\frac{1}{2} - \frac{\sin\left(2\pi - \frac{\pi}{6}\right)}{2\sin\frac{\pi}{6}} = 6.$$

DEVELOPMENT IN TRIGONOMETRICAL SERIES.

The primary individual from the last condition is

$$\frac{2\pi}{6}\sin\frac{\pi}{6} + 2\frac{2\pi}{6}\sin\frac{2\pi}{6} + 2\frac{3\pi}{6}\sin\frac{3\pi}{6} + 2\frac{4\pi}{6}\sin\frac{4\pi}{6} + 2\frac{5\pi}{6}\sin\frac{5\pi}{6}. \quad \text{Hence}$$

$$a_1 = \frac{2}{6}\sum_{k=1}^{k=5}\frac{k\pi}{6}\sin\frac{k\pi}{6} = \frac{\pi}{6}(2+\sqrt{3}) = 2 \quad \text{approximately.}$$

In the event that we duplicate the main condition of (6) Art. 19 by $2\sin\frac{2\pi}{6}$, the second by $2\sin\frac{4\pi}{6}$, the third by $2\sin\frac{6\pi}{6}$, the fourth by $2\sin\frac{8\pi}{6}$, the fifth by $2\sin\frac{10\pi}{6}$, add and lessen as before we will find

$$a_2 = \frac{2}{6}\sum_{k=1}^{k=5}\frac{k\pi}{6}\sin\frac{2k\pi}{6} = -\frac{\pi}{6}\sqrt{3} = -0.9;$$

and in this way we get

$$a_3 = \frac{2}{6}\sum_{k=1}^{k=5}\frac{k\pi}{6}\sin\frac{3k\pi}{6} = \frac{\pi}{6} = 0.5$$

$$a_4 = \frac{2}{6}\sum_{k=1}^{k=5}\frac{k\pi}{6}\sin\frac{4k\pi}{6} = -\frac{\pi\sqrt{3}}{18} = -0.3$$

$$a_5 = \frac{2}{6}\sum_{k=1}^{k=5}\frac{k\pi}{6}\sin\frac{5k\pi}{6} = \frac{\pi}{6}(2-\sqrt{3}) = 0.1.$$

Thusly
$$y = 2\sin x - 0.9\sin 2x + 0.5\sin 3x - 0.3\sin 4x + 0.1\sin 5x \qquad (1)$$

cuts the bend $y = x$ at the five focuses whose abscissas are $\frac{\pi}{6}$, $\frac{2\pi}{6}$, $\frac{3\pi}{6}$, $\frac{4\pi}{6}$, and $\frac{5\pi}{6}$.

22. The conditions (4) Art. 19 can be settled by the very same gadget. To find any coefficient a_m increase the principal condition by $2\sin m\Delta x$, the second by $2\sin 2m\Delta x$, the third by $2\sin 3m\Delta x$, &c. and add.

The coefficient of some other a as a_k in the subsequent condition will be

$$2\sin k\Delta x \sin m\Delta x + 2\sin 2k\Delta x \sin 2m\Delta x + 2\sin 3k\Delta x \sin 3m\Delta x + \cdots$$
$$+ 2\sin nk\Delta x \sin nm\Delta x$$
$$= \cos(m-k)\Delta x + \cos 2(m-k)\Delta x + \cos 3(m-k)\Delta x + \cdots + \cos n(m-k)\Delta x$$
$$- \cos(m+k)\Delta x - \cos 2(m+k)\Delta x - \cos 3(m+k)\Delta x - \cdots - \cos n(m+k)\Delta x$$
$$= \frac{\sin\frac{2n+1}{2}(m-k)\Delta x}{2\sin\frac{(m-k)\Delta x}{2}} - \frac{\sin\frac{2n+1}{2}(m+k)\Delta x}{2\sin\frac{(m+k)\Delta x}{2}}; \quad \text{by (1) Art. 20.}$$

$$\frac{2n+1}{2} = n+1-\frac{1}{2} \quad \text{and} \quad (n+1)\Delta x = \pi.$$

Consequently the coefficient of a_k might be composed

$$\frac{\sin\left[(m-k)\pi - \frac{(m-k)\Delta x}{2}\right]}{2\sin\frac{(m-k)\Delta x}{2}} - \frac{\sin\left[(m+k)\pi - \frac{(m+k)\Delta x}{2}\right]}{2\sin\frac{(m+k)\Delta x}{2}}$$

be that as it may, this is equivalent to $\frac{1}{2} - \frac{1}{2}$ or $-\frac{1}{2} + \frac{1}{2}$ proportionately as $m-k$ is odd or even as is zero regardless.

DEVELOPMENT IN TRIGONOMETRICAL SERIES.

The coefficient of a_m will be

$$2\sin^2 m\Delta x + 2\sin^2 2m\Delta x + 2\sin^2 3m\Delta x + \cdots + 2\sin^2 nm\Delta x$$
$$= \quad 1 \quad + \quad 1 \quad + \quad 1 \quad + \cdots + \quad 1$$
$$- \cos 2m\Delta x - \cos 4m\Delta x - \cos 6m\Delta x - \cdots - \cos 2nm\Delta x$$
$$= n + \frac{1}{2} - \frac{\sin(2n+1)m\Delta x}{2\sin m\Delta x}, \text{ by (1) Art. 20.}$$

Yet $and(2n+1)m\Delta x = 2m(nand+1)\Delta x - m\Delta x = 2m\pi - m\Delta x$,
therefore $and \dfrac{\sin(2n+1)m\Delta x}{2\sin m\Delta x} and = \dfrac{\sin(2m\pi - m\Delta x)}{2\sin m\Delta x} = -\dfrac{1}{2}$,
also, the coefficient of a_m is $n+1$.

The primary individual from our last condition will be

$$2\sum_{k=1}^{k=n} f(k\Delta x)\sin km\Delta x.$$

Thus

$$a_m = \frac{2}{n+1}\sum_{k=1}^{k=n} f(k\Delta x)\sin km\Delta x, \tag{1}$$

what's more, the bend

$$y = a_1\sin x + a_2\sin 2x + \cdots + a_n\sin nx, \tag{2}$$

where the coefficients are given by (1) will go through the n points of the bend $y = f(x)$ whose abscissas are $\Delta x, 2\Delta x, 3\Delta x, \cdots n\Delta x$. Δx being $\frac{\pi}{n+1}$.

It ought to be noticed that since the n conditions (4) Art. 19 are the entirety of the first degree there will exist just a single bunch of values for the n amounts $a_1, a_2, a_3, \cdots a_n$ that can fulfill these conditions. Thus the arrangement which we have acquired is the main arrangement conceivable.

23. The outcome just got clearly holds great regardless of how extraordinary a worth of n might be taken.

Assuming now we assume n endlessly expanded the two bends (2) Art. 22 and $y = f(x)$ will draw ever closer to agreeing all through the entire of their parts among $x = 0$ and $x = \pi$, and subsequently the restricting structure that condition (2) Art. 22 approaches as n is endlessly expanded will address a bend totally harmonizing between the upsides of x being referred to with $y = f(x)$.

Allow us to see what restricting worth a_m approaches as n is endlessly expanded.

$$a_m = \frac{2}{n+1}\sum_{k=1}^{k=n} f(k\Delta x)\sin km\Delta x \qquad (1) \text{ Art. 22.}$$

$$= \frac{2\Delta x}{\pi}\sum_{k=1}^{k=n} f(k\Delta x)\sin km\Delta x$$

$$= \frac{2}{\pi}\left[f(\Delta x)\sin m\Delta x.\Delta x + f(2\Delta x)\sin 2m\Delta x.\Delta xand + \cdots \atop \qquad\qquad\qquad\qquad\qquad\qquad + f(n\Delta x)\sin nm\Delta x.\Delta x\right]$$

$$= \frac{2}{\pi}\left[f(\Delta x)\sin m\Delta x.\Delta x + f(2\Delta x)\sin 2m\Delta x.\Delta x + \cdots \atop \qquad\qquad\qquad\qquad\qquad + f(\pi - \Delta x)\sin m(\pi - \Delta x).\Delta x\right]$$

DEVELOPMENT IN TRIGONOMETRICAL SERIES.   30

since $\Delta x = \frac{\pi}{n+1}$.

As n is expanded endlessly Δx approaches zero as a cutoff. Consequently the restricting worth of a_m as n increments endlessly is

$$\frac{2}{\pi} \operatorname*{limit}_{\Delta x \doteq 0} \left[\begin{array}{l} f(\Delta x)\sin m\Delta x.\Delta x + f(2\Delta x)\sin 2m\Delta x.\Delta x + \cdots \\ \qquad\qquad + f(\pi - \Delta x)\sin m(\pi - \Delta x).\Delta x \end{array}\right] \quad {}^{1}$$

$$= \frac{2}{\pi} \int_0^\pi f(x)\sin mx.dx. \quad \text{[v. Int. Cal. Arts. 80, 81.]}$$

Hence and $f(x) = a_1 \sin x + a_2 \sin 2x + a_3 \sin 3x + \cdots$, and   (2)

where any coefficient a_m is given by the recipe

$$a_m = \frac{2}{\pi} \int_0^\pi f(x) \sin mx.dx, \qquad (3)$$

is a genuine improvement of $f(x)$ for all upsides of x among $x = 0$ and $x = \pi$ *provided that the series* (2) *is convergent*, for it is all things considered just that we would be able accept that the restricting worth of the second individual from (2) Art. 22 can be acquired by adding the restricting upsides of the few terms.

When $x = 0$ and when $x = \pi$ each term in the second individual from (2) is zero, and the subsequent part is zero and won't be equivalent to $f(x)$ except if $f(x)$ is itself zero when $x = 0$ and $x = \pi$; however in any event, when $f(x)$ isn't zero for $x = 0$ and $x = \pi$ the advancement given above holds great for any worth of x among nothing and π regardless of how close to it very well might be taken to both of these values.

24. Rather than really playing out the end in conditions (4) Art. 19 furthermore, getting an equation for a_m as far as n, and afterward letting n increment endlessly, we could have saved work by the accompanying strategy.

Get back to conditions (4) Art. 19 and duplicate the first by $\Delta x \sin m\Delta x$, the second by $\Delta x \sin 2m\Delta x$, etc, that is duplicate every condition by Δx times the coefficient of a_m in that situation, and afterward add the conditions.

We get as the coefficient of a_k

$$\sin k\Delta x \sin m\Delta x.\Delta x + \sin 2k\Delta x \sin 2m\Delta x.\Delta x + \cdots + \sin nk\Delta x \sin nm\Delta x.\Delta x.$$

Allow us to track down its restricting worth as n is endlessly expanded. It very well might be composed, since $(n+1)\Delta x = \pi$,

$$\operatorname*{limit}_{\Delta x \doteq 0} \left[\begin{array}{l} \sin k\Delta x \sin m\Delta x.\Delta x + \sin 2k\Delta x \sin 2m\Delta x.\Delta x + \cdots \\ \qquad\qquad + \sin k(\pi - \Delta x)\sin m(\pi - \Delta x).\Delta x \end{array}\right]$$

$$= \int_0^\pi \sin kx \sin mx.dx;$$

but and $\int_0^\pi \sin kx \sin mx.dx = \frac{1}{2}\int_0^\pi [\cos(m-k)x - \cos(m+k)x]dx$

$$= 0 \text{ if } m \text{ and } k \text{ are not equal.}$$

---

[1] We will utilize the sign $\doteq$ for *approaches*. $\Delta x \doteq 0$ is perused Δx approaches zero.

The coefficient of a_m is

$$\Delta x(\sin^2 m\Delta x + \sin^2 2m\Delta x + \sin^2 3m\Delta x + \cdots + \sin^2 nm\Delta x).$$

Its restricting worth

$$\lim_{\Delta x \doteq 0} \left[\sin^2 m\Delta x.\Delta x + \sin^2 2m\Delta x.\Delta x + \cdots + \sin^2 m(\pi - \Delta x)\Delta x \right]$$
$$= \int_0^\pi \sin^2 mx.dx = \frac{\pi}{2}.$$

The principal part is

$$f(\Delta x)\sin m\Delta x.\Delta x + f(2\Delta x)\sin 2m\Delta x.\Delta x + \cdots + f(n\Delta x)\sin mn\Delta x.\Delta x$$

also, its restricting worth is

$$\int_0^\pi f(x)\sin mx.dx.$$

Consequently the restricting structure moved toward by the last condition as n is expanded is

$$\int_0^\pi f(x)\sin mx.dx = \frac{\pi}{2}a_m.$$

Whence $and a_m = \dfrac{2}{\pi}\int_0^\pi f(x)\sin mx.dx$ as before.

This strategy is for all intents and purposes equivalent to *multiplying the equation*

$$f(x) = a_1 \sin x + a_2 \sin 2x + a_3 \sin 3x + \cdots \tag{1}$$

by $\sin mx.dx$ and incorporating the two individuals from zero to π.

It is extremely critical to understand that the short strategy for deciding any coefficient a_m of the series (1) which has recently been portrayed in the emphasized passage, is basically equivalent to that of getting a_m by genuine end from the situations (4) Art. 19, and afterward assuming n to increment endlessly, in this way making the bends (3) Art. 19 and (2) Art. 19 totally correspond between the upsides of x which are taken as the constraints of the distinct joining.

25. We see, then, at that point, that any capability of x which is single-esteemed, limited, and constant between $x = 0$ and $x = \pi$, or on the other hand if spasmodic has just limited discontinuities every one of which is gone before and prevailed by ceaseless parts, can likely be formed into a progression of the structure

$$f(x) = a_1 \sin x + a_2 \sin 2x + a_3 \sin 3x + \cdots \tag{1}$$

where
$$a_m = \frac{2}{\pi}\int_0^\pi f(x)\sin mx.dx = \frac{2}{\pi}\int_0^\pi f(\alpha)\sin m\alpha.d\alpha; \tag{2}$$

what's more, the series and the capability will be indistinguishable for all upsides of x between $x = 0$ and $x = \pi$, excluding the qualities $x = 0$ and $x = \pi$ except if the given capability is equivalent to zero for those qualities.

An intricate examination of the topic of the combination of the series (1), for which we have not space, altogether affirms the outcome figured out above[2] furthermore, shows moreover that at

---

[2] Given the capability has not a limitless number of maxima and minima in the area of a point. v. Arts. 37- - 38.

DEVELOPMENT IN TRIGONOMETRICAL SERIES.

a place of limited intermittence the series has a worth equivalent to a portion of the amount of the two qualities which the capability approaches as we approach the point being referred to from inverse sides.

The examination which we have made in the first areas lays out the way that the bend addressed by $y = f(x)$ need not follow something similar numerical regulation all through its length, yet might be comprised of parts of completely various bends. For instance, a wrecked line or a locus comprising of limited pieces of a few unique and separated straight lines can be addressed completely well by $y =$ a sine series.

26. Let us get a couple of sine improvements.

(a) Let
$$f(x) = x. \tag{1}$$

We have
$$x = a_1 \sin x + a_2 \sin 2x + a_3 \sin 3x + \cdots \tag{2}$$

where
$$a_m = \frac{2}{\pi} \text{and} \int_0^\pi x \sin mx \, dx \tag{3}$$

$$\int x \sin mx \, dx = \frac{1}{m^2}(\sin mx - mx \cos mx),$$

$$\int_0^\pi x \sin mx \, dx = -\frac{(-1)^m \pi}{m},$$

and $andx = 2\left(\dfrac{\sin x}{1} - \dfrac{\sin 2x}{2} + \dfrac{\sin 3x}{3} - \dfrac{\sin 4x}{4} + \cdots\right)$ and (4)

(b) Let $andf(x) = 1.and$ (1)

$$a_m = \frac{2}{\pi} \int_0^\pi \sin mx \, dx; \tag{2}$$

$$\int \sin mx \, dx = -\frac{\cos mx}{m},$$

$$\int_0^\pi \sin mx \, dx = \frac{1}{m}(1 - \cos m\pi) = \frac{1}{m}[1 - (-1)^m]$$

$= 0$ in the event that m is even

$= \dfrac{2}{m}$ assuming m is odd.

Hence $and1 = \dfrac{4}{\pi}\left(\dfrac{\sin x}{1} + \dfrac{\sin 3x}{3} + \dfrac{\sin 5x}{5} + \dfrac{\sin 7x}{7} + \cdots\right).and$ (3)

It is to be seen that (3) gives without a moment's delay a sine improvement for any consistent c. It is,
$$c = \frac{4c}{\pi}\left(\frac{\sin x}{1} + \frac{\sin 3x}{3} + \frac{\sin 5x}{5} + \cdots\right). \tag{4}$$

On the off chance that we substitute $x = \frac{\pi}{2}$ in (4)(a) or (3)(b) we obtain a natural outcome, in particular
$$\frac{\pi}{4} = \frac{1}{1} - \frac{1}{3} + \frac{1}{5} - \frac{1}{7} + \cdots, \tag{5}$$

an equation typically determined by subbing $x = 1$ in the power series for $\tan^{-1} x$. (v. Dif. Cal. Art. 135.)

(4)(a) doesn't hold great when $x = \pi$, and (3)(b) fizzles when $x = 0$ and when $x = \pi$, for in this large number of cases the series lessens to zero.

DEVELOPMENT IN TRIGONOMETRICAL SERIES.

(c) Let $f(x) = x$ from $x = 0$ to $x = \frac{\pi}{2}$ also, $f(x) = \pi - x$ from $x = \frac{\pi}{2}$ to $x = \pi$. That is, let $y = f(x)$ address the messed up line in the figure.

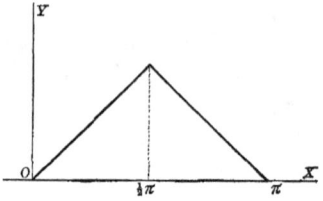

As the numerical articulation for $f(x)$ is different in the two parts of the bend we should separate

$$\int_0^\pi f(x) \sin mx.dx \quad \text{into} \quad \int_0^{\frac{\pi}{2}} f(x) \sin mx.dx + \int_{\frac{\pi}{2}}^\pi f(x) \sin mx.dx.$$

We have, then,

$$a_m = \frac{2}{\pi} \int_0^{\frac{\pi}{2}} x \sin mx.dx + \frac{2}{\pi} \int_{\frac{\pi}{2}}^\pi (\pi - x) \sin mx.dx \tag{1}$$

$$= \frac{4}{m^2 \pi} \sin m \frac{\pi}{2}.$$

But
$$\begin{aligned}
\sin m\frac{\pi}{2} &= 1 \quad \text{if} \quad m = 1 \quad \text{or} \quad 4k + 1 \\
&= 0 \quad " \quad m = 2 \quad " \quad 4k + 2 \\
&= -1 \quad " \quad m = 3 \quad " \quad 4k + 3 \\
&= 0 \quad " \quad m = 4 \quad " \quad 4k.
\end{aligned}$$

Thus if $y = f(x)$ addresses our wrecked line,

$$f(x) = \frac{4}{\pi} \left(\frac{\sin x}{1^2} - \frac{\sin 3x}{3^2} + \frac{\sin 5x}{5^2} - \frac{\sin 7x}{7^2} + \cdots \right). \tag{2}$$

When $x = \frac{\pi}{2}$ $f(x) = \frac{\pi}{2}$ and we have

$$\frac{\pi^2}{8} = \frac{1}{1^2} + \frac{1}{3^2} + \frac{1}{5^2} + \frac{1}{7^2} + \cdots \tag{3}$$

(d) As a situation where the capability has a limited intermittence, let

$$f(x) = 1 \quad \text{from} \quad x = 0 \quad \text{to} \quad x = \frac{\pi}{2} \quad \text{and}$$
$$f(x) = 0 \quad " \quad x = \frac{\pi}{2} \quad " \quad x = \pi.$$

$y = f(x)$ will for this situation address the locus in the figure.

As in the past

$$\int_0^\pi f(x) \sin mx.dx\, and = \int_0^{\frac{\pi}{2}} f(x) \sin mx.dx$$

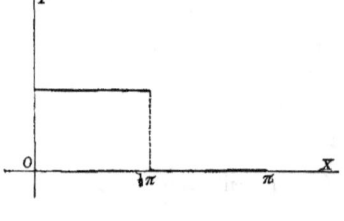

$$+ \int_{\frac{\pi}{2}}^\pi f(x) \sin mx.dx.$$

$$a_m = \frac{2}{\pi} \int_0^{\frac{\pi}{2}} \sin mx.dx + \frac{2}{\pi} \int_{\frac{\pi}{2}}^\pi 0. \sin mx.dx. \tag{1}$$

$$a_m = \frac{2}{\pi} \int_0^{\frac{\pi}{2}} \sin mx.dx = \frac{2}{\pi} \frac{1}{m} \left(1 - \cos m\frac{\pi}{2} \right).$$

DEVELOPMENT IN TRIGONOMETRICAL SERIES.

But
$$\cos m\frac{\pi}{2} = 0 \text{ if } m = 1 \text{ or } 4k+1$$
$$= -1 \text{ '' } m = 2 \text{ '' } 4k+2$$
$$= 0 \text{ '' } m = 3 \text{ '' } 4k+3$$
$$= 1 \text{ '' } m = 4 \text{ '' } 4k.$$

Consequently
$$f(x) = \frac{2}{\pi}\left(\frac{\sin x}{1} + \frac{2\sin 2x}{2} + \frac{\sin 3x}{3} + \frac{\sin 5x}{5} + \frac{2\sin 6x}{6} + \frac{\sin 7x}{7} + \cdots\right). \qquad (2)$$

If $x = \frac{\pi}{2}$ the second individual from (2) decreases to $\frac{1}{2}$, for
$$\frac{2}{\pi}\left(\frac{1}{1} - \frac{1}{3} + \frac{1}{5} - \frac{1}{7} + \cdots\right) = \frac{1}{2} \quad \text{by (5) (b);}$$

what's more, we see that the series addresses the capability totally for all upsides of x among $x = 0$ and $x = \pi$ with the exception of $x = \frac{\pi}{2}$ and there it has a esteem which is the mean of the qualities moved toward by the capability as x approaches $\frac{\pi}{2}$ from inverse sides.

EXAMPLES.

Acquire the accompanying turns of events:- - -

(1) $\quad x^2 = \dfrac{2}{\pi}\left[\left(\dfrac{\pi^2}{1} - \dfrac{4}{1^3}\right)\sin x - \dfrac{\pi^2}{2}\sin 2x + \left(\dfrac{\pi^2}{3} - \dfrac{4}{3^3}\right)\sin 3x - \dfrac{\pi^2}{4}\sin 4x\right.$
$\left. + \left(\dfrac{\pi^2}{5} - \dfrac{4}{5^3}\right)\sin 5x - \cdots\right].$

(2) $\quad x^3 = \dfrac{2}{\pi}\left[\left(\dfrac{\pi^3}{1} - \dfrac{6\pi}{1^3}\right)\sin x - \left(\dfrac{\pi^3}{2} - \dfrac{6\pi}{2^3}\right)\sin 2x + \left(\dfrac{\pi^3}{3} - \dfrac{6\pi}{3^3}\right)\sin 3x\right.$
$\left. - \left(\dfrac{\pi^3}{4} - \dfrac{6\pi}{4^3}\right)\sin 4x + \cdots\right].$

(3) $\quad f(x) = \dfrac{2}{\pi}\left[\dfrac{\sin x}{1^2} + \dfrac{\pi}{2^2}\sin 2x - \dfrac{\sin 3x}{3^2} - \dfrac{2\pi}{4^2}\sin 4x + \dfrac{\sin 5x}{5^2}\right.$
$\left. + \dfrac{3\pi}{6^2}\sin 6x - \cdots\right],$

in the event that $f(x) = x$ from $x = 0$ to $x = \frac{\pi}{2}$ and $f(x) = 0$ from $x = \frac{\pi}{2}$ to $x = \pi$.

(4) $\quad \sin \mu x = \dfrac{2}{\pi}\sin \mu\pi \left[\dfrac{\sin x}{1^2 - \mu^2} - \dfrac{2\sin 2x}{2^2 - \mu^2} + \dfrac{3\sin 3x}{3^2 - \mu^2} - \dfrac{4\sin 4x}{4^2 - \mu^2} + \cdots\right]$

if μ is a fraction.

(5) $\quad e^x = \dfrac{2}{\pi}\left[\dfrac{1}{2}(1 + e^\pi)\sin x + \dfrac{2}{5}(1 - e^\pi)\sin 2x + \dfrac{3}{10}(1 + e^\pi)\sin 3x\right.$
$\left. + \dfrac{4}{17}(1 - e^\pi)\sin 4x + \cdots\right].$

(6) $\quad \sinh x = \dfrac{2\sinh \pi}{\pi}\left[\dfrac{1}{2}\sin x - \dfrac{2}{5}\sin 2x + \dfrac{3}{10}\sin 3x - \dfrac{4}{17}\sin 4x + \cdots\right].$

(7) $\quad \cosh x = \dfrac{2}{\pi}\left[\dfrac{1}{2}(1 + \cosh \pi)\sin x + \dfrac{2}{5}(1 - \cosh \pi)\sin 2x\right.$
$\left. + \dfrac{3}{10}(1 + \cosh \pi)\sin 3x + \cdots\right].$

DEVELOPMENT IN TRIGONOMETRICAL SERIES.

27. Let us presently attempt to foster a given capability of x in a progression of cosines.

As before assume that $f(x)$ has a solitary incentive for each worth of x between $x = 0$ and $x = \pi$, that it doesn't become boundless among $x = 0$ and $x = \pi$, and that assuming irregular it has just limited discontinuities.

Expect
$$f(x) = b_0 + b_1 \cos x + b_2 \cos 2x + b_3 \cos 3x + \cdots \qquad (1)$$

To decide any coefficient b_m duplicate (1) by $\cos mx.dx$ and coordinate each term from 0 to π.

$$\int_0^\pi b_0 \cos mx.dx = 0.$$

$$\int_0^\pi b_k \cos kx \cos mx.dx = \frac{b_k}{2} \int_0^\pi [\cos(m-k)x + \cos(m+k)x]dx$$

$$= 0 \text{ if } m \text{ and } k \text{ are not equal.}$$

$$\int b_m \cos^2 mx.dx = \frac{b_m}{2m}(mx + \cos mx \sin mx),$$

$$\int_0^\pi b_m \cos^2 mx.dx = \frac{\pi}{2}b_m, \qquad \text{if } m \text{ isn't zero.}$$

Hence
$$b_m = \frac{2}{\pi} \int_0^\pi f(x) \cos mx.dx = \frac{2}{\pi} \int_0^\pi f(\alpha) \cos m\alpha.d\alpha, \text{ and} \qquad (2)$$

assuming m isn't zero.

To get b_0 increase (1) by dx and incorporate from zero to π.

$$\int_0^\pi b_0 dx = b_0 \pi,$$

$$\int_0^\pi b_k \cos kx.dx = 0.$$

Hence and $b_0 = \frac{1}{\pi}$
$$\int_0^\pi f(x)dx = \frac{1}{\pi} \int_0^\pi f(\alpha)d\alpha, \text{ and} \qquad (3)$$

which is simply around 50fill in for m.

To save a different equation (1) is generally composed
$$f(x) = \tfrac{1}{2}b_0 + b_1 \cos x + b_2 \cos 2x + b_3 \cos 3x + \cdots \qquad (4)$$

and afterward the recipe
$$b_m = \frac{2}{\pi} \int_0^\pi f(x) \cos mx.dx = \frac{2}{\pi} \int_0^\pi f(\alpha) \cos m\alpha.d\alpha \qquad (2)$$

will give b_0 as well as different coefficients.

It is vital to see plainly that what we have quite recently finished in deciding the coefficients of (1) is comparable to taking $n+1$ terms of (4), subbing in

$$y = \tfrac{1}{2}b_0 + b_1 \cos x + b_2 \cos 2x + \cdots + b_n \cos nx \qquad (5)$$

thus the coÃ¶rdinates of the $n+1$ points of the bend

$$y = f(x)$$

DEVELOPMENT IN TRIGONOMETRICAL SERIES.

whose projections on the hub of X are equidistant, deciding $b_0, b_1, b_2, \cdots b_n$ by disposal from the $n+1$ coming about conditions, and afterward taking the restricting values they approach as n is endlessly increased. (v. Art. 24.)

In the event that $\Delta x = \frac{\pi}{n+1}$ the abscissas of the $n+1$ focuses utilized are $0, \Delta x, 2\Delta x, 3\Delta x, \cdots n\Delta x$, so we ought to anticipate that our cosine improvement should hold for $x = 0$ as well concerning upsides of x among nothing and π.

28. Let us take a couple of models:

(a) Let and $f(x) = x$. and (1)

$$b_0 = \frac{2}{\pi} \int_0^\pi x\, dx = \frac{2}{\pi} \frac{\pi^2}{2} = \pi.$$

$$b_m = \frac{2}{\pi} \int_0^\pi x \cos mx \, dx = \frac{2}{m^2 \pi}(\cos m\pi - 1) = \frac{2}{m^2 \pi}[(-1)^m - 1].$$

Hence and x
$$= \frac{\pi}{2} - \frac{4}{\pi}\left(\cos x + \frac{\cos 3x}{3^2} + \frac{\cos 5x}{5^2} + \frac{\cos 7x}{7^2} + \cdots\right). \tag{2}$$

(2) holds great not just for upsides of x among nothing and π except for $x = 0$ also, $x = \pi$ too, since for these qualities we have

$$0 = \frac{\pi}{2} - \frac{4}{\pi}\left(1 + \frac{1}{3^2} + \frac{1}{5^2} + \frac{1}{7^2} + \cdots\right) \tag{3}$$

and
$$\pi = \frac{\pi}{2} + \frac{4}{\pi}\left(1 + \frac{1}{3^2} + \frac{1}{5^2} + \frac{1}{7^2} + \cdots\right) \tag{4}$$

which are valid by Art. 26 (c)(3).

(b) Let $\hspace{3em} f(x) = x \sin x.$ \hfill (1)

$$b_0 = \frac{2}{\pi} \int_0^\pi x \sin x \, dx = \frac{2}{\pi}\pi = 2,$$

$$b_1 = \frac{2}{\pi} \int_0^\pi x \sin x \cos x \, dx = \frac{1}{\pi} \int_0^\pi x \sin 2x \, dx = -\frac{1}{2},$$

$$b_m = \frac{2}{\pi} \int_0^\pi x \sin x \cos mx \, dx = \frac{1}{\pi} \int_0^\pi [x \sin(m+1)x - x \sin(m-1)x] dx$$

$$= \frac{2}{(m-1)(m+1)} \text{ if } m \text{ is odd}$$

$$= -\frac{2}{(m-1)(m+1)} \text{ if } m \text{ is even.}$$

Subsequently

$$x \sin x = 1 - \frac{\cos x}{2} - \frac{2 \cos 2x}{1.3} + \frac{2 \cos 3x}{2.4} - \frac{2 \cos 4x}{3.5} + \cdots \tag{2}$$

If $x = \frac{\pi}{2}$ we have

$$\frac{\pi}{4} = \frac{1}{2} + \frac{1}{1.3} - \frac{1}{3.5} + \frac{1}{5.7} - \cdots. \tag{3}$$

DEVELOPMENT IN TRIGONOMETRICAL SERIES. 37

EXAMPLES.

Acquire the accompanying turns of events:

(1) $f(x) = \dfrac{\pi}{4} - \dfrac{2}{\pi}\left[\dfrac{\cos 2x}{1^2} + \dfrac{\cos 6x}{3^2} + \dfrac{\cos 10x}{5^2} + \dfrac{\cos 14x}{7^2} + \cdots\right]$

if $f(x) = x$ from $x = 0$ to $x = \dfrac{\pi}{2}$ and $f(x) = \pi - x$ from $x = \dfrac{\pi}{2}$ to $x = \pi$.

(2) $f(x) = \dfrac{1}{2} + \dfrac{2}{\pi}\left[\dfrac{\cos x}{1} - \dfrac{\cos 3x}{3} + \dfrac{\cos 5x}{5} - \dfrac{\cos 7x}{7} + \cdots\right],$

if $f(x) = 1$ from $x = 0$ to $x = \dfrac{\pi}{2}$ and $f(x) = 0$ from $x = \dfrac{\pi}{2}$ to $x = \pi$.

(3) $x^2 = \dfrac{\pi^2}{3} - 4\left[\dfrac{\cos x}{1^2} - \dfrac{\cos 2x}{2^2} + \dfrac{\cos 3x}{3^2} - \dfrac{\cos 4x}{4^2} + \cdots\right],$

(4) $x^3 = \dfrac{\pi^3}{4} - \dfrac{6}{\pi}\left[\left(\dfrac{\pi^2}{1^2} - \dfrac{4}{1^4}\right)\cos x - \dfrac{\pi^2}{2^2}\cos 2x + \left(\dfrac{\pi^2}{3^2} - \dfrac{4}{3^4}\right)\cos 3x\right.$

$\left. - \dfrac{\pi^2}{4^2}\cos 4x + \left(\dfrac{\pi^2}{5^2} - \dfrac{4}{5^4}\right)\cos 5x - \cdots\right],$

(5) $f(x) = \dfrac{\pi}{8} + \dfrac{2}{\pi}\left[\left(\dfrac{\pi}{2} - 1\right)\cos x - \dfrac{2}{2^2}\cos 2x - \dfrac{1}{3^2}\left(\dfrac{3\pi}{2} + 1\right)\cos 3x\right.$

$\left. + \dfrac{1}{5^2}\left(\dfrac{5\pi}{2} - 1\right)\cos 5x - \dfrac{2}{6^2}\cos 6x - \cdots\right],$

if $f(x) = x$ from $x = 0$ to $x = \dfrac{\pi}{2}$ and $f(x) = 0$ from $x = \dfrac{\pi}{2}$ to $x = \pi$.

(6) $e^x = \dfrac{2}{\pi}\left[\dfrac{1}{2}(e^\pi - 1) - \dfrac{1}{1+1^2}(e^\pi + 1)\cos x + \dfrac{1}{1+2^2}(e^\pi - 1)\cos 2x\right.$

$\left. - \dfrac{1}{1+3^2}(e^\pi + 1)\cos 3x + \cdots\right],$

(7) $\cosh x = \dfrac{2\sinh \pi}{\pi}\left[\dfrac{1}{2} - \dfrac{1}{2}\cos x + \dfrac{1}{5}\cos 2x - \dfrac{1}{10}\cos 3x\right.$

$\left. + \dfrac{1}{17}\cos 4x - \cdots\right],$

(8) $\sinh x = \dfrac{2}{\pi}\left[\dfrac{1}{2}(\cosh \pi - 1) - \dfrac{1}{2}(\cosh \pi + 1)\cos x\right.$

$\left. + \dfrac{1}{5}(\cosh \pi - 1)\cos 2x - \dfrac{1}{10}(\cosh \pi + 1)\cos 3x + \cdots\right],$

(9) $\cos \mu x = \dfrac{2\mu \sin \mu \pi}{\pi}\left[\dfrac{1}{2\mu^2} - \dfrac{\cos x}{\mu^2 - 1^2} + \dfrac{\cos 2x}{\mu^2 - 2^2} - \dfrac{\cos 3x}{\mu^2 - 3^2}\right.$

$\left. + \dfrac{\cos 4x}{\mu^2 - 4^2} - \cdots\right],$

if μ is a fraction.

29. Albeit any capability can be communicated both as a sine series and as a cosine series, and the capability and either series will be equivalent for all upsides of x among nothing and π, there is a settled contrast in the two series for other upsides of x.

Both series are occasional elements of x having the period 2π. In the event that, we give y equivalent the series access question and build the piece of the comparing bend which lies between the qualities $x = -\pi$ and $x = \pi$ the entirety bend will comprise of redundancies of this piece.

Since $\sin mx = -\sin(-mx)$ the ordinate relating to any worth of x between $-\pi$ and zero in the sine bend; will be the negative of the ordinate relating to a similar worth of x with the positive sign. At the end of the day the bend

$$y = a_1 \sin x + a_2 \sin 2x + a_3 \sin 3x + \cdots \tag{1}$$

is balanced as for the beginning.

Since $\cos mx = \cos(-mx)$ the ordinate comparing to any worth of x between $-\pi$ and zero in the cosine bend will be equivalent to the ordinate having a place with the comparing positive worth of x. At the end of the day the bend

$$y = \tfrac{1}{2}b_0 + b_1 \cos x + b_2 \cos 2x + b_3 \cos 3x + \cdots \tag{2}$$

is even concerning the hub of Y.

On the off chance that, $f(x) = -f(-x)$, that is assuming $f(x)$ is a *odd* capability the sine series comparing to it will be equivalent to it for all upsides of x between $-\pi$ and π, aside from maybe for the worth $x = 0$ for which the series will essentially be zero.

If $f(x) = f(-x)$, that is assuming $f(x)$ is a *even* capability the cosine series relating to it will be equivalent to it for all upsides of x between $x = -\pi$ and $x = \pi$, not with the exception of the worth $x = 0$.

To act as an illustration of the contrast between the sine and cosine advancements of a similar capability let us take the series for x.

$$y = 2\left[\sin x - \frac{\sin 2x}{2} + \frac{\sin 3x}{3} - \frac{\sin 4x}{4} + \cdots\right] \tag{3}$$

$$y = \frac{\pi}{2} - \frac{4}{\pi}\left[\cos x + \frac{\cos 3x}{3^2} + \frac{\cos 5x}{5^2} + \frac{\cos 7x}{7^2} + \cdots\right] \tag{4}$$

[v. Art. 26(a) and Art. 28(a)]. (3) addresses the bend

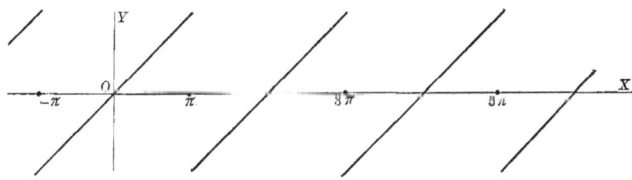

DEVELOPMENT IN TRIGONOMETRICAL SERIES.

and (4) the bend

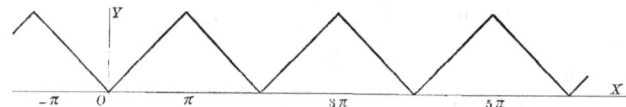

Both concur with $y = x$ from $x = 0$ to $x = \pi$, (3) harmonizes with $y = x$ from $x = -\pi$ to $x = \pi$, and neither concurs with $y = x$ for upsides of x not exactly $-\pi$ or more prominent than π. Besides (3), notwithstanding the ceaseless segments of the locus addressed in the figure, gives the separated focuses $(-\pi, 0)$ $(\pi, 0)$ $(3\pi, 0)$ &c.

30. We have seen that assuming $f(x)$ is a *odd* capability its improvement in sine series holds for all upsides of x from $-\pi$ to π, as does the improvement of $f(x)$ in cosine series on the off chance that $f(x)$ is a *even* capability.

Subsequently the improvements of Art. 26(*a*), Art. 26 Exs. (2), (4), (6); Art. 28(*b*), Art. 28 Exs. (3), (7), (9) are legitimate for all upsides of x between $-\pi$ and π.

Any capability of x can be formed into a Geometrical series to which it is equivalent for all upsides of x between $-\pi$ and π.

Let $f(x)$ be the given capability of x. It very well may be communicated as the amount of an indeed, even capability of x and an odd capability of x by the accompanying gadget.

$$f(x) = \frac{f(x) + f(-x)}{2} + \frac{f(x) - f(-x)}{2} \qquad (1)$$

indistinguishably; yet $\frac{f(x)+f(-x)}{2}$ isn't changed by turning around the indication of x and is thusly a *even* capability of x; and when we switch the indication of x, $\frac{f(x)-f(-x)}{2}$ is impacted exclusively to the degree of having its sign turned around and is thus a *odd* capability of x.

Hence for all upsides of x between $-\pi$ and π

$$\frac{f(x) + f(-x)}{2} = \frac{1}{2}b_0 + b_1 \cos x + b_2 \cos 2x + b_3 \cos 3x + \cdots$$

where
$$b_m = \frac{2}{\pi} \int_0^\pi \frac{f(x) + f(-x)}{2} \cos mx \, dx;$$
and

$$\frac{f(x) - f(-x)}{2} = a_1 \sin x + a_2 \sin 2x + a_3 \sin 3x + \cdots$$

where
$$a_m = \frac{2}{\pi} \int_0^\pi \frac{f(x) - f(-x)}{2} \sin mx \, dx.$$

DEVELOPMENT IN TRIGONOMETRICAL SERIES.

b_m and a_m can be improved on a bit.

$$b_m = \frac{2}{\pi} \int_0^\pi \frac{f(x) + f(-x)}{2} \cos mx . dx$$

$$= \frac{1}{\pi} \left[\int_0^\pi f(x) \cos mx . dx + \int_0^\pi f(-x) \cos mx . dx \right],$$

however, on the off chance that we supplant x by $-x$, we get

$$\int_0^\pi f(-x) \cos mx . dx = -\int_0^{-\pi} f(x) \cos mx . dx = \int_{-\pi}^0 f(x) \cos mx . dx,$$

and we have
$$b_m = \frac{1}{\pi} \int_{-\pi}^\pi f(x) \cos mx . dx.$$

Similarly we can decrease the worth of a_m to

$$\frac{1}{\pi} \int_{-\pi}^\pi f(x) \sin mx . dx.$$

Subsequently

$$\left\{ \begin{array}{l} f(x) = \frac{1}{2} b_0 + b_1 \cos x + b_2 \cos 2x + b_3 \cos 3x + \cdots \\ \qquad\qquad + a_1 \sin x + a_2 \sin 2x + a_3 \sin 3x + \cdots \end{array} \right\} \qquad (2)$$

where
$$b_m = \frac{1}{\pi} \int_{-\pi}^\pi f(x) \cos mx . dx = \frac{1}{\pi} \int_{-\pi}^\pi f(\alpha) \cos m\alpha . d\alpha. \qquad (3)$$

and
$$a_m = \frac{1}{\pi} \int_{-\pi}^\pi f(x) \sin mx . dx = \frac{1}{\pi} \int_{-\pi}^\pi f(\alpha) \sin m\alpha . d\alpha. \qquad (4)$$

also, this advancement holds for all upsides of x between $-\pi$ and π.

The second individual from (2) is known as a Fourier's Series.

EXAMPLES.

1. Get the accompanying turns of events, which are all legitimate from $x = -\pi$ to $x = \pi$:- - -

(1) $$e^x = \frac{2 \sinh \pi}{\pi} \left[\frac{1}{2} - \frac{1}{2} \cos x + \frac{1}{5} \cos 2x - \frac{1}{10} \cos 3x + \frac{1}{17} \cos 4x + \cdots \right]$$
$$+ \frac{2 \sinh \pi}{\pi} \left[\frac{1}{2} \sin x - \frac{2}{5} \sin 2x + \frac{3}{10} \sin 3x - \frac{4}{17} \sin 4x + \cdots \right].$$

(2) $$f(x) = \frac{\pi}{4} - \frac{2}{\pi} \left[\cos x + \frac{\cos 3x}{3^2} + \frac{\cos 5x}{5^2} + \frac{\cos 7x}{7^2} + \cdots \right]$$
$$+ \frac{\sin x}{1} - \frac{\sin 2x}{2} + \frac{\sin 3x}{3} - \frac{\sin 4x}{4} + \cdots ,$$

where $f(x) = 0$ from $x = -\pi$ to $x = 0$ and $f(x) = x$ from $x = 0$ to $x = \pi$.

(3) $$f(x) = -\frac{3\pi}{16} + \frac{1}{\pi} \left[\frac{1}{1^2} \cos x + \frac{2}{2^2} \cos 2x + \frac{1}{3^2} \cos 3x + \frac{1}{5^2} \cos 5x \right.$$
$$\left. + \frac{2}{6^2} \cos 6x + \cdots \right]$$
$$+ \frac{1}{\pi} \left[\left(\frac{3\pi}{2} - 1 \right) \sin x - \frac{3\pi}{4} \sin 2x + \left(\frac{3\pi}{6} + \frac{1}{3^2} \right) \sin 3x \right.$$
$$\left. - \frac{3\pi}{8} \sin 4x + \left(\frac{3\pi}{10} - \frac{1}{5^2} \right) \sin 5x - \cdots \right],$$

where $\quad f(x) = x$ from $x = -\pi$ to $x = 0$, $\quad f(x) = 0$ from $x = 0$ to $x = \dfrac{\pi}{2}$,

and $\quad f(x) = x - \dfrac{\pi}{2}$ from $x = \dfrac{\pi}{2}$ to $x = \pi$.

2. Show that formula (2) Art. 30 can be composed

$$f(x) = \frac{1}{2}c_0 \cos\beta_0 + c_1 \cos(x - \beta_1) + c_2 \cos(2x - \beta_2) + c_3 \cos(3x - \beta_3) + \cdots$$

where $\quad c_m = (a_m^2 + b_m^2)^{\frac{1}{2}}$ and $\beta_m = \tan^{-1}\dfrac{a_m}{b_m}$.

3. Show that formula (2) Art. 30 can be composed

$$f(x) = \frac{1}{2}c_0 \sin\beta_0 + c_1 \sin(x + \beta_1) + c_2 \sin(2x + \beta_2) + c_3 \sin(3x + \beta_3) + \cdots$$

where $\quad c_m = (a_m^2 + b_m^2)^{\frac{1}{2}}$ and $\beta_m = \tan^{-1}\dfrac{b_m}{a_m}$.

31. In fostering an element of x into a Mathematical series it is frequently badly designed to be held inside the limited limits $x = -\pi$ and $x = \pi$. Allow us to check whether we can't broaden them.

Allow it to be expected to foster a component of x into a Mathematical series which will be equivalent to $f(x)$ for all upsides of x between $x = -c$ and $x = c$.

Present another variable
$$z = \frac{\pi}{c}x,$$
which is equivalent to $-\pi$ when $x = -c$ and to π when $x = c$.

$f(x) = f\left(\dfrac{c}{\pi}z\right)$ can be created as far as z by Art. 30 (2), (3), and (4).

We have
$$\left. f\left(\frac{c}{\pi}z\right) = \frac{1}{2}b_0 + b_1 \cos z + b_2 \cos 2z + b_3 \cos 3z + \cdots \atop + a_1 \sin z + a_2 \sin 2z + a_3 \sin 3z + \cdots \right\} \quad (1)$$

where $\quad b_m = \dfrac{1}{\pi}\displaystyle\int_{-\pi}^{\pi} f\left(\dfrac{c}{\pi}z\right) \cos mz\, dz.$ \hfill (2)

and $\quad a_m = \dfrac{1}{\pi}\displaystyle\int_{-\pi}^{\pi} f\left(\dfrac{c}{\pi}z\right) \sin mz\, dz.$ and \hfill (3)

furthermore (1) holds great from $z = -\pi$ to $z = \pi$.

Supplant z by its worth as far as x and (1) becomes

$$\left. f(x) = \frac{1}{2}b_0 + b_1 \cos\frac{\pi x}{c} + b_2 \cos\frac{2\pi x}{c} + b_3 \cos\frac{3\pi x}{c} + \cdots \atop + a_1 \sin\frac{\pi x}{c} + a_2 \sin\frac{2\pi x}{c} + a_3 \sin\frac{3\pi x}{c} + \cdots \right\} \quad (4)$$

The coefficients in (4) are equivalent to in (1), and (4) holds great from $x = -c$ to $x = c$.

Recipes (2) and (3) can be placed into more advantageous shape.

$$b_m = \frac{1}{\pi}\int_{-\pi}^{\pi} f\left(\frac{c}{\pi}z\right) \cos mz\, dz = \frac{1}{\pi}\int_{-c}^{c} f(x) \cos\frac{m\pi x}{c}\frac{\pi}{c}dx$$

or $\quad b_m = \dfrac{1}{c}\displaystyle\int_{-c}^{c} f(x) \cos\dfrac{m\pi x}{c}dx = \dfrac{1}{c}\displaystyle\int_{-c}^{c} f(\lambda) \cos\dfrac{m\pi\lambda}{c}d\lambda.$ \hfill (5)

DEVELOPMENT IN TRIGONOMETRICAL SERIES.

In this way we can change (3) into

$$a_m = \frac{1}{c}\int_{-c}^{c} f(x)\sin\frac{m\pi x}{c}dx = \frac{1}{c}\int_{-c}^{c} f(\lambda)\sin\frac{m\pi\lambda}{c}d\lambda. \tag{6}$$

By treating in like design equations (1) and (2) Art. 25 and recipes (4) what's more (2) Art. 27 we get

$$f(x) = a_1\sin\frac{\pi x}{c} + a_2\sin\frac{2\pi x}{c} + a_3\sin\frac{3\pi x}{c} + \cdots \tag{7}$$

where
$$a_m = \frac{2}{c}\int_0^c f(x)\sin\frac{m\pi x}{c}dx = \frac{2}{c}\int_0^c f(\lambda)\sin\frac{m\pi\lambda}{c}d\lambda. \tag{8}$$

and
$$f(x) = \frac{1}{2}b_0 + b_1\cos\frac{\pi x}{c} + b_2\cos\frac{2\pi x}{c} + b_3\cos\frac{3\pi x}{c} + \cdots \tag{9}$$

where
$$b_m = \frac{2}{c}\int_0^c f(x)\cos\frac{m\pi x}{c}dx = \frac{2}{c}\int_0^c f(\lambda)\cos\frac{m\pi\lambda}{c}d\lambda. \tag{10}$$

what's more (7) and (9) hold great from $x = 0$ to $x = c$.

EXAMPLES.

1. Get the accompanying turns of events:

(1) $$1 = \frac{4}{\pi}\left[\sin\frac{\pi x}{c} + \frac{1}{3}\sin\frac{3\pi x}{c} + \frac{1}{5}\sin\frac{5\pi x}{c} + \cdots\right]$$
from $x = 0$ to $x = c$.

(2) $$x = \frac{2c}{\pi}\left[\sin\frac{\pi x}{c} - \frac{1}{2}\sin\frac{2\pi x}{c} + \frac{1}{3}\sin\frac{3\pi x}{c} - \frac{1}{4}\sin\frac{4\pi x}{c} + \cdots\right]$$
from $x = -c$ to $x = c$.

$$x = \frac{c}{2} - \frac{4c}{\pi^2}\left[\cos\frac{\pi x}{c} + \frac{1}{3^2}\cos\frac{3\pi x}{c} + \frac{1}{5^2}\cos\frac{5\pi x}{c} + \frac{1}{7^2}\cos\frac{7\pi x}{c} + \cdots\right]$$
from $x = -c$ to $x = c$.

(3) $$x^2 = \frac{2c^2}{\pi^3}\left[\left(\frac{\pi^2}{1} - \frac{4}{1^3}\right)\sin\frac{\pi x}{c} - \frac{\pi^2}{2}\sin\frac{2\pi x}{c} + \left(\frac{\pi^2}{3} - \frac{4}{3^2}\right)\sin\frac{3\pi x}{c}\right.$$
$$\left. - \frac{\pi^2}{4}\sin\frac{4\pi x}{c} + \left(\frac{\pi^2}{5} - \frac{4}{5^3}\right)\sin\frac{5\pi x}{c} + \cdots\right]$$
from $x = 0$ to $x = c$.

$$x^2 = \frac{c^2}{3} - \frac{4c^2}{\pi^2}\left[\cos\frac{\pi x}{c} - \frac{1}{2^2}\cos\frac{2\pi x}{c} + \frac{1}{3^2}\cos\frac{3\pi x}{c} - \frac{1}{4^2}\cos\frac{4\pi x}{c} + \cdots\right]$$
from $x = -c$ to $x = c$.

(4) $$e^x = 2\pi\left[\frac{1+e^c}{c^2+\pi^2}\sin\frac{\pi x}{c} + \frac{2(1-e^c)}{c^2+4\pi^2}\sin\frac{2\pi x}{c}\right.$$
$$\left. + \frac{3(1+e^c)}{c^2+9\pi^2}\sin\frac{3\pi x}{c} + \frac{4(1-e^c)}{c^2+16\pi^2}\sin\frac{4\pi x}{c} + \cdots\right],$$

$$e^x = 2c\left[\frac{1}{2}\frac{e^c-1}{c^2} - \frac{e^c+1}{c^2+\pi^2}\cos\frac{\pi x}{c} + \frac{e^c-1}{c^2+4\pi^2}\cos\frac{2\pi x}{c}\right.$$
$$\left. - \frac{e^c+1}{c^2+9\pi^2}\cos\frac{3\pi x}{c} + \cdots\right]$$
from $x = 0$ to $x = c$.

and $f(x) = \dfrac{4c}{\pi^2}\left[\sin\dfrac{\pi x}{c} - \dfrac{1}{3^2}\sin\dfrac{3\pi x}{c} + \dfrac{1}{5^2}\sin\dfrac{5\pi x}{c} + \cdots\right]$

from $x=0$ to $x=c$,

where $f(x)=x$ from $x=0$ to $x=\tfrac{c}{2}$ and $f(x)=c-x$ from $x=\tfrac{c}{2}$ to $x=c$.

2. Show that equation (4) Art. 31 can be composed

$$f(x) = \tfrac{1}{2}c_0 \cos\beta_0 + c_1\cos\left(\dfrac{\pi x}{c}-\beta_1\right) + c_2\cos\left(\dfrac{2\pi x}{c}-\beta_2\right)$$
$$+ c_3\cos\left(\dfrac{3\pi x}{c}-\beta_3\right) + \cdots$$

where $\qquad c_m = (a_m^2+b_m^2)^{\frac{1}{2}} \quad \text{and} \quad \beta_m = \tan^{-1}\dfrac{a_m}{b_m}.$

3. Show that equation (4) Art. 31 can be composed

$$f(x) = \tfrac{1}{2}c_0 \sin\beta_0 + c_1\sin\left(\dfrac{\pi x}{c}+\beta_1\right) + c_2\sin\left(\dfrac{2\pi x}{c}+\beta_2\right)$$
$$+ c_3\sin\left(\dfrac{3\pi x}{c}+\beta_3\right) + \cdots$$

where $\qquad c_m = (a_m^2+b_m^2)^{\frac{1}{2}} \quad \text{and} \quad \beta_m = \tan^{-1}\dfrac{a_m}{b_m}.$

32. In the recipes of Art. 31 c may have as extraordinary a worth however we see fit, so we can get a Geometrical Series for $f(x)$ that will address the given capability through as extraordinary a span as we might decide to take. If, then, at that point, we can get the restricting structure moved toward by the series (4) Art. 31 as c is endlessly expanded the articulation being referred to should be equivalent to the given capability of x for all upsides of x. Condition (4) Art. 31 can be composed as follows on the off chance that we supplant $b_0, b_1, b_2, \cdots a_1, a_2, \cdots$ by their qualities surrendered Art. 31 (5) and (6).

$$f(x) = \dfrac{1}{c}\left[\dfrac{1}{2}\int_{-c}^{c} f(\lambda)\,d\lambda\right.$$
$$+ \int_{-c}^{c} f(\lambda)\cos\dfrac{\pi\lambda}{c}\cos\dfrac{\pi x}{c}d\lambda + \int_{-c}^{c} f(\lambda)\cos\dfrac{2\pi\lambda}{c}\cos\dfrac{2\pi x}{c}d\lambda + \cdots$$
$$\left.+ \int_{-c}^{c} f(\lambda)\sin\dfrac{\pi\lambda}{c}\sin\dfrac{\pi x}{c}d\lambda + \int_{-c}^{c} f(\lambda)\sin\dfrac{2\pi\lambda}{c}\sin\dfrac{2\pi x}{c}d\lambda + \cdots\right]$$
$$= \dfrac{1}{c}\int_{-c}^{c} f(\lambda)\,d\lambda\left[\dfrac{1}{2} + \cos\dfrac{\pi\lambda}{c}\cos\dfrac{\pi x}{c} + \sin\dfrac{\pi\lambda}{c}\sin\dfrac{\pi x}{c}\right.$$
$$\left.+ \cos\dfrac{2\pi\lambda}{c}\cos\dfrac{2\pi x}{c} + \sin\dfrac{2\pi\lambda}{c}\sin\dfrac{2\pi x}{c} + \cdots\right]$$

$$f(x) = \dfrac{1}{c}\int_{-c}^{c} f(\lambda)\,d\lambda\left[\dfrac{1}{2} + \cos\dfrac{\pi}{c}(\lambda-x) + \cos\dfrac{2\pi}{c}(\lambda-x) + \cdots\right]$$
$$= \dfrac{1}{2c}\int_{-c}^{c} f(\lambda)\,d\lambda\left[1 + \cos\dfrac{\pi}{c}(\lambda-x) + \cos\dfrac{2\pi}{c}(\lambda-x) + \cdots\right.$$
$$\left.+ \cos\left(-\dfrac{\pi}{c}\right)(\lambda-x) + \cos\left(-\dfrac{2\pi}{c}\right)(\lambda-x) + \cdots\right]$$

DEVELOPMENT IN TRIGONOMETRICAL SERIES.

since $\cos(-\phi) = \cos\phi$.

$$f(x) = \frac{1}{2\pi} \int_{-c}^{c} f(\lambda)\, d\lambda \left[\cdots + \frac{\pi}{c}\cos\left(-\frac{2\pi}{c}\right)(\lambda - x) + \frac{\pi}{c}\cos\left(-\frac{\pi}{c}\right)(\lambda - x) \right. \\
\left. + \frac{\pi}{c}\cos\frac{0\pi}{c}(\lambda - x) + \frac{\pi}{c}\cos\frac{\pi}{c}(\lambda - x) \right. \\
\left. + \frac{\pi}{c}\cos\frac{2\pi}{c}(\lambda - x) + \cdots \right] \tag{1}$$

As c is endlessly expanded the restricting worth drew nearer by the bracket in (1) is

$$\int_{-\infty}^{\infty} \cos\alpha(\lambda - x)\, .d\alpha.$$

Consequently the restricting structure drew closer by (1) is

$$f(x) = \frac{1}{2\pi} \int_{-\infty}^{\infty} f(\lambda)\, d\lambda \int_{-\infty}^{\infty} \cos\alpha(\lambda - x)\, .d\alpha, \tag{2}$$

what's more, the second individual from (2) should be equivalent to $f(x)$ for all upsides of x.

The twofold essential in (2) is known as *Fourier's Integral*, and since it is a restricting type of *Fourier's Series* it is dependent upon similar constraints as the series.

That is, all together that (2) ought to be valid $f(x)$ should be limited, ceaseless, and single esteemed for all upsides of x, or on the other hand if intermittent, should have just limited discontinuities.[3]

(2) is here and there given in a marginally unique structure.

Since $$\int_{-\infty}^{\infty} \cos\alpha(\lambda - x)\, .d\alpha = \int_{-\infty}^{0} \cos\alpha(\lambda - x)\, .d\alpha + \int_{0}^{\infty} \cos\alpha(\lambda - x)\, .d\alpha$$

what's more,

$$\int_{-\infty}^{0} \cos\alpha(\lambda - x)\, .d\alpha = \int_{\infty}^{0} \cos(-\alpha)(\lambda - x)\, .d(-\alpha) = -\int_{\infty}^{0} \cos\alpha(\lambda - x)\, .d\alpha$$

$$\int_{-\infty}^{\infty} \cos\alpha(\lambda - x)\, .d\alpha = 2\int_{0}^{\infty} \cos\alpha(\lambda - x)\, .d\alpha$$

what's more (2) might be composed

$$f(x) = \frac{1}{\pi} \int_{-\infty}^{\infty} f(\lambda)\, d\lambda \int_{0}^{\infty} \cos\alpha(\lambda - x)\, .d\alpha. \tag{3}$$

Assuming $f(x)$ is a *even* capability or a *odd* capability (3) can be even additionally rearranged.
Let $$f(x) = -f(-x).$$

Since the constraints of reconciliation in (3) don't contain α or λ the mixes might be acted in whichever request we pick. That is

$$\int_{-\infty}^{\infty} f(\lambda)\, d\lambda \int_{0}^{\infty} \cos\alpha(\lambda - x)\, .d\alpha = \int_{0}^{\infty} d\alpha \int_{-\infty}^{\infty} f(\lambda)\cos\alpha(\lambda - x)\, .d\lambda.$$

---

[3] See note on page 31.

DEVELOPMENT IN TRIGONOMETRICAL SERIES.

Presently

$$\int_{-\infty}^{\infty} f(\lambda)\cos\alpha(\lambda-x).d\lambda = \int_{-\infty}^{0} f(\lambda)\cos\alpha(\lambda-x).d\lambda + \int_{0}^{\infty} f(\lambda)\cos\alpha(\lambda-x).d\lambda.$$

$$\int_{-\infty}^{0} f(\lambda)\cos\alpha(\lambda-x).d\lambda = \int_{\infty}^{0} f(-\lambda)\cos\alpha(-\lambda-x).d(-\lambda)$$

$$= -\int_{0}^{\infty} f(\lambda)\cos\alpha(\lambda+x).d\lambda$$

what's more (3) becomes

$$f(x) = \frac{1}{\pi}\int_{0}^{\infty} d\alpha \int_{0}^{\infty} f(\lambda)[\cos\alpha(\lambda-x) - \cos\alpha(\lambda+x).d\lambda$$

$$= \frac{2}{\pi}\int_{0}^{\infty} d\alpha \int_{0}^{\infty} f(\lambda)\sin\alpha\lambda\sin\alpha x.d\lambda$$

or
$$f(x) = \frac{2}{\pi}\int_{0}^{\infty} f(\lambda)d\lambda \int_{0}^{\infty}\sin\alpha\lambda\sin\alpha x.d\alpha. \qquad (4)$$

In the event that $f(x) = f(-x)$ (3) can be diminished similarly to

$$f(x) = \frac{2}{\pi}\int_{0}^{\infty} f(\lambda)\,d\lambda \int_{0}^{\infty}\cos\alpha\lambda\cos\alpha x.d\alpha. \qquad (5)$$

Albeit (4) holds for all upsides of x just on the off chance that $f(x)$ is a *odd* capability, what's more (5) just on the off chance that $f(x)$ is a *even*, capability, both (4) and (5) hold for all *positive* upsides of x on account of any capability.

EXAMPLE.

(1) Get recipes (4) and (5) straightforwardly from (7) and (9) Art. 31.

Chapter 3

CONVERGENCE OF FOURIER SERIES

33. The topic of the *convergence* of a Fourier's Series is out and out as well huge to be totally dealt with in a rudimentary composition. We will, be that as it may, consider at some length one of the most significant of the series we have gotten, in particular

$$\frac{4}{\pi}\left[\sin x + \frac{\sin 3x}{3} + \frac{\sin 5x}{5} + \frac{\sin 7x}{7} + \cdots\right], \qquad \text{[v. (3) Art. 26}(b).\text{]}$$

what's more, demonstrate that for all upsides of x among nothing and π its total is totally equivalent to solidarity; that will be, that the cutoff moved toward by the amount of n terms of the series

$$\frac{2}{\pi}\left[\sin x \int_0^\pi \sin\alpha.d\alpha + \sin 2x \int_0^\pi \sin 2\alpha.d\alpha + \sin 3x \int_0^\pi \sin 3\alpha.d\alpha + \cdots\right],$$

as n is endlessly expanded, is 1, gave that x lies among nothing and π.

Let

$$S_n = \frac{2}{\pi}\left[\sin x \int_0^\pi \sin\alpha.d\alpha + \sin 2x \int_0^\pi \sin 2\alpha.d\alpha + \sin 3x \int_0^\pi \sin 3\alpha.d\alpha + \cdots \right.$$
$$\left. + \sin nx \int_0^\pi \sin n\alpha.d\alpha\right]. \qquad (1)$$

Then, at that point,

$$S_n = \frac{2}{\pi}\int_0^\pi [\sin\alpha\sin x + \sin 2\alpha\sin 2x + \sin 3\alpha\sin 3x + \cdots + \sin n\alpha\sin nx]d\alpha$$

$$= \frac{1}{\pi}\int_0^\pi [\cos(\alpha - x) - \cos(\alpha + x) + \cos 2(\alpha - x) - \cos 2(\alpha + x) + \cdots$$
$$+ \cos n(\alpha - x) - \cos n(\alpha + x)]\,d\alpha$$

$$= \frac{1}{\pi}\int_0^\pi [\cos(\alpha - x) + \cos 2(\alpha - x) + \cos 3(\alpha - x) + \cdots + \cos n(\alpha - x)]d\alpha$$

$$- \frac{1}{\pi}\int_0^\pi [\cos(\alpha + x) + \cos 2(\alpha + x) + \cos 3(\alpha + x) + \cdots + \cos n(\alpha + x)]d\alpha.$$

CONVERGENCE OF FOURIER'S SERIES.

Along these lines by Art. 20 (1)

$$S_n = \frac{1}{\pi} \int_0^\pi \left[-\frac{1}{2} + \frac{1}{2} \frac{\sin(2n+1)\frac{\alpha-x}{2}}{\sin\frac{\alpha-x}{2}} \right] d\alpha$$

$$- \frac{1}{\pi} \int_0^\pi \left[-\frac{1}{2} + \frac{1}{2} \frac{\sin(2n+1)\frac{\alpha+x}{2}}{\sin\frac{\alpha+x}{2}} \right] d\alpha.$$

$$S_n = \frac{1}{2\pi} \int_0^\pi \frac{\sin(2n+1)\frac{\alpha-x}{2}}{\sin\frac{\alpha-x}{2}} d\alpha - \frac{1}{2\pi} \int_0^\pi \frac{\sin(2n+1)\frac{\alpha+x}{2}}{\sin\frac{\alpha+x}{2}} d\alpha.$$

In the primary fundamental substitute β for $\frac{\alpha-x}{2}$, and in the subsequent necessary substitute β for $\frac{\alpha+x}{2}$.

We get

$$S_n = \frac{1}{\pi} \int_{-\frac{x}{2}}^{\frac{\pi}{2}-\frac{x}{2}} \frac{\sin(2n+1)\beta}{\sin\beta} d\beta - \frac{1}{\pi} \int_{\frac{x}{2}}^{\frac{\pi}{2}+\frac{x}{2}} \frac{\sin(2n+1)\beta}{\sin\beta} d\beta. \qquad (2)$$

It stays to find the cutoff drew closer by S_n as n is endlessly expanded.

34.
$$\int_0^{\frac{\pi}{2}} \frac{\sin(2n+1)\beta}{\sin\beta} d\beta = \frac{\pi}{2}. \qquad (1)$$

For

$$\frac{\sin(2n+1)\beta}{2\sin\beta} = \tfrac{1}{2} + \cos 2\beta + \cos 4\beta + \cdots + \cos 2n\beta, \qquad \text{by Art. 20.}$$

and

$$\int_0^{\frac{\pi}{2}} \cos 2k\beta \, d\beta = 0.$$

Allow us to develop the bend

$$y = \frac{\sin(2n+1)x}{\sin x}.$$

We have just to draw the bend $y = \sin(2n+1)x$ and afterward to separate the length of each ordinate by the worth of the sine of the relating abscissa.

In $y = \sin(2n+1)x$ the progressive curves into which the bend is isolated by the pivot of X are equivalent, and subsequently their regions are equivalent.

Each curve has for its height solidarity and for its base $\frac{\pi}{2n+1}$ and is balanced concerning the ordinate of its most elevated or absolute bottom.

Assuming that now we structure the bend $y = \frac{\sin(2n+1)x}{\sin x}$ from the bend $y = \sin(2n+1)x$, obviously, since $\sin x$ increments as x increments from 0 to $\frac{\pi}{2}$, the ordinate of any place of the new bend will be more limited than the ordinate of the comparing point in the going before curve, and that subsequently the area of each curve $y = \frac{\sin(2n+1)x}{\sin x}$ will be not exactly that of the curve before it.

If $a_0, a_1, a_2, \cdots a_{n-1}$ are the region of the progressive curves and a_n that of the deficient curve ended by the ordinate comparing to $x = \frac{\pi}{2}$

$$\int_0^{\frac{\pi}{2}} \frac{\sin(2n+1)x}{\sin x} dx = a_0 - a_1 + a_2 - a_3 + \cdots.$$

However

$$\int_0^{\frac{\pi}{2}} \frac{\sin(2n+1)x}{\sin x} dx = \int_0^{\frac{\pi}{2}} \frac{\sin(2n+1)\beta}{\sin\beta} d\beta = \frac{\pi}{2} \qquad \text{by (1).}$$

CONVERGENCE OF FOURIER'S SERIES.

Consequently

$$\frac{\pi}{2} = a_0 - a_1 + a_2 - a_3 + a_4 - \cdots + a_n \quad \text{if } n \text{ is even,}$$

or on the other hand

$$\frac{\pi}{2} = a_0 - a_1 + a_2 - a_3 + a_4 - \cdots - a_n \quad \text{if } n \text{ is odd.}$$

These conditions can be composed

$$\frac{\pi}{2} = a_0 + (-a_1 + a_2) + (-a_3 + a_4)$$
$$+ (-a_5 + a_6) + \cdots + (-a_{n-1} + a_n)$$

assuming n is even, and

$$\frac{\pi}{2} = a_0 + (-a_1 + a_2) + (-a_3 + a_4)$$
$$+ (-a_5 + a_6) + \cdots + (-a_{n-2} + a_{n-1}) + (-a_n)$$

assuming n is odd.

Regardless every bracket is a negative amount since

$$a_0 > a_1 > a_2 > a_3 \cdots > a_n,$$

also, it follows that a_0 is more prominent than $\frac{\pi}{2}$.

Once more

$$\frac{\pi}{2} = a_0 - a_1 + (a_2 - a_3) + (a_4 - a_5) + \cdots + (a_{n-2} - a_{n-1}) + a_n$$

if n is even and

$$\frac{\pi}{2} = a_0 - a_1 + (a_2 - a_3) + (a_4 - a_5) + \cdots + (a_{n-1} - a_n)$$

assuming n is odd.

Regardless every enclosure is positive and it follows that $a_0 - a_1$ is is under $\frac{\pi}{2}$.

Since

$$a_0 > \frac{\pi}{2} > a_0 - a_1,$$

a_0 and $a_0 - a_1$ vary from $\frac{\pi}{2}$ by short of what they contrast from one another, that is, by under a_1.

Likewise we can show that $a_0 - a_1$ and $a_0 - a_1 + a_2$ vary from $\frac{\pi}{2}$ by under a_2; and overall that $a_0 - a_1 + a_2 - a_3 + \cdots \pm a_k$ contrasts from $\frac{\pi}{2}$ by not exactly a_k; or even that

$$a_0 - a_1 + a_2 - a_3 + \cdots \pm \frac{a_k}{p}$$

varies from $\frac{\pi}{2}$ by not exactly a_k regardless of the worth of p, if p is more noteworthy than solidarity.

35. From what has been demonstrated in the last article it follows that

$$\int_0^b \frac{\sin(2n+1)x}{\sin x} dx,$$

where b is some worth somewhere in the range of $\frac{\pi}{2n+1}$ and $\frac{\pi}{2}$, contrasts from $\frac{\pi}{2}$ by not exactly the region of the curve wherein the ordinate of $y = \frac{\sin(2n+1)x}{\sin x}$ relating to $x = b$ falls assuming

CONVERGENCE OF FOURIER'S SERIES.

this ordinate isolates a curve, or by not exactly the region of the curve next past the point $(b, 0)$ assuming that the bend crosses the pivot of X at that point.

The region of the curve being referred to is under $\frac{\pi}{2n+1}$, its base, duplicated by $\frac{1}{\sin\left(b - \frac{\pi}{2n+1}\right)}$, a worth more prominent than the length of its longest ordinate.

Thusly
$$\int_0^b \frac{\sin(2n+1)x}{\sin x} dx$$
contrasts from $\frac{\pi}{2}$ by under $\frac{\pi}{2n+1} \frac{1}{\sin\left(b - \frac{\pi}{2n+1}\right)}$.

On the off chance that now n is endlessly expanded $\frac{\pi}{2n+1} \frac{1}{\sin\left(b - \frac{\pi}{2n+1}\right)}$ approaches zero as its cutoff, and we come by the vital outcome

$$\lim_{n=\infty}\left[\int_0^b \frac{\sin(2n+1)x}{\sin x} dx\right] = \frac{\pi}{2} \tag{1}$$

if $0 < b < \frac{\pi}{2}$.

36.
$$S_n = \frac{1}{\pi}\int_{-\frac{x}{2}}^{\frac{\pi}{2}-\frac{x}{2}} \frac{\sin(2n+1)\beta}{\sin\beta} d\beta - \frac{1}{\pi}\int_{\frac{x}{2}}^{\frac{\pi}{2}+\frac{x}{2}} \frac{\sin(2n+1)\beta}{\sin\beta} d\beta. \quad [\text{Art. 33. (2)}]$$

$$= \frac{1}{\pi}\int_{-\frac{x}{2}}^{0} \frac{\sin(2n+1)\beta}{\sin\beta} d\beta + \frac{1}{\pi}\int_0^{\frac{\pi}{2}-\frac{x}{2}} \frac{\sin(2n+1)\beta}{\sin\beta} d\beta$$

$$- \frac{1}{\pi}\int_0^{\frac{x}{2}} \frac{\sin(2n+1)\beta}{\sin\beta} d\beta + \frac{1}{\pi}\int_0^{\frac{x}{2}} \frac{\sin(2n+1)\beta}{\sin\beta} d\beta$$

$$- \frac{1}{\pi}\int_{\frac{\pi}{2}}^{\frac{\pi}{2}+\frac{x}{2}} \frac{\sin(2n+1)\beta}{\sin\beta} d\beta.$$

This last incentive for S_n can be to some degree rearranged.

Subbing $\gamma = -\beta$ we get

$$\int_{-\frac{x}{2}}^{0} \frac{\sin(2n+1)\beta}{\sin\beta} d\beta = -\int_{\frac{x}{2}}^{0} \frac{\sin(2n+1)\gamma}{\sin\gamma} d\gamma = \int_0^{\frac{x}{2}} \frac{\sin(2n+1)\beta}{\sin\beta} d\beta.$$

Subbing $\gamma = \pi - \beta$ in
$$\int_{\frac{\pi}{2}}^{\frac{\pi}{2}+\frac{x}{2}} \frac{\sin(2n+1)\beta}{\sin\beta} d\beta \qquad\qquad \text{we have}$$

$$\int_{\frac{\pi}{2}}^{\frac{\pi}{2}+\frac{x}{2}} \frac{\sin(2n+1)\beta}{\sin\beta} d\beta = -\int_{\frac{\pi}{2}}^{\frac{\pi}{2}-\frac{x}{2}} \frac{\sin(2n+1)\gamma}{\sin\gamma} d\gamma = \int_{\frac{\pi}{2}-\frac{x}{2}}^{\frac{\pi}{2}} \frac{\sin(2n+1)\beta}{\sin\beta} d\beta$$

$$= \int_0^{\frac{\pi}{2}} \frac{\sin(2n+1)\beta}{\sin\beta} d\beta - \int_0^{\frac{\pi}{2}-\frac{x}{2}} \frac{\sin(2n+1)\beta}{\sin\beta} d\beta.$$

CONVERGENCE OF FOURIER'S SERIES. 50

Consequently

$$S_n = \frac{2}{\pi} \int_0^{\frac{x}{2}} \frac{\sin(2n+1)\beta}{\sin \beta} d\beta + \frac{2}{\pi} \int_0^{\frac{\pi}{2} - \frac{x}{2}} \frac{\sin(2n+1)\beta}{\sin \beta} d\beta - \frac{2}{\pi} \int_0^{\frac{\pi}{2}} \frac{\sin(2n+1)\beta}{\sin \beta} d\beta.$$

$$\int_0^{\frac{\pi}{2}} \frac{\sin(2n+1)\beta}{\sin \beta} d\beta = \frac{\pi}{2} \qquad \text{by (1) Art. 34.}$$

$$\lim_{n=\infty} \left[\int_0^{\frac{x}{2}} \frac{\sin(2n+1)\beta}{\sin \beta} d\beta \right] = \frac{\pi}{2} \quad \text{if} \quad 0 < x < \pi \qquad \text{by (1) Art. 35}$$

and

$$\lim_{n=\infty} \left[\int_0^{\frac{\pi}{2} - \frac{x}{2}} \frac{\sin(2n+1)\beta}{\sin \beta} d\beta \right] = \frac{\pi}{2} \quad \text{if} \quad 0 < x < \pi \qquad \text{by (1) Art. 35.}$$

In this manner $\quad \lim_{n=\infty}[S_n] = 1 + 1 - 1 = 1 \quad \text{if} \quad 0 < x < \pi \quad$ and

$$\frac{4}{\pi}\left[\sin x + \frac{\sin 3x}{3} + \frac{\sin 5x}{5} + \frac{\sin 7x}{7} + \cdots\right] = 1$$

for all upsides of x among nothing and π.

37. By a fairly lengthy however not particularly troublesome expansion of the thinking just given it very well may be shown that assuming $f(x)$ is *single-valued* and *finite* between $x = -\pi$ and $x = \pi$, and has just a *finite number* of *discontinuities* furthermore, of *maxima* and *minima* between $x = -\pi$ and $x = \pi$ the *Fourier's Series*

$$\frac{1}{2}b_0 + b_1 \cos x + b_2 \cos 2x + b_3 \cos 3x + \cdots$$
$$+ a_1 \sin x + a_2 \sin 2x + a_3 \sin 3x + \cdots$$

where
$$a_m = \frac{1}{\pi} \int_{-\pi}^{\pi} f(\alpha) \sin m\alpha \, d\alpha$$

and
$$b_m = \frac{1}{\pi} \int_{-\pi}^{\pi} f(\alpha) \cos m\alpha \, d\alpha,$$

and that Fourier's Series only is equivalent to $f(x)$ for all upsides of x between $x = -\pi$ and $x = \pi$, excepting the upsides of x relating to the discontinuities of $f(x)$, and the qualities π and $-\pi$ in the event that $f(\pi)$ isn't equivalent to $f(-\pi)$; furthermore, that assuming c is a worth of x comparing to a brokenness of $f(x)$, the worth of the series when $x = c$ is

$$\frac{1}{2} \lim_{\epsilon \doteq 0}[f(c-\epsilon) + f(c+\epsilon)];$$

furthermore, that assuming $f(\pi)$ isn't equivalent to $f(-\pi)$ the worth of the series when $x = -\pi$ what's more, when $x = \pi$ is

$$\frac{1}{2}[f(-\pi) + f(\pi)].$$

In the event that $f(x)$ while fulfilling the circumstances named in the first section with the exception of a limited number of upsides of x, becomes endless for those qualities, the series is equivalent to the capability with the exception of the upsides of x being referred to gave that $\int_{-\pi}^{\pi} f(x)\,dx$ is limited and determinate. (v. Int. Cal. Arts. 83 and 84.)

CONVERGENCE OF FOURIER'S SERIES.

38. The topic of the convergency of a Fourier's Series and the circumstances under which a capability might be created in such a series was first gone after effectively by Dirichlet in 1829, and his decisions have been censured and stretched out by later mathematicians, remarkably by Riemann, Heine, Lipschitz, and du Bois Reymond. It could be noticed that the reactions relate not to the adequacy but rather to the need of Dirichlet's circumstances.

A fantastic list of references of the writing of the subject is given by Arnold Sachse in a short thesis distributed by Gauthier- - Villars, Paris, 1880, named "Essai Historique sur la ReprÃ©sentation d'une Fonction Arbitraire d'une seule variable standard une SÃ©rie TrigonomÃ©trique."

39. A fair plan of light is tossed on the eccentricities of mathematical series by the endeavor to develop roughly the bends comparing to them.

In the event that we develop $y = a_1 \sin x$ and $y = a_2 \sin 2x$ and add the ordinates of the focuses having similar abscissas we will get focuses on the bend

$$y = a_1 \sin x + a_2 \sin 2x.$$

In the event that now we develop $y = a_3 \sin 3x$ and add the ordinates to those of $y = a_1 \sin x + a_2 \sin 2x$ we will get the bend

$$y = a_1 \sin x + a_2 \sin 2x + a_3 \sin 3x.$$

By proceeding with this cycle we get progressive approximations to

$$y = a_1 \sin x + a_2 \sin 2x + a_3 \sin 3x + a_4 \sin 4x + \cdots$$

Allow us to apply this strategy to a couple of the series which we have gotten in Chapter II.

Take

$$y = \sin x + \frac{1}{3} \sin 3x + \frac{1}{5} \sin 5x + \cdots \qquad (1)$$
$$= 0 \text{ when } x = 0, \ \frac{\pi}{4} \text{ from } x = 0 \text{ to } x = \pi, \text{ and } 0 \text{ when } x = \pi,$$

v. Art. 26 $[b](3)$.

$$y = 2\left(\sin x - \frac{1}{2}\sin 2x + \frac{1}{3}\sin 3x - \frac{1}{4}\sin 4x + \cdots\right) \qquad (2)$$
$$= x \text{ from } x = 0 \text{ to } x = \pi, \text{ and } 0 \text{ when } x = \pi,$$

Art. 26 $[a](4)$.

$$y = \frac{4}{\pi}\left[\frac{1}{1^2}\sin x - \frac{1}{3^2}\sin 3x + \frac{1}{5^2}\sin 5x - \frac{1}{7^2}\sin 7x + \cdots\right] \qquad (3)$$
$$= x \text{ from } x = 0 \text{ to } x = \frac{\pi}{2}, \text{ and } \pi - x \text{ from } x = \frac{\pi}{2} \text{ to } x = \pi,$$

Art. 26 $[c](2)$.

$$y = \frac{1}{1}\sin x + \frac{2}{2}\sin 2x + \frac{1}{3}\sin 3x + \frac{1}{5}\sin 5x - \frac{2}{6}\sin 6x + \frac{1}{7}\sin 7x + \cdots \qquad (4)$$
$$= 0 \text{ when } x = 0, \ \frac{\pi}{2} \text{ from } x = 0 \text{ to } x = \frac{\pi}{2}, \text{ and } 0 \text{ from } x = \frac{\pi}{2} \text{ to } x = \pi,$$

v. Art. 26 $[d](2)$.

It should be borne as a top priority that every one of these bends is intermittent having the period 2π, and is even regarding the beginning.

CONVERGENCE OF FOURIER'S SERIES. 52

The accompanying figures I, II, III, and IV address the initial four approximations to every one of these bends.

In each figure the bend $y =$ the series, and the estimation being referred to are defined in persistent boundaries, and the first estimation and the bend comparing to the term to be added are defined in spotted boundaries.

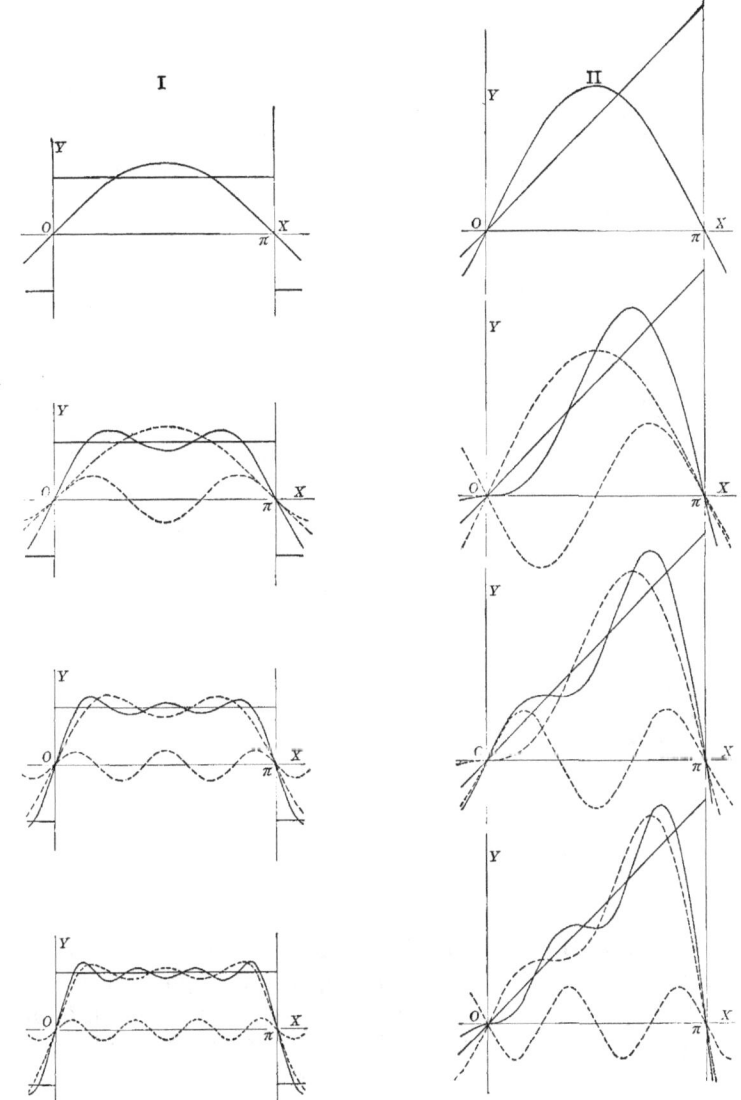

Figs. I, II, III, and IV promptly recommend the accompanying realities:

(a) The bend addressing every guess is consistent in any event, when the bend addressing the series is broken.

(b) When the bend addressing the series is broken the part of each progressive estimated bend in the neighborhood of the point whose abscissa is a worth of x for which the series bend is spasmodic methodologies increasingly more almost a straight line opposite to the pivot of X and interfacing the different parts of the series bend.

CONVERGENCE OF FOURIER'S SERIES.

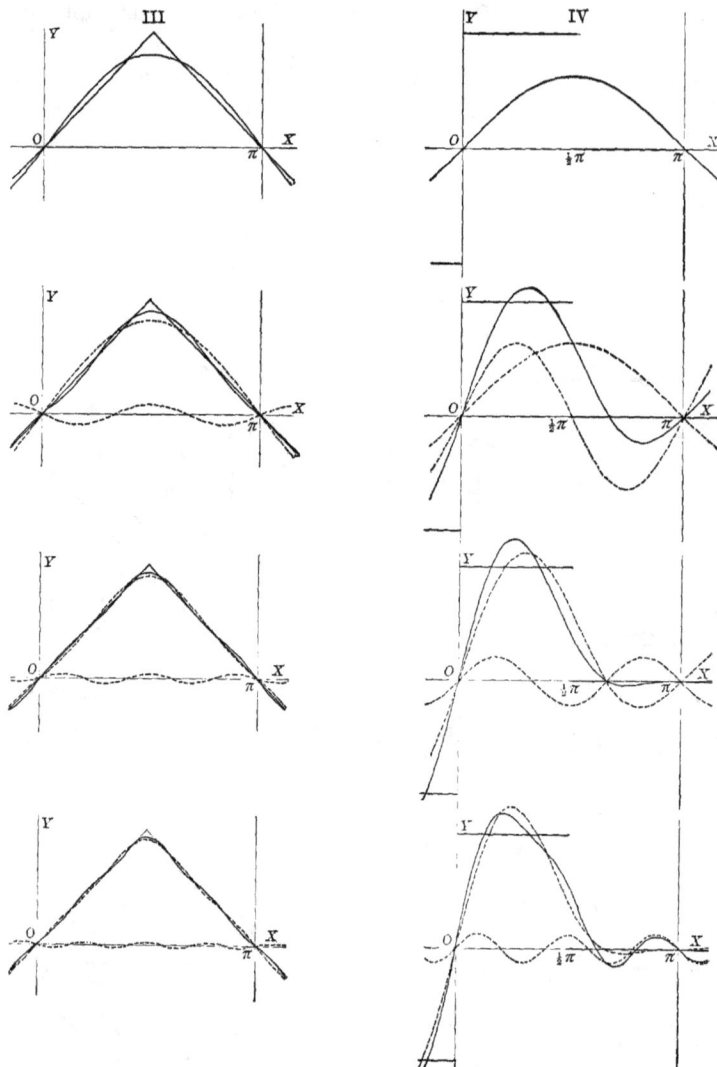

(c) The bends addressing progressive approximations don't really will generally lose their wavy person, since each is acquired from the former one by superposing upon it a wave line whose waves are more limited each time yet try not to fundamentally lose their sharpness of pitch. This is the situation in Figures I, II, and IV. In Fig. III the floods of the superposed bends develop quickly compliment.

It follows from this that in such cases as those addressed in Figures I, II, what's more, IV the bearing of the estimated bend at a point having guaranteed abscissa doesn't overall methodology the heading of the series bend at the comparing point, or for sure, move toward any restricting worth, as the estimate is made consistently nearer; and that the length of any part of the estimated bend won't in everyday methodology the length of the relating piece of the series bend.

Logically this adds up to saying that the subsidiary of a component of x can't overall be acquired by separating term by term the Fourier's Series which addresses the capability.

(d) The region limited by a given ordinate, the estimated bend, the hub of X, and any subsequent ordinate will approach as its breaking point the comparing area of the series bend on the off chance that the series bend is constant between the ordinates in question; and will move toward the area limited by the given ordinate, the series bend, the hub of X, any subsequent ordinate, and a line opposite to the pivot of X, and joining the different bits of the series bend in the event that the last option has a irregularity between the ordinates being referred to.

CONVERGENCE OF FOURIER'S SERIES. 54

Scientifically this adds up to saying that the Fourier's Series comparing to some random capability can be coordinated term by term and the subsequent series will address the indispensable of the capability in any event, when the capability is intermittent (v. Int. Cal. Art. 83).

We might note in passing that assuming that the capability bend is constant a bend addressing the essential of the capability will be nonstop and will not shift its course suddenly anytime; while on the off chance that the capability bend is irregular the bend addressing the necessary will in any case be constant yet will shift its course unexpectedly at focuses comparing to the discontinuities of the given capability.

40. The realities that the subordinate of a Fourier's Series can't overall be acquired by separating the series term by term and that its basic can be gotten by coordinating the series term by term are vital to the point that it is worth while to intently check the matter somewhat more out. Allow us to consider the separation of the series addressed in Art. 39 Figure I.

Let
$$S_n = \sin x + \frac{1}{3}\sin 3x + \frac{1}{5}\sin 5x + \cdots + \frac{1}{2n+1}\sin(2n+1)x.$$

Then
$$\frac{dS_n}{dx} = \cos x + \cos 3x + \cos 5x + \cdots + \cos(2n+1)x.$$

If $x = \frac{\pi}{2}$
$$\frac{dS_n}{dx} = 0$$

also, the bend is lined up with the pivot of X for $x = \frac{\pi}{2}$ regardless of what the worth of n.

On the off chance that $x = 0$ or $x = \pi$
$$\frac{dS_n}{dx} = 1 + 1 + 1 + 1 + \cdots + 1 = n + 1$$

furthermore, the bend $y = S_n$ turns out to be all the more almost opposite to the hub of X at the beginning and for $x = \pi$ as we increment n.

If $x = \frac{\pi}{3}$
$$\frac{dS_n}{dx} = \frac{1}{2} - 1 + \frac{1}{2} + \frac{1}{2} - 1 + \frac{1}{2} + \cdots$$

That is
$$\frac{dS_n}{dx} = \frac{1}{2} \quad \text{if} \quad n = 0 \quad \text{or} \quad n = 3k$$
$$= -\frac{1}{2} \quad \text{"} \quad and\, n = 1\, and\text{"}\quad and\, n = 3k+1$$
$$= 0\, and\text{"}\quad and\, n = 2\, and\text{"}\quad and\, n = 3k+2.$$

Subsequently when $x = \frac{\pi}{3}$ $\frac{dS_n}{dx}$ moves toward no restricting worth as n is endlessly expanded. To be sure, in the progressive approximations the point whose abscissa is $\frac{\pi}{3}$ is progressively on the back, on the front, and on the peak or on the other hand in the box of a wave, and albeit the waves are getting more modest they do not lose their sharpness of pitch.

If x has some other worth among 0 and π $\frac{dS_n}{dx}$ will change suddenly as n is changed and won't move toward any restricting worth as n is expanded.

41. overall in the event that we separate a Fourier's Series
$$S = \frac{1}{2}b_0 + b_1 \cos x + b_2 \cos 2x + b_3 \cos 3x + \cdots$$
$$+ a_1 \sin x + a_2 \sin 2x + a_3 \sin 3x + \cdots$$

we get
$$- b_1 \sin x - 2b_2 \sin 2x - 3b_3 \sin 3x - \cdots$$
$$+ a_1 \cos x + 2a_2 \cos 2x + 3a_3 \cos 3x + \cdots.$$

CONVERGENCE OF FOURIER'S SERIES.

Differentiate once more and we get

$$-b_1 \cos x - 2^2 b_2 \cos 2x - 3^2 b_3 \cos 3x - \cdots$$
$$-a_1 \sin x - 2^2 a_2 \sin 2x - 3^2 b_3 \sin 3x - \cdots.$$

We see that each time we separate we duplicate the coefficient of $\sin kx$ furthermore, of $\cos kx$ by k while the term actually includes $\cos kx$ or $\sin kx$.

Since the series

$$\cos x + \cos 2x + \cos 3x + \cdots$$
$$+ \sin x + \sin 2x + \sin 3x + \cdots$$

isn't joined, and a Fourier's Series meets simply because its coefficients decline as we advance in the series, the separation of a Fourier's Series should make its intermingling less quick in the event that it doesn't really obliterate it, and reiterations of the cycle will typically at last make the inferred series veer.

It is to be seen that the inferred series are Fourier's Series, yet of fairly exceptional structure, that is they miss the mark on steady term. (v. Art. 30.)

On the off chance that now we coordinate a Fourier's Series

$$\frac{1}{2}b_0 + b_1 \cos x + b_2 \cos 2x + b_3 \cos 3x + \cdots$$
$$+ a_1 \sin x + a_2 \sin 2x + a_3 \sin 3x + \cdots$$

we get

$$C + \frac{1}{2}b_0 x + b_1 \sin x + \frac{1}{2}b_2 \sin 2x + \frac{1}{3}b_3 \sin 3x + \cdots$$
$$- a_1 \cos x - \frac{1}{2}a_2 \cos 2x - \frac{1}{3}a_3 \cos 3x - \cdots,$$

a Geometrical Series which merges more quickly than the given series.

It is to be seen that the series got by incorporating a Fourier's Series isn't overall a Fourier's Series inferable from the presence of the term $\frac{1}{2}b_0 x$. (v. Art. 30.)

42. We are currently prepared to consider the circumstances under which an element of x can be formed into a Fourier's Series whose term by term subordinate will be equivalent to the subsidiary of the capability.

Let the capability $f(x)$ fulfill the circumstances expressed in Art. 37. Then, at that point, there is one Fourier's Series and yet one which is equivalent to it. Call this series S.

Let the subordinate $f'(x)$[1] of the given capability additionally fulfill the circumstances expressed in Art. 37. Then, at that point, $f'(x)$ can be communicated as a Fourier's Series. By Art. 39 (d) the vital of this last series will be equivalent to the essential of $f'(x)$, that is to $f(x)$ in addition to a consistent, and one vital will be equivalent to $f(x)$.

On the off chance that this vital which is fundamentally a Mathematical Series is a Fourier's Series it should be indistinguishable with S. It will be a Fourier's Series just on the off chance that the Fourier's Series for $f'(x)$ misses the mark on steady term $\frac{1}{2}b_0$.

But $$b_0 = \frac{1}{\pi} \int_{-\pi}^{\pi} and f'(x) dx \text{ by (3) Art. 30.}$$

Therefore $$b_0 = \frac{1}{\pi}[f(\pi) and - f(-\pi)];$$

furthermore, will be zero if $f(\pi) = f(-\pi)$.

All together that $f'(x)$ will fulfill the circumstances expressed in Art. 37 $f(x)$ while fulfilling similar circumstances must moreover be limited and nonstop between $x = -\pi$ and $x = \pi$.

---
[1] We will consistently utilize the documentation $f'(x)$ for $\frac{df(x)}{dx}$. v. Dif. Cal. Art. 124.

CONVERGENCE OF FOURIER'S SERIES. 56

In the event that, $f(x)$ is *single-esteemed, limited, and nonstop, and has just a limited number of maxima and minima*, between $x = -\pi$ and $x = \pi$, (the qualities $x = -\pi$ and $x = \pi$ being incorporated), *and if* $f(\pi) = f(-\pi)$ $f(x)$ can be formed into a Fourier's Series whose term by term subordinate will be equivalent to the subordinate of the capability.

It will be seen that for this situation the intermittent bend $y = S$ is constant all through its entire degree.

43. Since a Fourier's Basic is a restricting instance of a Fourier's Series the ends expressed in this section hold, *mutatis mutandis* for a Fourier's Basic.

For instance if an element of x is limited and single-esteemed for all upsides of x furthermore, has not a limitless number of discontinuities or of maxima and minima in the neighborhood valuable of x it will be equivalent to Fourier's Necessary

$$\frac{1}{\pi} \int_0^\infty d\alpha \int_{-\infty}^\infty f(\lambda) \cos \alpha (\lambda - x) . d\lambda$$

also, to that Fourier's Vital just, and the indispensable as for x of this Fourier's Essential will be equivalent to $\int f(x) dx$.

On the off chance that moreover $f(x)$ is limited and ceaseless for all upsides of x the subordinate of the Fourier's Indispensable concerning x will be equivalent to $\frac{df(x)}{dx}$.

Chapter 4

PHYSICS APPLICATIONS BY FOURIER INTEGRAL

44. In Art. 7 we have previously considered at some length an issue in Heat Conduction which required the utilization of a Fourier's Series. We will start the current part with an issue intently practically equivalent to in its treatment to that of Art. 7, however requiring the utilization of a Fourier's Basic.

Assume that power is streaming in a meager plane sheet of limitless degree furthermore, that the worth of the potential capability is given for each point in some straight line in the sheet, required the worth of the expected capability at any place of the sheet.

Allow us to accept the line as the pivot of X and consider at first just those places for which y is positive:

We have, then, at that point, to fulfill the condition

$$D_x^2 V + D_y^2 V = 0 \qquad (1)$$

dependent upon the circumstances

$$V = 0 \quad \text{when} \quad y = \infty \qquad (2)$$
$$V = f(x) \quad "and\, y = 0 \qquad (3)$$

where $f(x)$ is a given capability, and we are not worried about negative upsides of y.

As in Art. 7 we have $e^{-\alpha y}\sin\alpha x$ and $e^{-\alpha y}\cos\alpha x$ as specific upsides of V which fulfill (1) and (2). We should duplicate them by steady coefficients thus join them as to fulfill condition (3).

By (3) Art. 32

$$f(x) = \frac{1}{\pi}\int_0^\infty d\alpha \int_{-\infty}^\infty f(\lambda)\cos\alpha(\lambda - x).d\lambda. \qquad (4)$$

We wish to develop a worth of V which will decrease to (4) when $y = 0$. This requires a little consideration however not much resourcefulness.

Take $e^{-\alpha y}\cos\alpha x$ and $e^{-\alpha y}\sin\alpha x$ and duplicate the first by $\cos\alpha\lambda$, and the second by $\sin\alpha\lambda$; they are still upsides of V which fulfill (1). Add these and we get

$$e^{-\alpha y}\cos\alpha(\lambda - x),$$

still a worth of V which fulfills (1), regardless of what the upsides of α and λ. Duplicate by $f(\lambda)d\lambda$ and we have

$$e^{-\alpha y}f(\lambda)\cos\alpha(\lambda - x).d\lambda \qquad (5)$$

PHYSICS APPLICATIONS

as a worth of V which fulfills (1).

$$V = \int_{-\infty}^{\infty} e^{-\alpha y} f(\lambda) \cos \alpha(\lambda - x).d\lambda \tag{6}$$

is as yet an answer of (1) since it is the constraint of the amount of terms covered by the structure (5); lastly

$$V = \frac{1}{\pi} \int_0^{\infty} d\alpha \int_{-\infty}^{\infty} e^{-\alpha y} f(\lambda) \cos \alpha(\lambda - x).d\lambda \tag{7}$$

is an answer of (1) as it is $\frac{1}{\pi}$ duplicated by the constraint of the amount of terms framed by duplicating the second individual from (6) by $d\alpha$ and giving unique values to α.

However, (7) should be our expected arrangement since while it fulfills (1) and (2), it diminishes to (4) when $y = 0$ and hence fulfills condition (3).

Assuming that $f(x)$ is a *even* capability we can lessen (7) to the structure

$$V = \frac{2}{\pi} \int_0^{\infty} d\alpha \int_0^{\infty} e^{-\alpha y} f(\lambda) \cos \alpha x \cos \alpha \lambda.d\lambda \tag{8}$$

furthermore, on the off chance that $f(x)$ is a *odd* capability to the structure

$$V = \frac{2}{\pi} \int_0^{\infty} d\alpha \int_0^{\infty} e^{-\alpha y} f(\lambda) \sin \alpha x \sin \alpha \lambda.d\lambda. \tag{9}$$

(7), (8), and (9) are substantial just for positive upsides of y, yet as the issue is clearly balanced as for the hub of X, (7), (8), and (9) empower us to get the worth of the possible capability anytime of the plane.

EXAMPLES.

1. Get structures (8) and (9) straight by the guide of (5) and (4) Art. 32.

2. Express an issue in statical power of which the arrangement surrendered Art. 44 is the arrangement.

45. As an exceptional case under Art. 44 let us think about the issue:- - - To find the worth of the possible capability anytime of a slender plane sheet of boundless degree where all places of a given line which lie to one side of the beginning are kept at likely zero, and all focuses which lie to one side of the beginning are kept at likely solidarity.

Here $f(x) = 0$ if $x < 0$ and $f(x) = 1$ if $x > 0$.

(7) Art. 44 gives us the expected arrangement. It is

$$V = \frac{1}{\pi} \int_0^{\infty} d\alpha \int_0^{\infty} e^{-\alpha y} \cos \alpha(\lambda - x).d\lambda; \tag{1}$$

yet, this can be tremendously rearranged.

We have

$$V = \frac{1}{\pi} \int_0^{\infty} d\lambda \int_0^{\infty} e^{-\alpha y} \cos \alpha(\lambda - x).d\alpha.$$

Now

$$\int_0^{\infty} e^{-ax} \cos mx.dx = \frac{a}{a^2 + m^2}$$

PHYSICS APPLICATIONS

in the event that $a > 0$. (Int. Cal. Art. 82, Ex. 8.)

Hence and $\int_0^\infty e^{-\alpha y} \cos\alpha(\lambda - x).d\alpha = \dfrac{y}{y^2 + (\lambda - x)^2}$,

and
$$V = \frac{1}{\pi}\int_0^\infty \frac{y\,d\lambda}{y^2 + (\lambda-x)^2} = \frac{1}{\pi}\left(\frac{\pi}{2} + \tan^{-1}\frac{x}{y}\right).$$

$$\tan\left(\frac{\pi}{2} - \tan^{-1}\frac{x}{y}\right) = \operatorname{ctn}\left(\tan^{-1}\frac{x}{y}\right) = \frac{y}{x};$$

what's more, subsequently

$$V = \frac{1}{\pi}\left(\frac{\pi}{2} + \tan^{-1}\frac{x}{y}\right) = 1 - \frac{1}{\pi}\tan^{-1}\frac{y}{x}. \tag{2}$$

Since $\log z = \log(x+yi) = \tfrac{1}{2}\log(x^2+y^2) + i\tan^{-1}\frac{y}{x}$, [Int. Cal. Art. 33 (2)],

$$I - \frac{1}{\pi}\log z = I - \frac{1}{\pi}\log(x+yi) = -\frac{1}{2\pi}\log(x^2+y^2) + i\left(1 - \frac{1}{\pi}\tan^{-1}\frac{y}{x}\right)$$

also, $1 - \frac{1}{\pi}\tan^{-1}\frac{y}{x}$ and $-\frac{1}{2\pi}\log(x^2+y^2)$ are *conjugate functions*. (v. Int. Cal. Arts. 209 and 210.) Consequently

$$V_1 = -\frac{1}{2\pi}\log(x^2+y^2) \tag{3}$$

is an answer of the situation

$$D_x^2 V_1 + D_y^2 V_1 = 0; \tag{4}$$

also, the bends

$$\frac{1}{\pi}\left(\frac{\pi}{2} + \tan^{-1}\frac{x}{y}\right) = a \tag{5}$$

and
$$and \frac{1}{2\pi}\log(x^2+y^2) = b \tag{6}$$

cut each other at right points.

Assuming we develop the bends got by giving various qualities to a in (5) we get a bunch of *equipotential lines* for the directing sheet depicted toward the start of this article, and the bends acquired by giving various qualities to (b) in (6) will be the *lines of flow*.

Also since

$$V_1 = -\frac{1}{2\pi}\log(x^2+y^2) \tag{3}$$

is an answer of Laplace's Situation (4), the lines of stream just referenced will be equipotential lines for a specific conveyance of potential, for which the equipotential lines previously mentioned will be lines of stream.

$V = a$, that is

$$\frac{1}{\pi}\left(\frac{\pi}{2} + \tan^{-1}\frac{x}{y}\right) = a, \tag{5}$$

reduces to *and y* $\qquad = -x\tan a\pi. \tag{7}$

In the event that now we provide for a values varying by a steady sum we get a bunch of straight lines emanating from the beginning and at equivalent rakish spans.

$V_1 = b$, that is

$$-\frac{1}{2\pi}\log(x^2+y^2) = b, \tag{6}$$

reduces to

$$x^2 + y^2 = e^{-2\pi b}. \tag{8}$$

PHYSICS APPLICATIONS

On the off chance that we provide for b a bunch of values varying by a steady sum we get a set of circles whose focuses are at the beginning and whose radii structure a mathematical movement. They are the equipotential lines for a dainty plane sheet of boundless degree where the potential capability is kept equivalent to given different steady values on the peripheries of two given concentric circles or where we have a *source* at the beginning; and for this framework the lines (7) are lines of stream, what's more (3) is the finished arrangement.

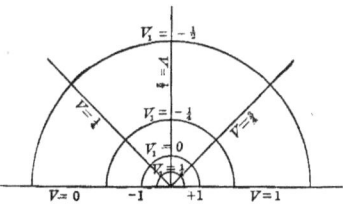

The figure gives the equipotential endlessly lines of stream for one or the other framework, in any case, just for positive upsides of y. The total figure has the pivot of X as a pivot of evenness.

EXAMPLES.

1. Tackle the issue of Art. 44 for the situation where
$$f(x) = -1 \text{ if } x < 0 \text{ and } f(x) = 1 \text{ if } x > 0.$$
Ans., $V = \frac{2}{\pi} \tan^{-1} \frac{x}{y}$.

2. Tackle the issue of Art. 44 for the situation where
$$f(x) = a \text{ if } x < 0 \text{ and } f(x) = b \text{ if } x > 0.$$
Ans., $V = \frac{1}{2}(a+b) + \frac{1}{\pi}(b-a) \tan^{-1} \frac{x}{y}$.

3. Lessen (7), (8), and (9) Art. 44 to the structures
$$V = \frac{1}{\pi} \int_{-\infty}^{\infty} \frac{yf(\lambda)d\lambda}{y^2 + (\lambda - x)^2},$$
$$V = \frac{1}{\pi} \int_0^{\infty} yf(\lambda)d\lambda \left[\frac{1}{y^2 + (\lambda - x)^2} + \frac{1}{y^2 + (\lambda + x)^2} \right],$$
$$V = \frac{1}{\pi} \int_0^{\infty} yf(\lambda)d\lambda \left[\frac{1}{y^2 + (\lambda - x)^2} - \frac{1}{y^2 + (\lambda + x)^2} \right],$$
separately.

46. A particularly fascinating instance of Art. 44 is the accompanying where
$$f(x) = 0 \text{ if } x < -1, \quad f(x) = 1 \text{ if } -1 < x < 1, \quad \text{and} \quad f(x) = 0 \text{ if } x > 1.$$

Here and $V = \frac{1}{\pi} \left[\tan^{-1} \frac{1+x}{y} + \tan^{-1} \frac{1-x}{y} \right].$ \hfill (1)

Now and $\frac{1}{\pi} \log[(1-z)i] = \frac{1}{\pi} \log[(1-x-yi)i] = \frac{1}{\pi} \log[y + (1-x)i]$
$$= \frac{1}{2\pi} \log[(1-x)^2 + y^2] + \frac{i}{\pi} \tan^{-1} \frac{1-x}{y},$$

also,
$$-\frac{1}{\pi} \log[(-1-z)i] = -\frac{1}{\pi} \log[(-1-x-yi)i] = -\frac{1}{\pi} \log[y - (1+x)i]$$
$$= -\frac{1}{2\pi} \log[(1+x)^2 + y^2] + \frac{i}{\pi} \tan^{-1} \frac{1+x}{y}.$$

PHYSICS APPLICATIONS

$$\frac{1}{\pi}\log\frac{1-z}{-1-z} = \frac{1}{2\pi}\log\frac{(1-x)^2+y^2}{(1+x)^2+y^2} + \frac{i}{\pi}\left[\tan^{-1}\frac{1+x}{y}+\tan^{-1}\frac{1-x}{y}\right].$$

Consequently

$$\frac{1}{\pi}\left(\tan^{-1}\frac{1+x}{y}+\tan^{-1}\frac{1-x}{y}\right) \quad \text{and} \quad \frac{1}{2\pi}\log\frac{(1-x)^2+y^2}{(1+x)^2+y^2}$$

are *conjugate functions*:[1] what's more,

$$\frac{1}{\pi}\left(\tan^{-1}\frac{1+x}{y}+\tan^{-1}\frac{1-x}{y}\right) = a \tag{2}$$

is any equipotential line, and

$$\frac{1}{2\pi}\log\frac{(1-x)^2+y^2}{(1+x)^2+y^2} = b \tag{3}$$

any line of stream for the framework portrayed toward the start of this article; and

$$V_1 = \frac{1}{2\pi}\log\frac{(1-x)^2+y^2}{(1+x)^2+y^2} \tag{4}$$

is the arrangement of another issue for which (3) addresses any equipotential line and (2) any line of stream.

(2) lessens to

$$\frac{2y}{x^2+y^2-1} = \tan a\pi$$

or and

$$x^2 + (y - \operatorname{ctn} a\pi)^2 = \csc^2 a\pi; \tag{5}$$

and (3) to and

$$x^2 + y^2 + 2\frac{e^{2b\pi}+1}{e^{2b\pi}-1}x + 1 = 0$$

---

[1] The capability form to

$$\frac{1}{\pi}\left[\tan^{-1}\frac{1+x}{y}+\tan^{-1}\frac{1-x}{y}\right]$$

could have been found as follows. Assuming that ϕ is the necessary capability and ψ the given capability we have by Int. Cal. Arts. 211, 212, and 213 the relations

$$D_x\phi = D_y\psi \quad \text{and} \quad D_y\phi = -D_x\psi.$$

Here

$$D_y\psi = -\frac{1}{\pi}\left[\frac{1+x}{(1+x)^2+y^2}+\frac{1-x}{(1-x)^2+y^2}\right]$$

and and $-D_x\psi = -\frac{1}{\pi}\left[\frac{y}{(1+x)^2+y^2}-\frac{y}{(1-x)^2+y^2}\right].$

In the event that now we coordinate $D_y\psi$ regarding x treating y as a consistent and add an erratic capability of y we will have ϕ. So that

$$\phi = -\frac{1}{2\pi}\left\{\log[(1+x)^2+y^2]-\log[(1-x)^2+y^2]\right\} + f(y).$$

$$D_y\phi = -\frac{1}{\pi}\left[\frac{y}{(1+x)^2+y^2}-\frac{y}{(1-x)^2+y^2}\right] + \frac{df(y)}{dy}$$

Contrasting this and its equivalent $-D_x\psi$ above we find $\frac{df(y)}{dy} = 0$ and $f(y) = C$ a steady therefore

$$\frac{1}{2\pi}\log\frac{(1-x)^2+y^2}{(1+x)^2+y^2} + C,$$

where C might be taken at delight, is our necessary form function.

PHYSICS APPLICATIONS

or
$$\left(x + \frac{e^{b\pi} + e^{-b\pi}}{e^{b\pi} - e^{-b\pi}}\right)^2 + y^2 = \left(\frac{e^{b\pi} + e^{-b\pi}}{e^{b\pi} - e^{-b\pi}}\right)^2 - 1$$

or $and (x + \ctnh b\pi)^2 + y^2 = \csch^2 b\pi.$ (6)

(5) and (6) are circles. The circles (5) have their focuses in the pivot of Y, furthermore, go through the focuses $(-1, 0)$ and $(1, 0)$; and the circles (6) have their focuses in the hub of X.

(4) is the finished arrangement, (6) is any equipotential line and (5) any line of stream for a plane sheet in which the focuses in the circuits of two given circles whose focuses are further separated than the amount of their radii are kept at different steady possibilities, or where a source and a sink of equivalent force are set at the places $(-1, 0)$ and $(1, 0)$. A significant useful model is where two wires associated with the shafts of a battery are put with their free finishes in touch with a slight plane sheet of leading material. The figure shows the equipotential endlessly lines of stream of one or the other framework.

The total figure would have the hub of X for a pivot of evenness.

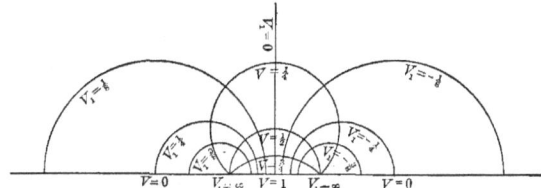

EXAMPLES.

1. Show that if $f(x) = a_1$ when $x < -b$, $f(x) = a_2$ when $-b < x < b$, $f(x) = a_3$ when $x > b$,
$$V = \frac{a_1 + a_3}{2} + \frac{1}{\pi}\left[(a_2 - a_1)\tan^{-1}\frac{b+x}{y} + (a_2 - a_3)\tan^{-1}\frac{b-x}{y}\right].$$

2. Show that if $f(x) = 0$ if $x < 0$, $f(x) = a_1$ if $0 < x < b_1$, $f(x) = a_2$ if $b_1 < x < b_2$, $f(x) = a_3$ if $b_2 < x < b_3$, &c.,
$$V = \frac{1}{\pi}\left[a_1 \tan^{-1}\frac{x}{y} + (a_1 - a_2)\tan^{-1}\frac{b_1 - x}{y} + (a_2 - a_3)\tan^{-1}\frac{b_2 - x}{y} \right.$$
$$\left. + (a_3 - a_4)\tan^{-1}\frac{b_3 - x}{y} + \cdots\right].$$

3. Show that if $f(x) = -1$ if $x < -1$, $f(x) = x$ if $-1 < x < 1$, $f(x) = 1$ if $x > 1$,
$$V = \frac{1}{\pi}\left[(1+x)\tan^{-1}\frac{1+x}{y} - (1-x)\tan^{-1}\frac{1-x}{y} + \frac{y}{2}\log\frac{(1-x)^2 + y^2}{(1+x)^2 + y^2}\right].$$

4. Show that if $f(x) = -1$ if $x < -1$, $f(x) = 0$ if $-1 < x < 1$, $f(x) = 1$ if $x > 1$,
$$V = \frac{1}{\pi}\left[\tan^{-1}\frac{1+x}{y} - \tan^{-1}\frac{1-x}{y}\right].$$

Show that the equipotential lines are symmetrical hyperbolas going through the focuses $(-1, 0)$ and $(1, 0)$, and that the lines of stream are Cassinian ovals having $(-1, 0)$ and $(1, 0)$ as foci. The lines of stream are equipotential lines furthermore, the equipotential lines will be lines of stream for the situation where the focuses $(-1, 0)$ and $(1, 0)$ are kept at a similar boundless potential, or where tiny ovals encompassing these focuses are kept at a similar limited potential. The case is roughly that of a couple of wires associated with a similar shaft of a battery whose other post is grounded, and afterward positioned with their finishes in touch with a flimsy plane directing sheet.

5. Show that if $f(x) = 0$ if $x < 0$, $f(x) = -1$ if $0 < x < a$, $f(x) = 0$ in the event that $a < x < b$, and $f(x) = 1$ if $x > b$,

$$V = \frac{1}{\pi}\left[\frac{\pi}{2} - \tan^{-1}\frac{a-x}{y} - \tan^{-1}\frac{b-x}{y} - \tan^{-1}\frac{x}{y}\right].$$

The form capability

$$V = \frac{1}{2\pi}\log\frac{x^2+y^2}{[(a-x)^2+y^2][(b-x)^2+y^2]}$$

is the answer for the situation where a sink and two wellsprings of equivalent power lie on the hub of X, the sink at the beginning and the sources at the distances a and b to one side of the beginning. One of the lines of stream is effortlessly seen to be the circle $x^2 + y^2 = ab$.

47. Assuming that the plane directing sheet has two straight edges at right points with one another and one is kept at possible zero while the worth of the potential capability is given at each place of the second, that is if $V = 0$ when $x = 0$ and $V = f(x)$ when $y = 0$, the arrangement is promptly gotten. It is

$$V = \frac{2}{\pi}\int_0^\infty d\alpha \int_0^\infty e^{-\alpha y} f(\lambda)\sin\alpha x \sin\alpha\lambda \, d\lambda. \tag{1}$$

v. (9) Art. 44.

This diminishes to

$$V = \frac{1}{\pi}\int_0^\infty f(\lambda)d\lambda\left[\frac{y}{y^2+(\lambda-x)^2} - \frac{y}{y^2+(\lambda+x)^2}\right]. \tag{2}$$

v. Ex. 3 Art. 45.

EXAMPLES.

1. If $V = 0$ when $y = 0$ and $V = F(y)$ when $x = 0$ show that

$$V = \frac{2}{\pi}\int_0^\infty d\alpha \int_0^\infty e^{-\alpha x}F(\lambda)\sin\alpha y\sin\alpha\lambda \, d\lambda$$

$$= \frac{1}{\pi}\int_0^\infty F(\lambda)d\lambda\left[\frac{x}{x^2+(\lambda-y)^2} - \frac{x}{x^2+(\lambda+y)^2}\right].$$

2. If $V = f(x)$ when $y = 0$ and $V = F(y)$ when $x = 0$ show that

$$V = \frac{1}{\pi}\int_0^\infty and\left[f(\lambda)\left(\frac{y}{y^2+(\lambda-x)^2} - \frac{y}{y^2+(\lambda+x)^2}\right)\right.$$

$$\left. + F(\lambda)\left(\frac{x}{x^2+(\lambda-y)^2} - \frac{x}{x^2+(\lambda+y)^2}\right)\right]d\lambda.$$

3. On the off chance that $F(y) = b$ the aftereffect of Ex. 2 diminishes to

$$V = \frac{2b}{\pi}\tan^{-1}\frac{y}{x} + \frac{1}{\pi}\int_0^\infty f(\lambda)d\lambda\left[\frac{y}{y^2+(\lambda-x)^2} - \frac{y}{y^2+(\lambda+x)^2}\right].$$

PHYSICS APPLICATIONS

4. If $F(y) = 1$ for $0 < y < 1$ and $F(y) = 0$ for $y > 1$ while $f(x) = 1$ for $0 < x < 1$ and $f(x) = 0$ for $x > 1$

$$V = \frac{1}{\pi} and \left[\tan^{-1}\frac{1-x}{y} - \tan^{-1}\frac{1+x}{y} + 2\tan^{-1}\frac{y}{x}\right.$$
$$\left. + \tan^{-1}\frac{1-y}{x} - \tan^{-1}\frac{1+y}{x} + 2\tan^{-1}\frac{x}{y}\right].$$

5. Assuming one edge of the leading sheet treated in Art. 47 is protected, so that $D_x V = 0$ if $x = 0$ and $V = f(x)$ when $y = 0$

$$V = \frac{2}{\pi}\int_0^\infty d\alpha \int_0^\infty e^{-\alpha y} f(\lambda)\cos\alpha x \cos\alpha\lambda.d\lambda$$
$$= \frac{1}{\pi}\int_0^\infty f(\lambda)d\lambda \left[\frac{y}{y^2+(\lambda+x)^2} + \frac{y}{y^2+(\lambda-x)^2}\right].$$

48. On the off chance that the directing sheet is a long strip with equal edges one of which is at expected zero while the worth of the potential capability is given by any stretch of the imagination marks of the other, that is if $V = 0$ when $y = 0$ and $V = F(x)$ when $y = b$ the issue is definitely not a truly challenging one.

Since we are not generally worried about the worth of V when $y = \infty$ $V = e^{\alpha y}\sin\alpha x$ and $V = e^{\alpha y}\cos\alpha x$ are accessible as specific arrangements of the condition

$$D_x^2 V + D_y^2 V = 0 \tag{1}$$

as well as $V = e^{-\alpha y}\sin\alpha x$ and $V = e^{-\alpha y}\cos\alpha x$.

Consequently $\qquad\dfrac{e^{\alpha y}+e^{-\alpha y}}{2}\sin\alpha x = \cosh\alpha y\sin\alpha x\, and$[Int. Cal. Art. 43 (2)]

and $\qquad\dfrac{e^{\alpha y}-e^{-\alpha y}}{2}\sin\alpha x = \sinh\alpha y\sin\alpha x\, and$[Int. Cal. Art. 43 (1)]

and $\qquad\cosh\alpha y\cos\alpha x\quad$ and $\quad\sinh\alpha y\cos\alpha x$

are presently accessible upsides of V and can be utilized definitively as $e^{-\alpha y}\cos\alpha x$ and $e^{-\alpha y}\sin\alpha x$ are utilized in Art. 44.

Following a similar course as in Art. 44 we get

$$V = \frac{1}{\pi}\int_0^\infty d\alpha \int_{-\infty}^\infty \frac{\sinh\alpha y}{\sinh\alpha b} F(\lambda)\cos\alpha(\lambda-x).d\lambda \tag{2}$$

as an answer of (1) which will lessen to $V = F(x)$ when $y = b$

and to $\qquad V = 0\quad$ when $\quad y = 0$, since $\sinh 0 = \dfrac{1-1}{2} = 0$,

also (2) is consequently our expected arrangement.

Assuming V is to be equivalent to zero when $y = b$ and to $f(x)$ when $y = 0$ we have just to supplant y by $b - y$ and $F(x)$ by $f(x)$ in (2). We get

$$V = \frac{1}{\pi}\int_0^\infty d\alpha \int_{-\infty}^\infty \frac{\sinh\alpha(b-y)}{\sinh\alpha b} f(\lambda)\cos\alpha(\lambda-x).d\lambda. \tag{3}$$

PHYSICS APPLICATIONS

On the off chance that $V = f(x)$ when $y = 0$ and $V = F(x)$ when $y = b$ then, at that point,

$$V = \frac{1}{\pi}\int_0^\infty d\alpha \int_{-\infty}^\infty \frac{\sinh\alpha(b-y)}{\sinh\alpha b} f(\lambda)\cos\alpha(\lambda - x).d\lambda$$

$$+ \frac{1}{\pi}\int_0^\infty d\alpha \int_{-\infty}^\infty \frac{\sinh\alpha y}{\sinh\alpha b} F(\lambda)\cos\alpha(\lambda - x).d\lambda.$$

This can be extensively improved on by the guide of the equation

$$\int_0^\infty \frac{\sinh px}{\sinh qx}\cos rx.dx = \frac{\pi}{2q}\frac{\sin\frac{p\pi}{q}}{\cos\frac{p\pi}{q} + \cosh\frac{r\pi}{q}}$$

in the event that $p^2 < q^2$. [Bierens de Haan, Tables of Def. Int. (7) 265] and becomes

$$V = and\frac{1}{2b}\sin\frac{\pi}{b}(b-y)\int_{-\infty}^\infty f(\lambda)\frac{d\lambda}{\cos\frac{\pi(b-y)}{b} + \cosh\frac{\pi}{b}(\lambda - x)}$$

$$+ \frac{1}{2b}\sin\frac{\pi y}{b}\int_{-\infty}^\infty F(\lambda)\frac{d\lambda}{\cos\frac{\pi y}{b} + \cosh\frac{\pi}{b}(\lambda - x)} \quad \text{or}$$

$$V = \frac{1}{2b}\sin\frac{\pi y}{b}\int_{-\infty}^\infty \left[\frac{f\lambda}{\cosh\frac{\pi}{b}(\lambda - x) - \cos\frac{\pi y}{b}} + \frac{F\lambda}{\cosh\frac{\pi}{b}(\lambda - x) + \cos\frac{\pi y}{b}}\right]d\lambda. \tag{5}$$

EXAMPLES.

1. Given the equation

$$\int \frac{dx}{a + b\cosh x} = \frac{2}{\sqrt{b^2 - a^2}}\tan^{-1}\left(\sqrt{\frac{b-a}{b+a}}\tanh\frac{x}{2}\right) \quad \text{if } b > a,$$

show that if $V = 1$ when $y = 0$ and $V = 0$ when $y = b$ $V = \frac{1}{b}(b-y)$.

2. Show that if $V = 0$ when $y = b$, $V = -1$ when $y = 0$ and $x < 0$, and $V = 1$ when $y = 0$ and $x > 0$

$$V = \frac{2}{\pi}\tan^{-1}\left[\frac{\tanh\frac{\pi x}{2b}}{\tan\frac{\pi y}{2b}}\right]$$

The answer for the form framework, or at least, for a strip having a source at $(0,0)$ and a vastly far off sink is

$$V = -\frac{1}{\pi}\log\left[\cosh^2\frac{\pi x}{2b} - \cos^2\frac{\pi y}{2b}\right].$$

3. Show that if $V = -1$ when $y = 0$ and $x < 0$, $V = 1$ when $y = 0$ what's more, $x > 0$, $V = -1$ when $y = b$ and $x < 0$, and $V = 1$ when $y = b$ and $x > 0$,

$$V and = \frac{2}{\pi}\tan^{-1}\left(\tan\frac{\pi}{2b}(b-y)\tanh\frac{\pi x}{2b}\right) + \frac{2}{\pi}\tan^{-1}\left(\tan\frac{\pi}{2b}y\tanh\frac{\pi x}{2b}\right)$$

$$= \frac{2}{\pi}\tan^{-1}\left[\frac{\sinh\frac{\pi x}{b}}{\sin\frac{\pi y}{b}}\right].$$

The answer for the form framework, that is to say, for a strip having a source and a sink at the focuses $(0,0)$ and $(0,b)$ is

$$V = \frac{1}{\pi}\log\left[\frac{\cosh\frac{\pi x}{b} + \cos\frac{\pi y}{b}}{\cosh\frac{\pi x}{b} - \cos\frac{\pi y}{b}}\right].$$

PHYSICS APPLICATIONS   66

4. If $V = 0$ when $x = 0$, $V = f(x)$ when $y = 0$ and $x > 0$, and $V = 0$ when $y = b$ and $x > 0$.

$$V = \frac{1}{\pi} \int_0^\infty d\alpha \int_0^\infty \frac{\sinh\alpha(b-y)}{\sinh\alpha b}[\cos\alpha(\lambda - x) - \cos\alpha(\lambda + x)]f(\lambda)d\lambda$$

$$= \frac{1}{2b}\sin\frac{\pi y}{b}\int_0^\infty \left[\frac{1}{\cosh\frac{\pi}{b}(\lambda - x) - \cos\frac{\pi y}{b}} - \frac{1}{\cosh\frac{\pi}{b}(\lambda + x) - \cos\frac{\pi y}{b}}\right]f(\lambda)d\lambda$$

for positive upsides of x and for upsides of y among 0 and b.

5. If $V_1 = 0$ when $x = 0$, $V_1 = F(x)$ when $y = b$ and $x > 0$, and $V_1 = 0$ when $y = 0$ and $x > 0$

$$V_1 = \frac{1}{2b}\sin\frac{\pi y}{b}\int_0^\infty \left[\frac{1}{\cosh\frac{\pi}{b}(\lambda - x) + \cos\frac{\pi y}{b}} - \frac{1}{\cosh\frac{\pi}{b}(\lambda + x) - \cos\frac{\pi y}{b}}\right]F(\lambda)d\lambda$$

for positive upsides of x and upsides of y among 0 and b.

6. If $V_2 = 0$ when $x = 0$, $V_2 = f(x)$ when $y = 0$ and $x > 0$, and $V_2 = F(x)$ when $y = b$ and $x > 0$

$$V_2 = V + V_1 \quad \text{for} \quad x > 0 \quad \text{and} \quad 0 < y < b. \qquad \text{(v. Exs. 4 and 5)}$$

7. Assuming one edge of the strip depicted in Art. 48 is protected with the goal that we have $V = f(x)$ when $y = 0$ and $D_y V = 0$ when $y = b$ show that

$$V = \frac{1}{\pi}\int_0^\infty d\alpha \int_{-\infty}^\infty \frac{\cosh\alpha(b-y)}{\cosh\alpha b} f(\lambda)\cos\alpha(\lambda - x).d\lambda.$$

By the guide of the equation

$$\int_0^\infty \frac{\cosh px}{\cosh qx}\cos rx.dx = \frac{\pi}{q}\frac{\cosh\frac{r\pi}{2q}\cos\frac{p\pi}{2q}}{\cos\frac{p\pi}{q} + \cosh\frac{r\pi}{q}} \qquad \text{if } p < q,$$

[Bierens de Haan, Def. Int. Tables (6) 265],
lessen this to

$$V = \frac{1}{b}\sin\frac{\pi y}{2b}\int_{-\infty}^\infty \frac{f(\lambda)\cosh\frac{\pi}{2b}(\lambda - x)}{\cosh\frac{\pi}{b}(\lambda - x) - \cos\frac{\pi y}{b}}d\lambda.$$

8. If $V = 0$ when $y = 0$ or b and $x < -a$, $V = 1$ when $y = 0$ or b what's more, $-a < x < a$, and $V = 0$ when $y = 0$ or b and $x > a$

$$V = \frac{1}{\pi}\left[\tan^{-1}\frac{\sinh\frac{\pi(a-x)}{b}}{\sin\frac{\pi y}{b}} + \tan^{-1}\frac{\sinh\frac{\pi(a+x)}{b}}{\sin\frac{\pi y}{b}}\right].$$

9. If $V = 0$ when $y = 0$ or b and $x < -a$, $V = 1$ when $y = 0$ and $-a < x < a$, $V = 0$ when $y = 0$ or b and $x > a$, and $V = -1$ when $y = b$ and $-a < x < a$

$$V = \frac{1}{\pi}\left[\tan^{-1}\frac{\tanh\frac{\pi(a-x)}{b}}{\tan\frac{\pi y}{b}} + \tan^{-1}\frac{\tanh\frac{\pi(a+x)}{b}}{\tan\frac{\pi y}{b}}\right].$$

10. A framework form to that of Ex. 9 is $V = +\infty$ when $y = 0$ or b what's more, $x = -a$, $V = -\infty$ when $y = 0$ or b and $x = a$. For this situation

$$V = \frac{1}{2\pi}\log\frac{\sin^2\frac{\pi y}{b} + \sinh^2\frac{\pi(a-x)}{b}}{\sin^2\frac{\pi y}{b} + \sinh^2\frac{\pi(a+x)}{b}}.$$

PHYSICS APPLICATIONS

49. Let us take now an issue in the progression of intensity. Assume we have an endless strong in which intensity streams just in one heading, and that toward the beginning the temperature of each mark of the strong is given. Allow it to be expected to see as the temperature of any place of the strong toward the finish of the time t.

Here we need to address the condition

$$D_t u = a^2 D_x^2 u \tag{1}$$

[v. Art. 1 (II)] subject to the condition

$$u = f(x) \quad \text{when} \quad t = 0. \tag{2}$$

As the situation (1) is direct with steady coefficients we can get a specific arrangement by the gadget utilized in Arts. 7 and 8.

Let $u = e^{\beta t + \alpha x}$ and substitute in (1). We get

$$\beta = a^2 \alpha^2$$

as the main connection which need hold among β and α.

Hence
$$u = e^{\alpha x + a^2 \alpha^2 t} = e^{a^2 \alpha^2 t} e^{\alpha x} \tag{3}$$

is an answer of (1) regardless of what worth is given to α.

To get a mathematical structure supplant α by αi.

Then
$$u = e^{-a^2 \alpha^2 t} e^{\alpha x i}.$$

If in (3) we supplant α by $-\alpha i$ we get

$$u = e^{-a^2 \alpha^2 t} e^{-\alpha x i}.$$

As in Arts. 7 and 8 we get from these qualities

$$u = e^{-a^2 \alpha^2 t} \sin \alpha x \quad \text{and} \quad u = e^{-a^2 \alpha^2 t} \cos \alpha x$$

as specific arrangements of (1), α being entirely unlimited.

From these qualities we wish to develop a worth of u which will decrease to $f(x)$ when $t = 0$ and will in any case be an answer of (1).

We have
$$f(x) = \frac{1}{\pi} \int_0^\infty d\alpha \int_{-\infty}^\infty f(\lambda) \cos \alpha(\lambda - x).d\lambda \tag{4}$$

v. Art. 32 (3), and by continuing as in Art. 44 we get

$$u = \frac{1}{\pi} \int_0^\infty d\alpha \int_{-\infty}^\infty e^{-a^2 \alpha^2 t} f(\lambda) \cos \alpha(\lambda - x).d\lambda \tag{5}$$

as our expected worth of u.

This can be extensively rearranged.

Changing the request for combination

$$u = \frac{1}{\pi} \int_{-\infty}^\infty f(\lambda) d\lambda \int_0^\infty e^{-a^2 \alpha^2 t} \cos \alpha(\lambda - x).d\alpha. \tag{6}$$

$$\int_0^\infty e^{-a^2 \alpha^2 t} \cos \alpha(\lambda - x).d\alpha = \frac{1}{2a} \sqrt{\frac{\pi}{t}} \cdot e^{-\frac{(\lambda - x)^2}{4a^2 t}} \tag{7}$$

PHYSICS APPLICATIONS   68

by the recipe

$$\int_0^\infty e^{-a^2x^2} \cos bx \, dx = \frac{\sqrt{\pi}}{2a} e^{-\frac{b^2}{4a^2}} \qquad \text{[Int. Cal. Art. 94 (2)]}$$

Hence
$$u = \frac{1}{2a\sqrt{\pi t}} \int_{-\infty}^{\infty} f(\lambda) e^{-\frac{(\lambda-x)^2}{4a^2 t}} \, d\lambda. \qquad (8)$$

Let now
$$\beta = \frac{\lambda - x}{2a\sqrt{t}},$$

then
$$\lambda = x + 2a\sqrt{t}.\beta$$

and
$$u = \frac{1}{\sqrt{\pi}} \int_{-\infty}^{\infty} f(x + 2a\sqrt{t}.\beta) e^{-\beta^2} \, d\beta. \text{ and} \qquad (9)$$

EXAMPLES.

1. Let the strong be of limitless degree and allowed the temperature to be equivalent to a steady c at the time $t = 0$.

Then
$$u = \frac{c}{\sqrt{\pi}} \int_{-\infty}^{\infty} e^{-\beta^2} \, d\beta = \frac{2c}{\sqrt{\pi}} \int_0^\infty e^{-\beta^2} \, d\beta = c.$$

v. Int. Cal. Art. 92 (2).

2. Let $u = x$ when $t = 0$.

Then
$$u = \frac{1}{\sqrt{\pi}} \int_{-\infty}^{\infty} (x + 2a\sqrt{t}.\beta) e^{-\beta^2} \, d\beta = x.$$

3. Let $u = x^2$ when $t = 0$.

Then
$$u = x^2 + 2a^2 t.$$

4. Let $u = 0$ if $x < -b$, $u = 1$ if $-b < x < b$, and $u = 0$ if $x > b$, when $t = 0$.

Then

$$u = \frac{1}{\sqrt{\pi}} \int_{-\frac{b+x}{2a\sqrt{t}}}^{\frac{b-x}{2a\sqrt{t}}} e^{-\beta^2} \, d\beta = \frac{2}{\sqrt{\pi}} \left[\frac{b}{2a\sqrt{t}} - \frac{b^3 + 3bx^2}{3(2a\sqrt{t})^3} + \frac{b^5 + 10b^3 x^2 + 5bx^4}{5.2!(2a\sqrt{t})^5} - \cdots \right].$$

5. Let $u = 0$ if $x < 0$ and $u = 1$ if $x > 0$ when $t = 0$.
Then

$$u = \frac{1}{\sqrt{\pi}} \int_{-\frac{x}{2a\sqrt{t}}}^{\infty} e^{-\beta^2} \, d\beta = \frac{1}{\sqrt{\pi}} \left[\int_0^{\frac{x}{2a\sqrt{t}}} e^{-\beta^2} \, d\beta + \int_0^\infty e^{-\beta^2} \, d\beta \right] = \frac{1}{\sqrt{\pi}} \int_0^{\frac{x}{2a\sqrt{t}}} e^{-\beta^2} \, d\beta + \frac{1}{2}$$

$$= \frac{1}{2} + \frac{1}{\sqrt{\pi}} \left[\frac{x}{2a\sqrt{t}} - \frac{x^3}{3.(2a\sqrt{t})^3} + \frac{x^5}{5.2!(2a\sqrt{t})^5} - \frac{x^7}{7.3!(2a\sqrt{t})^7} + \cdots \right].$$

6. An iron section 10 c.m. thick is put between and in touch with two exceptionally thick iron chunks. The underlying temperature of the center piece is 100Â°, and of every one of the

PHYSICS APPLICATIONS   69

external pieces 0Â°. Required the temperature of a point in the center of the internal section fifteen minutes after the chunks have been assembled. Given $a^2 = 0.185$ in C.G.S. units. *Ans.*, 21Â°.6.

7. Two extremely thick iron chunks one of which is at the temperature 0Â° and the other at the temperature 100Â° all through are set together up close and personal. Find the temperature of every piece 10 c. m. from their familiar face fifteen minutes after they have been put together. *Ans.*, 70Â°.8, 29Â°.2.

8. Track down a specific arrangement of $D_t u = a^2 D_x^2 u$ with the understanding that it is of the structure $u = T.X$ where T is an element of t alone and X is a capability of x alone.

50. Assuming our strong has one plane face which is kept at the steady temperature zero, and we start with some random dispersion of intensity, the issue is to some degree adjusted.

Take the beginning of coÃ¶rdinates in the plane face. Then we have as in the past the condition

$$D_t u = a^2 D_x^2 u, \tag{1}$$

be that as it may, our circumstances are

$$u = 0 \quad \text{when} \quad x = 0 \tag{2}$$
$$u = f(x) \quad " \quad and\, t = 0 \tag{3}$$

also, we are concerned exclusively with positive upsides of x.

We may then utilize the structure (4) Art. 32

$$f(x) = \frac{2}{\pi} \int_0^\infty d\alpha \int_0^\infty f(\lambda) \sin \alpha x \sin \alpha \lambda . d\lambda, \tag{4}$$

also, continuing as in the last area we get

$$u = \frac{2}{\pi} \int_0^\infty d\alpha \int_0^\infty e^{-a^2 \alpha^2 t} f(\lambda) \sin \alpha x \sin \alpha \lambda . d\lambda \tag{5}$$

as our necessary arrangement. This might be decreased significantly.

$$u = \frac{1}{\pi} \int_0^\infty f(\lambda) d\lambda \int_0^\infty e^{-a^2 \alpha^2 t} [\cos \alpha (\lambda - x) - \cos \alpha (\lambda + x)] d\alpha,$$

$$\text{or} \quad u = \frac{1}{2a\sqrt{\pi t}} \int_0^\infty f(\lambda)(e^{-\frac{(\lambda-x)^2}{4a^2 t}} - e^{-\frac{(\lambda+x)^2}{4a^2 t}}) d\lambda \tag{6}$$

by (7) Art. 49, and this might be decreased to the structure

$$u = \frac{1}{\sqrt{\pi}} \left[\int_{-\frac{x}{2a\sqrt{t}}}^\infty e^{-\beta^2} f(x + 2a\sqrt{t}.\beta) d\beta - \int_{\frac{x}{2a\sqrt{t}}}^\infty e^{-\beta^2} f(-x + 2a\sqrt{t}.\beta) d\beta \right]. \tag{7}$$

EXAMPLES.

1. Let the underlying temperature be steady and equivalent to c.

PHYSICS APPLICATIONS

Then, at that point,

$$u = \frac{c}{\sqrt{\pi}}\left[\int_{\frac{x}{2a\sqrt{t}}}^{\infty} e^{-\beta^2}d\beta - \int_{\frac{x}{2a\sqrt{t}}}^{\infty} e^{-\beta^2}d\beta\right]$$

$$= \frac{2c}{\sqrt{\pi}}\int_{0}^{\frac{x}{2a\sqrt{t}}} e^{-\beta^2}d\beta$$

$$= \frac{2c}{\sqrt{\pi}}\left[\frac{x}{2a\sqrt{t}} - \frac{x^3}{3.(2a\sqrt{t})^3} + \frac{x^5}{5.2!.(2a\sqrt{t})^5} - \frac{x^7}{7.3!.(2a\sqrt{t})^7} + \cdots\right].$$

2. Accepting that the earth was initially at the temperature 7000Â° Fahrenheit all through, and that the surface was kept at the consistent temperature 0Â°, find (1) the temperature 10 miles underneath the surface 10,000,000 years after the cooling started; (2) the temperature 1 mile underneath the surface at something similar age; (3) the temperature 10 miles beneath the surface 100,000,000 years later the cooling started; (4) the temperature 1 mile beneath the surface at something similar age; (5) the rate at which the temperature was expanding with the distance from the surface at each point at every age.

Disregard the convexity of the world's surface and take Sir Wm. Thomson's worth of a^2(400) the foot, the Fahrenheit degree, and the year being taken as units. (Thomson and Tait's Nat. Phil. Vol. II. Informative supplement.)

Ans., (1) 3114Â°; (2) 329Â°.5; (3) 1036Â°; (4) 103Â°; (5) 1Â° for each 20 feet, 3Â° for each 50 feet, 1Â° for each 50 feet, 1Â° for each 50 feet.

3. Let the underlying temperature be steady and equivalent to $-b$, then by Ex. 1

$$u = -\frac{2b}{\sqrt{\pi}}\int_{0}^{\frac{x}{2a\sqrt{t}}} e^{-\beta^2}d\beta.$$

4. Let the temperature of the plane face be b rather than nothing, and let the introductory temperature be zero.

Then we have just to add b to the second individual from the arrangement in Ex. 3, as we may since $u = b$ is an answer of (1) Art. 49, and we get

$$u = b\left(1 - \frac{2}{\sqrt{\pi}}\int_{0}^{\frac{x}{2a\sqrt{t}}} e^{-\beta^2}d\beta\right).$$

5. Let $u = b$ when $x = 0$ and $u = f(x)$ when $t = 0$.
Then

$$u = b\left(1 - \frac{2}{\sqrt{\pi}}\int_{0}^{\frac{x}{2a\sqrt{t}}} e^{-\beta^2}d\beta\right) + \frac{1}{2a\sqrt{\pi t}}\int_{0}^{\infty} f(\lambda)[e^{-\frac{(\lambda-x)^2}{4a^2 t}} - e^{-\frac{(\lambda+x)^2}{4a^2 t}}]d\lambda$$

by (6) Art. 50.

6. Let $u = b$ when $x = 0$ and $u = c$ when $t = 0$.

Then
$$u = b + (c-b)\frac{2}{\sqrt{\pi}}\int_{0}^{\frac{x}{2a\sqrt{t}}} e^{-\beta^2}d\beta.$$

7. On the off chance that the earth has been cooling for a long time from a uniform temperature, demonstrate that the pace of cooling is most prominent at a profundity of around 76 miles, and that at a profundity of around 130 miles the pace of cooling has reached its most extreme

PHYSICS APPLICATIONS

incentive forever. Let $a^2 = 400$.

8. Show that assuming the plane essence of the strong considered in Art. 50 rather than being kept at temperature zero is impenetrable to warm

$$u = \frac{1}{2a\sqrt{\pi t}} \int_0^\infty f(\lambda)(e^{-\frac{(\lambda-x)^2}{4a^2 t}} + e^{-\frac{(\lambda+x)^2}{4a^2 t}})d\lambda. \qquad \text{v. (6) Art. 50.}$$

51. Assuming the temperature of the plane essence of the strong depicted in Art. 50 is a given capability of the time and the underlying temperature is zero, the arrangement of the issue can be gotten by an extremely brilliant technique because of Riemann.

Here we need to address the condition

$$D_t u = a^2 D_x^2 u \qquad (1)$$

dependent upon the circumstances

$$\left. \begin{array}{ll} u = F(t) & \text{when} \quad x = 0 \\ u = 0 & \text{''} \qquad t = 0. \end{array} \right\} \qquad (2)$$

That's what we know

$$u = \frac{2}{\sqrt{\pi}} \int_0^{\frac{x}{2a\sqrt{t}}} e^{-\beta^2} d\beta$$

is an answer of (1), v. Ex. 1 Art. 50. It is effortlessly shown that

$$u = \frac{2}{\sqrt{\pi}} \int_0^{\frac{x}{2a\sqrt{t-c}}} e^{-\beta^2} d\beta, \qquad (3)$$

where c is any steady, is an answer of (1).

For

$$D_t u = -\frac{2}{\sqrt{\pi}} \frac{x}{2a} \frac{1}{2(t-c)^{\frac{3}{2}}} e^{-\frac{x^2}{4a^2(t-c)}} = -\frac{x}{2a\sqrt{\pi}}(t-c)^{-\frac{3}{2}} e^{-\frac{x^2}{4a^2(t-c)}}$$

$$D_x u = \frac{2}{\sqrt{\pi}} \frac{1}{2a\sqrt{t-c}} e^{-\frac{x^2}{4a^2(t-c)}}$$

$$D_x^2 u = -\frac{2}{\sqrt{\pi}} \frac{1}{2a\sqrt{t-c}} \frac{2x}{4a^2(t-c)} e^{-\frac{x^2}{4a^2(t-c)}} = -\frac{x}{2a^3\sqrt{\pi}}(t-c)^{-\frac{3}{2}} e^{-\frac{x^2}{4a^2(t-c)}}$$

and
$$D_t u = a^2 D_x^2 u.$$

Let $\phi(x,t)$ be an element of x and t which will be equivalent to nothing in the event that t is negative and will be equivalent to

$$1 - \frac{2}{\sqrt{\pi}} \int_0^{\frac{x}{2a\sqrt{t}}} e^{-\beta^2} d\beta$$

assuming t is equivalent to or more prominent than nothing; so that if $x = 0$ $\phi(x,t) = 1$ and if $t = 0$ $\phi(x,t) = 0$. We will currently tackle the accompanying issue, to settle condition (1) dependent upon the circumstances

$$\begin{array}{lll} u = 0 & \text{if} \quad t = 0 \\ u = F(0) & \text{''} \quad x = 0 & \text{and} \quad 0 < t < \tau \\ u = F(k\tau) & \text{''} \quad x = 0 & \text{''} \qquad k\tau < t < (k+1)\tau, \end{array}$$

PHYSICS APPLICATIONS

where k is any entire number and τ is any with no obvious end goal in mind picked time frame.

On the off chance that we structure the worth

$$u = F(k\tau)[\phi(x, t - k\tau) - \phi(x, t - (k+1)\tau)] \tag{4}$$

u will fulfill condition (1) beginning around nothing, solidarity and

$$\frac{2}{\sqrt{\pi}} \int_0^{\frac{x}{2a\sqrt{t-k\tau}}} e^{-\beta^2} d\beta$$

are upsides of u which fulfill (1). u will be zero if $t < k\tau$ by the definition of the capability $\phi(x, t)$; if $x = 0$ $u = 0$ if $t > (k+1)\tau$ and $u = F(k\tau)$ if $k\tau < t < (k+1)\tau$.

In this way

$$u = \sum_{k=0}^{k=\infty} F(k\tau)[\phi(x, t - k\tau) - \phi(x, t - (k+1)\tau)] \tag{5}$$

is the arrangement of the issue expressed previously.

(5) can be rearranged to some degree from the thought that for a given worth of t $\phi(x, t-k\tau) = 0$ if $k\tau > t$. On the off chance that, $n\tau$ is the best entire different of τ not surpassing t,

$$u = \sum_{k=0}^{k=n} F(k\tau)[\phi(x, t - k\tau) - \phi(x, t - (k+1)\tau)]. \tag{6}$$

Assuming now we decline τ endlessly the restricting type of (6) will be the arrangement of the issue expressed toward the start of this article.

(6) might be composed

$$u = \sum_{k=0}^{k=n} F(k\tau) \left[\frac{\phi(x, t - k\tau) - \phi(x, t - (k+1)\tau)}{\tau} \right] \tau \tag{7}$$

also, assuming τ is endlessly diminished the restricting type of (7) is

$$u = -\int_0^t F(\lambda) D_\lambda \phi(x, t - \lambda) d\lambda. \tag{8}$$

Since $t - \lambda$ is positive between the constraints of combination

$$\phi(x, t - \lambda) = 1 - \frac{2}{\sqrt{\pi}} \int_0^{\frac{x}{2a\sqrt{t-\lambda}}} e^{-\beta^2} d\beta,$$

and and $D_\lambda \phi(x, t - \lambda) = \qquad -\frac{x}{2a\sqrt{\pi}} e^{-\frac{x^2}{4a^2(t-\lambda)}} (t - \lambda)^{-\frac{3}{2}};$

also (8) might be composed

$$u = \frac{x}{2a\sqrt{\pi}} \int_0^t F(\lambda) e^{-\frac{x^2}{4a^2(t-\lambda)}} (t - \lambda)^{-\frac{3}{2}} d\lambda, \tag{9}$$

or on the other hand assuming we let $\qquad \beta = \frac{x}{2a\sqrt{t-\lambda}}$

$$u = \frac{2}{\sqrt{\pi}} \int_{\frac{x}{2a\sqrt{t}}}^{\infty} e^{-\beta^2} F\left(t - \frac{x^2}{4a^2\beta^2}\right) d\beta. \tag{10}$$

PHYSICS APPLICATIONS

EXAMPLES.

1. If $u = nt$ when $x = 0$ and $u = 0$ when $t = 0$

$$u = n\left(t + \frac{x^2}{2a^2}\right)\left[1 - \frac{2}{\sqrt{\pi}} \int_0^{\frac{x}{2a\sqrt{t}}} e^{-\beta^2} d\beta\right] - \frac{nx\sqrt{t}}{a\sqrt{\pi}} e^{-\frac{x^2}{4a^2 t}}.$$

2. A thick iron chunk is at the temperature zero all through, one of its plane faces is then kept at the temperature 100Â° Centigrade for 5 minutes, then at the temperature zero for the following 5 minutes, then at the temperature 100Â° for the following 5 minutes, and afterward at the temperature zero. Required the temperature of a point in the chunk 5 c.m. from the face at the termination of 18 minutes. Given; $a^2 = .185$. Ans., 20Â°.1.

3. On the off chance that $u = F(t)$ when $x = 0$ and $u = f(x)$ when $t = 0$,

$$u = \frac{2}{\sqrt{\pi}} \int_{\frac{x}{2a\sqrt{t}}}^{\infty} e^{-\beta^2} F\left(t - \frac{x^2}{4a^2\beta^2}\right) d\beta + \frac{1}{2a\sqrt{\pi t}} \int_0^{\infty} (e^{-\frac{(\lambda-x)^2}{4a^2 t}} - e^{-\frac{(\lambda+x)^2}{4a^2 t}}) f(\lambda) d(\lambda).$$

v. (6) Art. 50.

4. If in Art. (51) $F(t)$ is an occasional capability of the hour of period T it can be communicated by a Fourier's series of the structure

$$F(t) = \frac{1}{2}b_0 + \sum_{m=1}^{m=\infty} [a_m \sin m\alpha t + b_m \cos m\alpha t], \quad \text{where} \quad \alpha = \frac{2\pi}{T},$$

or and $F(t) = \frac{1}{2}b_0 + \sum_{m=1}^{m=\infty} \rho_m \sin(m\alpha t + \lambda_m),$

where $\rho_m \cos \lambda_m = a_m$ and $\rho_m \sin \lambda_m = b_m$. v. Art. 31 Ex. 3.

Show that with this worth of $F(t)$ (10) Art. 51. becomes

$$u = \frac{1}{\sqrt{\pi}} b_0 \int_{\frac{x}{2a\sqrt{t}}}^{\infty} e^{-\beta^2} d\beta + \frac{2}{\sqrt{\pi}} \sum_{m=1}^{m=\infty} \rho_m \left[\sin(m\alpha t + \lambda_m) \int_{\frac{x}{2a\sqrt{t}}}^{\infty} e^{-\beta^2} \cos \frac{m\alpha x^2}{4a^2 \beta^2} d\beta \right.$$
$$\left. - \cos(m\alpha t + \lambda_m) \int_{\frac{x}{2a\sqrt{t}}}^{\infty} e^{-\beta^2} \sin \frac{m\alpha x^2}{4a^2 \beta^2} d\beta \right]$$

also, that as t increments u approaches the worth

$$u = \frac{1}{2}b_0 + \sum_{m=1}^{m=\infty} \rho_m e^{-\frac{x}{a}\sqrt{\frac{m\alpha}{2}}} \sin(m\alpha t - \frac{x}{a}\sqrt{\frac{m\alpha}{2}} + \lambda_m).$$

Considering that

$$\int_0^{\infty} e^{-x^2} \sin \frac{b^2}{x^2} dx = \frac{\sqrt{\pi}}{2} e^{-b\sqrt{2}} \sin b\sqrt{2}; \quad \int_0^{\infty} e^{-x^2} \cos \frac{b^2}{x^2} dx = \frac{\sqrt{\pi}}{2} e^{-b\sqrt{2}} \cos b\sqrt{2}.$$

v. Riemann, Lin. par. dif. gl. Â§ 54.

5. In the event that we are managing a bar of little cross-segment where the intensity not just streams along the bar and yet escapes at the outer layer of the bar into air at the temperature zero we need to address the differential condition

$$D_t u = a^2 D_x^2 u - b^2 u. \qquad \text{v. Fourier, Intensity Â§ 105.}$$

PHYSICS APPLICATIONS   74

Show that for this case

$$u = e^{-(b^2+a^2\alpha^2)t} \sin \alpha x \quad \text{and} \quad u = e^{-(b^2+a^2\alpha^2)t} \cos \alpha x$$

are specific arrangements, and that if $u = f(x)$ when $t = 0$

$$u = \frac{e^{-b^2 t}}{2a\sqrt{\pi t}} \int_{-\infty}^{\infty} e^{-\frac{(\lambda-x)^2}{4a^2 t}} f(\lambda) d\lambda = \frac{e^{-b^2 t}}{\sqrt{\pi}} \int_{-\infty}^{\infty} e^{-\beta^2} f(x + 2a\sqrt{t}.\beta) d\beta.$$

cf. (8) and (9) Art. 49.

In the event that $u = 0$ when $x = 0$ and $u = f(x)$ when $t = 0$

$$u = \frac{e^{-b^2 t}}{\sqrt{\pi}} \left[\int_{\frac{x}{2a\sqrt{t}}}^{\infty} e^{-\beta^2} f(x + 2a\sqrt{t}.\beta) d\beta - \int_{\frac{x}{2a\sqrt{t}}}^{\infty} e^{-\beta^2} f(-x + 2a\sqrt{t}.\beta) d\beta \right].$$

cf. (7) Art. 50.

On the off chance that $u = -e^{-\frac{bx}{a}}$ when $t = 0$ and $u = 0$ when $x = 0$

$$u = \frac{1}{\sqrt{\pi}} \left[e^{\frac{bx}{a}} \int_{\frac{x}{2a\sqrt{t}}}^{\infty} e^{-(b\sqrt{t}+\beta)^2} d\beta - e^{-\frac{bx}{a}} \int_{-\frac{x}{2a\sqrt{t}}}^{\infty} e^{-(b\sqrt{t}+\beta)^2} d\beta \right],$$

what's more, if $u = 1$ when $x = 0$ and $u = 0$ when $t = 0$ we have just to add $e^{-\frac{bx}{a}}$ to the second individual from the last condition, since $u = e^{-\frac{bx}{a}}$ fulfills the condition

$$D_t u = a^2 D_x^2 u - b^2 u.$$

In the event that $u = F(t)$ when $x = 0$ and $u = 0$ when $t = 0$ we can utilize the technique for Art. 51.

$$\phi(x, t - \lambda) = e^{-\frac{bx}{a}} + \frac{1}{\sqrt{\pi}} \left[e^{\frac{bx}{a}} \int_{\frac{x}{2a\sqrt{t-\lambda}}}^{\infty} e^{-(b\sqrt{t-\lambda}+\beta)^2} d\beta - e^{-\frac{bx}{a}} \int_{-\frac{x}{2a\sqrt{t-\lambda}}}^{\infty} e^{-(b\sqrt{t-\lambda}+\beta)^2} d\beta \right],$$

$$-D_\lambda \phi(x, t - \lambda) = \frac{x(t-\lambda)^{-\frac{3}{2}}}{2a\sqrt{\pi}} e^{-b^2(t-\lambda) - \frac{x^2}{4a^2(t-\lambda)}};$$

and

$$u = \frac{x}{2a\sqrt{\pi}} \int_0^t (t-\lambda)^{-\frac{3}{2}} e^{-b^2(t-\lambda) - \frac{x^2}{4a^2(t-\lambda)}} F(\lambda) d\lambda,$$

cf. (9) Art. 51,

or

$$u = \frac{2}{\sqrt{\pi}} \int_{\frac{x}{2a\sqrt{t}}}^{\infty} e^{-\beta^2 - \frac{b^2 x^2}{4a^2 \beta^2}} F\left(t - \frac{x^2}{4a^2 \beta^2}\right) d\beta,$$

cf. (10) Art. 51.

Assuming $F(t)$ is occasional and has the worth taken in Ex. 4, show that the worth drawn nearer by u as t increments is

$$u = \frac{1}{2} b_0 e^{-\frac{bx}{a}} + \sum_{m=1}^{m=\infty} \rho_m e^{-\frac{x\sqrt{2}}{2a} p} \sin\left(m\alpha t - \frac{x\sqrt{2}}{2a} q + \lambda_m\right),$$

where
$$p = (b^2 + \sqrt{b^4 + m^2\alpha^2})^{\frac{1}{2}} \quad \text{and} \quad q = (-b^2 + \sqrt{b^4 + m^2\alpha^2})^{\frac{1}{2}}.$$

PHYSICS APPLICATIONS

Given *and* $\int_0^\infty e^{-x^2-\frac{a^2}{x^2}} dx = \frac{\sqrt{\pi}}{2} e^{-2a}$

$$\int_0^\infty e^{-x^2-\frac{a^2}{x^2}} \sin\frac{b^2}{x^2} dx = \frac{\sqrt{\pi}}{2} e^{-2c} \sin 2d$$

and *and* $\int_0^\infty e^{-x^2-\frac{a^2}{x^2}} \cos\frac{b^2}{x^2} dx = \frac{\sqrt{\pi}}{2} e^{-2c} \cos 2d,$

where

$$c = \frac{\sqrt{2}}{2}(a^2 + \sqrt{a^4+b^4})^{\frac{1}{2}} \quad \text{and} \quad d = \frac{\sqrt{2}}{2}(-a^2 + \sqrt{a^4+b^4})^{\frac{1}{2}}.$$

Ãngstrom's strategy for deciding the conductivity of a metal depends on the outcome recently given (v. Phil. Mag. Feb. 1863), and is portrayed by Sir Wm. Thomson (Encyc. Brit. Article "Intensity") as by a wide margin the best that has yet been conceived.

52. In the event that u is an occasional capability of when $x = 0$ as in Art. 51 Ex. 4 furthermore, we are worried about the restricting worth drew closer by u as t increments we can try not to assess a convoluted unequivocal vital on the off chance that we take the accompanying course.

Since as we have seen in Art. 49 $u = e^{\beta t + \alpha x}$ is an answer of

$$D_t u = a^2 D_x^2 u \tag{1}$$

given just that $\beta = a^2\alpha^2$ we have

$$u = e^{\beta t \hat{A} \pm \frac{x}{a}\sqrt{\beta}}$$

as an answer.

Supplanting β by $\hat{A} \pm \beta i$ this becomes

$$u = e^{\hat{A} \pm \beta t i \hat{A} \pm \frac{x}{a}\sqrt{\beta}\sqrt{\hat{A} \pm i}}$$

or *and* u $= e^{\hat{A} \pm \beta t i \hat{A} \pm \frac{x}{a}\sqrt{\frac{\beta}{2}}(1\hat{A} \pm i)}$

since $\sqrt{I} = \hat{A} \pm \frac{1}{2}\sqrt{2}(1+i)$

and *and* $\sqrt{-i}$ $= \hat{A} \pm \frac{1}{2}\sqrt{2}(1-i).$

Consequently

$$u = e^{-\frac{x}{a}\sqrt{\frac{\beta}{2}}} \sin\left(\beta t - \frac{x}{a}\sqrt{\frac{\beta}{2}}\right), \quad u = e^{-\frac{x}{a}\sqrt{\frac{\beta}{2}}} \cos\left(\beta t - \frac{x}{a}\sqrt{\frac{\beta}{2}}\right), \tag{2}$$

$$u = e^{\frac{x}{a}\sqrt{\frac{\beta}{2}}} \sin\left(\beta t + \frac{x}{a}\sqrt{\frac{\beta}{2}}\right), \quad u = e^{\frac{x}{a}\sqrt{\frac{\beta}{2}}} \cos\left(\beta t + \frac{x}{a}\sqrt{\frac{\beta}{2}}\right), \tag{3}$$

are specific arrangements of (1).

From these we get promptly

$$u = \rho_m e^{-\frac{x}{a}\sqrt{\frac{m\alpha}{2}}} \sin\left(m\alpha t - \frac{x}{a}\sqrt{\frac{m\alpha}{2}} + \lambda_m\right) \tag{4}$$

as an answer. (4) decreases to

$$u = \rho_m \sin(m\alpha t + \lambda_m) \quad \text{when} \quad x = 0$$

and to

$$u = \rho_m e^{-\frac{x}{a}\sqrt{\frac{m\alpha}{2}}} \sin\left(\lambda_m - \frac{x}{a}\sqrt{\frac{m\alpha}{2}}\right) \quad \text{when} \quad t = 0.$$

PHYSICS APPLICATIONS

On the off chance that we add a term which fulfills (1) and which is equivalent to zero when $x = 0$ what's more, to $-\rho_m e^{-\frac{x}{a}\sqrt{\frac{m\alpha}{2}}} \sin\left(\lambda_m - \frac{x}{a}\sqrt{\frac{m\alpha}{2}}\right)$ when $t = 0$ (v. Art. 50) we will have an answer of (1) which is zero when $t = 0$ and which is

$$\rho_m \sin(m\alpha t + \lambda_m) \quad \text{when} \quad x = 0.$$

The term being referred to approaches zero as t increments [v. (7) Art. 50] and we have immediately the arrangement given in Art. 51 Ex. 4, as our expected outcome.

EXAMPLE.

Show that $u = e^{\beta t + \alpha x}$ is an answer of $D_t u = a^2 D_x^2 u - b^2 u$ if $\beta = a^2 \alpha^2 - b^2$, also, subsequently that

$$u = e^{\beta t \pm \frac{x}{a}\sqrt{b^2 + \beta}}, \quad u = e^{\pm \beta t i \pm \frac{x}{a}\sqrt{b^2 \pm \beta i}}, \quad u = e^{\pm \beta t i \pm \frac{x}{a\sqrt{2}}(p \pm qi)},$$

$$u = e^{\pm \frac{px}{a\sqrt{2}}} \sin\left(\beta t \pm \frac{qx}{a\sqrt{2}}\right), \quad \text{and} \quad u = e^{\pm \frac{px}{a\sqrt{2}}} \cos\left(\beta t \pm \frac{qx}{a\sqrt{2}}\right),$$

where

$$p = [\sqrt{\beta^2 + b^4} + b^2]^{\frac{1}{2}} \quad \text{and} \quad q = [\sqrt{\beta^2 + b^4} - b^2]^{\frac{1}{2}},$$

are arrangements. Consequently

$$u = \rho_m e^{-\frac{px}{a\sqrt{2}}} \sin\left(\beta t - \frac{qx}{a\sqrt{2}} + \lambda_m\right)$$

is an answer.

In the event that $\beta = m\alpha$ this last outcome lessens to $u = \rho_m \sin(m\alpha t + \lambda_m)$ when $x = 0$ furthermore, by the thinking of Art. 52 it should be the worth u approaches as t increments in the event that we have similar circumstances as in the last piece of Art. 51 Ex. 5.

53. The entire issue of the progression of intensity is treated by Sir William Thomson (v. Math. and Phys. Papers, Vol. II), and other late journalists from an alternate what's more, unequivocally fascinating perspective, which we will momentarily draw regarding the issue of *Linear Flow*.

Assume we are managing a bar having a little cross-segment and an adiathermanous surface, and take as our unit of intensity the sum expected to raise by a unit the temperature of a unit of length of the bar. In the event that at a mark of the bar a amount Q of intensity is abruptly produced the point is called a *instantaneous heat source* of solidarity Q.

If the intensity as opposed to being unexpectedly created is produced step by step and at a rate that would give Q units of intensity per unit of time the point is known as a *permanent heat source* of solidarity Q.

The temperature anytime of the bar whenever due to a prompt wellspring of solidarity Q at the point $x = \lambda$ is handily tracked down by the guide of recipe (8) Art. 49 as follows:- - -

In the event that an amount of intensity Q is unexpectedly created along the piece of the bar from $x = \lambda$ to $x = \lambda + \Delta\lambda$, where $\Delta\lambda$ is any erratic length, the temperature of that piece will be abruptly raised to $\frac{Q}{\Delta\lambda}$, and we will have by (8) Art. 49

$$u = \frac{Q}{2a\sqrt{\pi t}} \frac{1}{\Delta\lambda} \int\limits_{\lambda}^{\lambda+\Delta\lambda} e^{-\frac{(\lambda-x)^2}{4a^2 t}} d\lambda \tag{1}$$

as the temperature of any place of the bar whenever t from that point.

In the event that now we compose u equivalent to the restricting worth moved toward continuously individual from (1) as $\Delta\lambda$ is made to move toward zero we get

$$u = \frac{Q}{2a\sqrt{\pi t}} e^{-\frac{(\lambda-x)^2}{4a^2 t}} \tag{2}$$

PHYSICS APPLICATIONS   77

as the answer for the situation where we have a momentary source at the point $x = \lambda$.

It is to be seen that in (2) $u = 0$ when $t = 0$ and $u = \frac{Q}{2a\sqrt{\pi t}}$ when $x = \lambda$ and $t > 0$.

Assuming we have a few sources we have just to add the temperatures due to the separate sources.

Recipe (8) Art. 49 may now be viewed as the answer for the situation where we start with a prompt intensity wellspring of solidarity $f(\lambda)d\lambda$ in each component of length of the bar.

A wellspring of solidarity $-Q$ is known as a sink of solidarity Q; and (6) Art. 50 might be viewed as the answer for the situation where we have toward the beginning an immediate wellspring of solidarity $f(\lambda)d\lambda$ in each component of the bar whose distance to one side of the beginning is λ, and a quick sink of solidarity $f(\lambda)d\lambda$ in each component of the bar whose distance to one side of the beginning is λ.

Assuming we have a quick source at the beginning (2) lessens to

$$u = \frac{Q}{2a\sqrt{\pi t}} e^{-\frac{x^2}{4a^2 t}} \tag{3}$$

For an extremely durable wellspring of consistent strength Q at the beginning (3) gives

$$u = \frac{Q}{2a\sqrt{\pi}} \int_0^t e^{-\frac{x^2}{4a^2(t-\tau)}} (t-\tau)^{-\frac{1}{2}} d\tau \tag{4}$$

what's more, for a long-lasting wellspring of variable strength $f(t)$

$$u = \frac{1}{2a\sqrt{\pi}} \int_0^t e^{-\frac{x^2}{4a^2(t-\tau)}} (t-\tau)^{-\frac{1}{2}} f(\tau) d\tau. \tag{5}$$

In (4) and (5) u clearly diminishes to zero when $t = 0$ and $x > 0$, yet its esteem when $x = 0$ not set in stone. We can keep away from the trouble by presenting the origination of a *doublet*.

54. In the event that a source and a sink of equivalent strength Q are made to move toward each other while Q increased by their distance separated is held equivalent to a consistent P the restricting situation is supposed to be expected to a *doublet* of solidarity P whose pivot is digression to the line of approach and focuses from sink to source. A *doublet* of solidarity $-P$ varies from a doublet of solidarity P just in that its hub has the other way.

Allow us to track down the temperature because of a momentary doublet of solidarity P set at the beginning. For a wellspring of solidarity Q at $x = \eta$ and an equivalent sink at $x = -\eta$ we have

$$u = \frac{Q}{2a\sqrt{\pi t}} (e^{-\frac{(\eta-x)^2}{4a^2 t}} - e^{-\frac{(\eta+x)^2}{4a^2 t}}),$$

or if $2\eta Q = P$,

$$u = \frac{P}{4a\eta\sqrt{\pi t}} e^{-\frac{(\eta^2+x^2)}{4a^2 t}} (e^{\frac{\eta x}{2a^2 t}} - e^{-\frac{\eta x}{2a^2 t}})$$

$$= \frac{P}{2a\eta\sqrt{\pi t}} e^{-\frac{(\eta^2+x^2)}{4a^2 t}} \sinh \frac{\eta x}{2a^2 t}.$$

In the event that η is made to move toward nothing

$$\lim \left[\frac{1}{\eta} \sinh \frac{\eta x}{2a^2 t} \right] = \frac{x}{2a^2 t},$$

and and $u = \dfrac{Px}{4a^3\sqrt{\pi t^3}} e^{-\frac{x^2}{4a^2 t}}$ \hfill (1)

PHYSICS APPLICATIONS

is the answer for the temperature whenever and place due to an immediate doublet of solidarity P set at the beginning. For a doublet at some other point $x = \lambda$ we have

$$u = \frac{P(x-\lambda)}{4a^3\sqrt{\pi t^3}} e^{-\frac{(x-\lambda)^2}{4a^2 t}}. \qquad (2)$$

For a long-lasting doublet of steady strength P put at the beginning we have

$$u = \frac{Px}{4a^3\sqrt{\pi}} \int_0^t e^{-\frac{x^2}{4a^2(t-\tau)}} (t-\tau)^{-\frac{3}{2}} d\tau; \qquad (3)$$

and for a long-lasting doublet of variable strength $f(t)$

$$u = \frac{x}{4a^3\sqrt{\pi}} \int_0^t e^{-\frac{x^2}{4a^2(t-\tau)}} (t-\tau)^{-\frac{3}{2}} f(\tau)d\tau, \text{and} \qquad (4)$$

or $u = \dfrac{1}{a^2\sqrt{\pi}} \displaystyle\int_{\frac{x}{2a\sqrt{t}}}^{\infty} e^{-\beta^2} f\left(t - \dfrac{x^2}{4a^2\beta^2}\right) d\beta$ and $\qquad (5)$

if $x > 0$, and

$$u = \frac{1}{a^2\sqrt{\pi}} \int_{\frac{x}{2a\sqrt{t}}}^{-\infty} e^{-\beta^2} f\left(t - \frac{x^2}{4a^2\beta^2}\right) d\beta \text{ and} \qquad (6)$$

in the event that $x < 0$, assuming we let $\beta = \frac{x}{2a\sqrt{t-\tau}}$.

From (5) and (6) we see promptly that $u = 0$ when $t = 0$ and that $u = \frac{f(t)}{2a^2}$ when $x = 0$ assuming we approach the beginning from the right and that $u = -\frac{f(t)}{2a^2}$ when $x = 0$ on the off chance that we approach the beginning from the left.

If the point $x = 0$ is kept at the consistent temperature b and we are concerned just with positive upsides of x we can get from (5) the arrangement surrendered Art. 50 Ex. 4 by assuming an extremely durable doublet of solidarity $2a^2 b$ set at the beginning.

To take care of the issue treated in Art. 51 we have just to assume a long-lasting doublet of solidarity $2a^2 F(t)$ put at $x = 0$ and from (5) we get without a moment's delay (10) Art. 51.

EXAMPLE.

Show that if $D_t u = a^2 D_x^2 u - b^2 u$ and a quick wellspring of solidarity Q is set at $x = \lambda$

$$u = \frac{Q}{2a\sqrt{\pi t}} e^{-b^2 t - \frac{(\lambda - x)^2}{4a^2 t}} \qquad \text{v. Art. 51, Ex. 5.}$$

Show that assuming a momentary doublet of solidarity P is put at the point $x = 0$

$$u = \frac{Px}{4a^3\sqrt{\pi t^3}} e^{-b^2 t - \frac{x^2}{4a^2 t}}.$$

If a long-lasting doublet of solidarity $f(t)$ is set at $x = 0$

$$u = \frac{x}{4a^3\sqrt{\pi}} \int_0^t e^{-b^2(t-\tau) - \frac{x^2}{4a^2(t-\tau)}} (t-\tau)^{-\frac{3}{2}} f(\tau) d\tau$$

$$= \frac{1}{a^2\sqrt{\pi}} \int_{\frac{x}{2a\sqrt{t}}}^{\hat{A}\pm\infty} e^{-\beta^2 - \frac{b^2 x^2}{4a^2\beta^2}} f\left(t - \frac{x^2}{4a^2\beta^2}\right) d\beta,$$

PHYSICS APPLICATIONS

whence $u = 0$ when $t = 0$ and $x > 0$ or $x < 0$ and $u = \hat{A} \pm \frac{f(t)}{2a^2}$ when $x = 0$.

Subsequently on the off chance that we place at $x = 0$ a super durable doublet of solidarity $2a^2 F(t)$ we get the arrangement given in Art. 51 Ex. 5 for the situation where $u = F(t)$ when $x = 0$ and $u = 0$ when $t = 0$ furnished we are concerned exclusively with positive upsides of x.

In the event that $F(t) = c$ this diminishes to

$$u = \frac{2c}{\sqrt{\pi}} \int_{\frac{x}{2a\sqrt{t}}}^{\infty} e^{-\beta^2 - \frac{b^2 x^2}{4a^2 \beta^2}} \, d\beta.$$

55. As one more illustration of the utilization of Fourier's Basic we will consider the transmission of an unsettling influence along an extended versatile string.

Assume we have an extended versatile string so lengthy that we want not consider what occurs at its finishes, that is long to such an extent that we might regard its length as limitless. Allow the string to be at first contorted into some given structure and afterward delivered; to explore its ensuing movement.

Allow us to take the place of harmony of the string as the hub of X and some random point as beginning.

We have, then, to tackle the differential condition

$$D_t^2 y = a^2 D_x^2 y \tag{1}$$

[v. (VIII) Art. 1] subject to the circumstances

$$y = f(x) \text{ and} \quad \text{when} \quad \text{and} t = 0 \tag{2}$$
$$D_t y = 0 \text{ and} \quad \text{''} \quad \text{and} t = 0. \tag{3}$$

As in Art. 8 we find

$$y = \cos\alpha(x\hat{A} \pm at) \quad \text{and} \quad y = \sin\alpha(x\hat{A} \pm at)$$

as specific arrangements of (1).

From these we should develop a worth that will lessen to

$$f(x) = \frac{1}{\pi} \int_0^{\infty} d\alpha \int_{-\infty}^{\infty} f(\lambda) \cos\alpha(\lambda - x) . d\lambda \tag{4}$$

when $t = 0$ and will simultaneously fulfill (3).

$$y = \cos\alpha\lambda \cos\alpha(x + at) + \sin\alpha\lambda \sin\alpha(x + at)$$
or
$$y = \cos\alpha(\lambda - x - at)$$

is an answer of (1).

Hence
$$y = \frac{1}{\pi} \int_0^{\infty} d\alpha \int_{-\infty}^{\infty} f(\lambda) \cos\alpha(\lambda - x - at) . d\lambda \text{ and} \tag{5}$$

is likewise an answer of (1).

(5) lessens to $y = f(x)$ when $t = 0$ however it gives

$$D_t y = \frac{a}{\pi} \int_0^{\infty} \alpha \, d\alpha \int_{-\infty}^{\infty} f(\lambda) \sin\alpha(\lambda - x) . d\lambda$$

when $t = 0$ and therefore doesn't fulfill condition (3).

PHYSICS APPLICATIONS   80

If in framing (5) we use $\cos\alpha(x-at)$ and $\sin\alpha(x-at)$ rather than $\cos\alpha(x+at)$ and $\sin\alpha(x+at)$ we get

$$y = \frac{1}{\pi}\int_0^\infty d\alpha \int_{-\infty}^\infty f(\lambda)\cos\alpha(\lambda - x + at).d\lambda \tag{6}$$

which is an answer of (1), and decreases to $y = f(x)$ when $t = 0$, however it gives

$$D_t y = -\frac{a}{\pi}\int_0^\infty \alpha d\alpha \int_{-\infty}^\infty f(\lambda)\sin\alpha(\lambda - x).d\lambda$$

when $t = 0$ and doesn't fulfill (3).

In the event that, nonetheless, we take one-around 50get

$$y = \frac{1}{2}\left[\frac{1}{\pi}\int_0^\infty d\alpha \int_{-\infty}^\infty f(\lambda)\cos\alpha(\lambda - x - at).d\lambda \right.$$
$$\left. + \frac{1}{\pi}\int_0^\infty d\alpha \int_{-\infty}^\infty f(\lambda)\cos\alpha(\lambda - x + at).d\lambda\right], \tag{7}$$

an answer of (1) which fulfills both (2) and (3), and is, thusly, our required arrangement.

This outcome can be especially rearranged.

Assuming we substitute $z = x + at$

$$\frac{1}{\pi}\int_0^\infty d\alpha \int_{-\infty}^\infty f(\lambda)\cos\alpha(\lambda - x - at).d\lambda$$
$$= \frac{1}{\pi}\int_0^\infty d\alpha \int_{-\infty}^\infty f(\lambda)\cos\alpha(\lambda - z).d\lambda = f(z) = f(x + at);$$

furthermore, in this way we can show that

$$\frac{1}{\pi}\int_0^\infty d\alpha \int_{-\infty}^\infty f(\lambda)\cos\alpha(\lambda - x + at).d\lambda = f(x - at).$$

Thus our answer becomes

$$y = \frac{1}{2}[f(x + at) + f(x - at)]. \tag{8}$$

This outcome is critical in the hypothesis of versatile strings and it shows that the underlying aggravation parts into two equivalent waves which run along the string, one to the right and the other to the left, with a uniform speed a, what's more, that there is nothing similar to an occasional movement or vibration of any kind except if the closures of the string produce some outcome.

56. In the event that the string isn't at first contorted yet begins from its place of harmony with a given beginning speed presented for each point we need to settle the condition

$$D_t^2 y = a^2 D_x^2 y \tag{1}$$

dependent upon the circumstances

$$y = 0 \quad\text{and when}\quad\text{and}\quad t = 0 \tag{2}$$
$$D_t y = F(x) \quad\text{and"}\quad\text{and}\quad t = 0. \tag{3}$$

PHYSICS APPLICATIONS

We get by the cycle utilized in Art. 55

$$y = \frac{1}{2\pi a} \int_0^\infty d\alpha \int_{-\infty}^\infty F(\lambda) \left[\frac{\sin \alpha(\lambda - x + at)}{\alpha} - \frac{\sin \alpha(\lambda - x - at)}{\alpha} \right] d\lambda$$

$$= \frac{1}{2\pi a} \int_{-\infty}^\infty F(\lambda) d\lambda \int_0^\infty \left[\frac{\sin \alpha(\lambda - x + at)}{\alpha} - \frac{\sin \alpha(\lambda - x - at)}{\alpha} \right] d\alpha;$$

however,
$$\int_0^\infty \frac{\sin \alpha(\lambda - x + at)}{\alpha} d\alpha - \int_0^\infty \frac{\sin \alpha(\lambda - x - at)}{\alpha} d\alpha = \pi$$

on the off chance that $x - at < \lambda < x + at$, and is equivalent to zero for any remaining upsides of λ; since

$$\int_0^\infty \frac{\sin mx}{x} dx = \frac{\pi}{2} \quad \text{if} \quad and\, m > 0$$
$$= -\frac{\pi}{2} \quad \text{if} \quad and\, m < 0$$
$$= 0 \quad \text{if} \quad and\, m = 0.$$

v. Int. Cal. Art. 92 (3).

Hence and $y = \dfrac{1}{2a} \displaystyle\int_{x-at}^{x+at} F(\lambda) d\lambda$ \hfill (4)

is our necessary arrangement.

EXAMPLES.

1. Assuming the string is at first contorted and begins with starting speed so that $y = f(x)$ and $D_t y = F(x)$ when $t = 0$

$$y = \frac{1}{2}[f(x+at) + f(x-at)] + \frac{1}{2a} \int_{x-at}^{x+at} F(\lambda) d\lambda.$$

2. In the event that the underlying unsettling influence is brought about by a blow, as from the mallet in a piano, which presents for every one of the places in a piece of the line of length c an equivalent cross over speed b show that the front of the wave which will be seen to race to one side along the string will be a straight line having a slant equivalent to $\frac{b}{2a}$ and a length equivalent to $\frac{c}{2a}\sqrt{4a^2 + b^2}$. Obviously a wave having a front of a similar length with an incline equivalent to $-\frac{b}{2a}$ will be believed to rush to the right along the string, and the impact of the two waves will be to lift the string real and forever to a distance $\frac{bc}{2a}$ over its unique position.

57. We will presently take up a couple of instances of the utilization of *Fourier's Series*.

In the issue of Art. 7 let the temperature of the foundation of the plate be a given capability of x, different circumstances staying unaltered.

Since and $f(x) = \displaystyle\sum_{m=1}^{m=\infty} (a_m \sin mx)$

where $a_m = \dfrac{2}{\pi} \displaystyle\int_0^\pi f(\alpha) \sin m\alpha \, d\alpha$

we have $u = \dfrac{2}{\pi} \displaystyle\sum_{m=1}^{m=\infty} \left[e^{-my} \sin mx \int_0^\pi f(\alpha) \sin m\alpha \, d\alpha \right]$ .and \hfill (1)

PHYSICS APPLICATIONS

In the event that the broadness of the plate is a rather than π

$$u = \frac{2}{a}\sum_{m=1}^{m=\infty}\left[e^{-\frac{m\pi y}{a}}\sin\frac{m\pi x}{a}\int_0^a f(\lambda)\sin\frac{m\pi\lambda}{a}d\lambda\right]. \qquad (2)$$

58. Assuming that the temperature of the base is solidarity and the broadness of the plate is π the arrangement is, as we have seen in Art. 7,

$$u = \frac{4}{\pi}\left[e^{-y}\sin x + \frac{1}{3}e^{-3y}\sin 3x + \frac{1}{5}e^{-5y}\sin 5x + \cdots\right]. \qquad (1)$$

This series can be added easily. We have the turn of events

$$\log(1+z) = \frac{z}{1} - \frac{z^2}{2} + \frac{z^3}{3} - \frac{z^4}{4} + \cdots$$

in the event that the modulus of z is under 1. Int. Cal. Art. 221 (4).

Hence and $\log(1-z) = -\dfrac{z}{1} - \dfrac{z^2}{2} - \dfrac{z^3}{3} - \dfrac{z^4}{4} - \cdots$

in the event that mod. $z < 1$.

and and $\dfrac{1}{2}[\log(1+z) - \log(1-z)] = \dfrac{z}{1} + \dfrac{z^3}{3} + \dfrac{z^5}{5} + \cdots \qquad (2)$

on the off chance that mod. $z < 1$.

However

$$\log(1+z) = \log[1 + r(\cos\phi + I\sin\phi)]$$
$$= \frac{1}{2}\log[(1+r\cos\phi)^2 + (r\sin\phi)^2] + I\tan^{-1}\frac{r\sin\phi}{1+r\cos\phi}$$
$$= \frac{1}{2}\log(1+2r\cos\phi+r^2) + I\tan^{-1}\frac{r\sin\phi}{1+r\cos\phi},$$

and

$$\log(1-z) = \frac{1}{2}\log(1-2r\cos\phi+r^2) - I\tan^{-1}\frac{r\sin\phi}{1-r\cos\phi},$$

[Int. Cal. Art. 33 (2)],
also (2) becomes

$$\frac{1}{2}\left[\frac{1}{2}\log\frac{1+2r\cos\phi+r^2}{1-2r\cos\phi+r^2} + I\tan^{-1}\frac{2r\sin\phi}{1-r^2}\right]$$
$$= \frac{r(\cos\phi + i\sin\phi)}{1} + \frac{r^3(\cos 3\phi + i\sin 3\phi)}{3} + \cdots \qquad (3)$$

From (3) we get two conditions

$$\frac{1}{4}\log\frac{1+2r\cos\phi+r^2}{1-2r\cos\phi+r^2} = \frac{r\cos\phi}{1} + \frac{r^3\cos 3\phi}{3} + \frac{r^5\cos 5\phi}{5} + \cdots \qquad (4)$$

$$\frac{1}{2}\tan^{-1}\frac{2r\sin\phi}{1-r^2} = \frac{r\sin\phi}{1} + \frac{r^3\sin 3\phi}{3} + \frac{r^5\sin 5\phi}{5} + \cdots \qquad (5)$$

PHYSICS APPLICATIONS

both legitimate for all upsides of ϕ gave $r < 1$.
e^{-y} is under 1 on the off chance that y is positive.
Subsequently from (5)

$$\frac{e^{-y}\sin x}{1} + \frac{e^{-3y}\sin 3x}{3} + \frac{e^{-5y}\sin 5x}{5} + \cdots = \frac{1}{2}\tan^{-1}\frac{2e^{-y}\sin x}{1-e^{-2y}}$$
$$= \frac{1}{2}\tan^{-1}\frac{2\sin x}{e^y - e^{-y}} = \frac{1}{2}\tan^{-1}\frac{\sin x}{\sinh y},$$

furthermore (1) might be written

$$u = \frac{2}{\pi}\tan^{-1}\frac{\sin x}{\sinh y}. \tag{6}$$

On the off chance that we supplant r by e^{-y} and ϕ by x in

$$\log[1 + r(\cos\phi + I\sin\phi)]$$

it becomes *and* $\log[1 + e^{-y} \quad \cos x + Ie^{-y}\sin x]$
or *and* $\log[1+ \quad \cos z + I\sin z]$
v. Int. Cal. Art. 35 (3) and (4)
a component of z overall; and

$$\log[1 - r(\cos\phi + I\sin\phi)]$$

becomes *and* $\log(1- \quad \cos z - i\sin z)$;
subsequently by Int. Cal. Arts. 212 and 213,

$$\frac{1}{4}\log\frac{1+2e^{-y}\cos x + e^{-2y}}{1-2e^{-y}\cos x + e^{-2y}} \quad \text{and} \quad \frac{1}{2}\tan^{-1}\frac{2e^{-y}\sin x}{1-e^{-2y}}$$

or $\quad \dfrac{1}{4}\log\dfrac{\cosh y + \cos x}{\cosh y - \cos x} \quad$ and $\quad \dfrac{1}{2}\tan^{-1}\dfrac{\sin x}{\sinh y}$

are form capabilities, and

$$u_1 = \frac{1}{\pi}\log\frac{\cosh y + \cos x}{\cosh y - \cos x} \tag{7}$$

is the answer for the issue where the isothermal lines are the lines of stream of the current issue and the lines of stream are the isothermal lines of the present issue.

For our concern, then, the isothermal lines are given by the situation

$$\frac{2}{\pi}\tan^{-1}\frac{\sin x}{\sinh y} = a$$

or *and* $\dfrac{\sin x}{\sinh y} \qquad = \tan\dfrac{a\pi}{2} \tag{8}$

furthermore, the lines of stream by

$$\frac{1}{\pi}\log\frac{\cosh y + \cos x}{\cosh y - \cos x} = b,$$

or $\qquad \dfrac{\cosh y + \cos x}{\cosh y - \cos x} = e^{\pi b}. \tag{9}$

EXAMPLES.

1. If $D_x^2 u + D_y^2 u = 0$, and $u = 1$ when $y = 0$, and $u = 0$ when $x = 0$ and when $x = a$,

$$u = \frac{4}{\pi}\left[e^{-\frac{\pi y}{a}}\sin\frac{\pi x}{a} + \frac{1}{3}e^{-\frac{3\pi y}{a}}\sin\frac{3\pi x}{a} + \frac{1}{5}e^{-\frac{5\pi y}{a}}\sin\frac{5\pi x}{a} + \cdots\right]$$
$$= \frac{2}{\pi}\tan^{-1}\frac{\sin\frac{\pi x}{a}}{\sinh\frac{\pi y}{a}}.$$

2. If $u = \phi(x)$ when $y = 0$, $u = f(y)$ when $x = 0$, and $u = F(y)$ when $x = a$

$$u = \frac{2}{a}\sum_{m=1}^{m=\infty} e^{-\frac{m\pi y}{a}} \sin\frac{m\pi x}{a} \int_0^a \phi(\lambda)\sin\frac{m\pi\lambda}{a}d\lambda$$

$$+\frac{1}{2a}\sin\frac{\pi x}{a}\int_0^\infty\left[\frac{1}{\cosh\frac{\pi}{a}(\lambda-y)-\cos\frac{\pi x}{a}}-\frac{1}{\cosh\frac{\pi}{a}(\lambda+y)-\cos\frac{\pi x}{a}}\right]f(\lambda)\,d\lambda$$

$$+\frac{1}{2a}\sin\frac{\pi x}{a}\int_0^\infty\left[\frac{1}{\cosh\frac{\pi}{a}(\lambda-y)+\cos\frac{\pi x}{a}}-\frac{1}{\cosh\frac{\pi}{a}(\lambda+y)+\cos\frac{\pi x}{a}}\right]F(\lambda)\,d\lambda$$

v. Art. 48, Exs. 4, 5, and 6.

59. In the event that three sides of a plane rectangular sheet of directing material be kept at likely zero and the worth of the expected capability at each mark of the fourth side be given; to track down the worth of this expected capability at any mark of the sheet.

To plan:- - -

$$D_x^2 V + D_y^2 V = 0. \tag{1}$$
$$V = 0 \quad \text{when} \quad x = 0. \tag{2}$$
$$V = 0 \quad " \quad x = a. \tag{3}$$
$$V = 0 \quad " \quad y = b. \tag{4}$$
$$V = f(x) \quad " \quad y = 0. \tag{5}$$

Filling in as in Art. 48 we get

$$\frac{\sinh\frac{m\pi}{a}(b-y)}{\sinh\frac{m\pi b}{a}}\sin\frac{m\pi x}{a}$$

as a worth of V which fulfills conditions (1), (2), (3), and (4) in the event that m is a number. Thusly

$$V = \frac{2}{a}\sum_{m=1}^{m=\infty}\left[\frac{\sinh\frac{m\pi}{a}(b-y)}{\sinh\frac{m\pi b}{a}}\sin\frac{m\pi x}{a}\int_0^a f(\lambda)\sin\frac{m\pi\lambda}{a}d\lambda\right] \tag{6}$$

is our expected arrangement.

EXAMPLES.

1. If $f(x) = 1$ Eq. (6) Art. 59 decreases to

$$V = \frac{4}{\pi}\left[\frac{\sinh\frac{\pi}{a}(b-y)}{\sinh\frac{\pi b}{a}}\sin\frac{\pi x}{a} + \frac{1}{3}\frac{\sinh\frac{3\pi}{a}(b-y)}{\sinh\frac{3\pi b}{a}}\sin\frac{3\pi x}{a}\right.$$
$$\left.+\frac{1}{5}\frac{\sinh\frac{5\pi}{a}(b-y)}{\sinh\frac{5\pi b}{a}}\sin\frac{5\pi x}{a}+\cdots\right]$$

2. If $V = 0$ when $x = 0$, $V = 0$ when $x = a$, $V = 0$ when $y = 0$, furthermore, $V = F(x)$ when $y = b$, then

$$V = \frac{2}{a}\sum_{m=1}^{m=\infty}\left[\frac{\sinh\frac{m\pi y}{a}}{\sinh\frac{m\pi b}{a}}\sin\frac{m\pi x}{a}\int_0^a F(\lambda)\sin\frac{m\pi\lambda}{a}d\lambda\right].$$

3. If $F(x) = 1$ the response of Ex. 2 lessens to

$$V = \frac{4}{\pi}\left[\frac{\sinh\frac{\pi y}{a}}{\sinh\frac{\pi b}{a}}\sin\frac{\pi x}{a}+\frac{1}{3}\frac{\sinh\frac{3\pi y}{a}}{\sinh\frac{3\pi b}{a}}\sin\frac{3\pi x}{a}+\frac{1}{5}\frac{\sinh\frac{5\pi y}{a}}{\sinh\frac{5\pi b}{a}}\sin\frac{5\pi x}{a}+\cdots\right].$$

PHYSICS APPLICATIONS 85

4. If $V = 0$ when $x = 0$, $V = 0$ when $x = a$, $V = f(x)$ when $y = 0$, also, $V = F(x)$ when $y = b$, then, at that point,

$$V = \frac{2}{a} \sum_{m=1}^{m=\infty} \left[\sin \frac{m\pi x}{a} \left(\frac{\sinh \frac{m\pi}{a}(b-y)}{\sinh \frac{m\pi b}{a}} \int_0^a f(\lambda) \sin \frac{m\pi \lambda}{a} d\lambda \right. \right.$$
$$\left. \left. + \frac{\sinh \frac{m\pi y}{a}}{\sinh \frac{m\pi b}{a}} \int_0^a F(\lambda) \sin \frac{m\pi \lambda}{a} d\lambda \right) \right].$$

5. If $f(x) = F(x)$ the response of Ex. 4 decreases to

$$V = \frac{2}{a} \sum_{m=1}^{m=\infty} \left[\frac{\cosh \frac{m\pi}{a} \left(\frac{b}{2} - y \right)}{\cosh \frac{m\pi b}{2a}} \sin \frac{m\pi x}{a} \int_0^a f(\lambda) \sin \frac{m\pi \lambda}{a} d\lambda \right].$$

6. If $f(x) = F(x) = 1$ the response of Ex. 5 decreases to

$$V = \frac{4}{\pi} \left[\frac{\cosh \frac{\pi}{a}\left(\frac{b}{2}-y\right)}{\cosh \frac{\pi b}{2a}} \sin \frac{\pi x}{a} + \frac{1}{3} \frac{\cosh \frac{3\pi}{a}\left(\frac{b}{2}-y\right)}{\cosh \frac{3\pi b}{2a}} \sin \frac{3\pi x}{a} \right.$$
$$\left. + \frac{1}{5} \frac{\cosh \frac{5\pi}{a}\left(\frac{b}{2}-y\right)}{\cosh \frac{5\pi b}{2a}} \sin \frac{5\pi x}{a} + \cdots \right].$$

7. If $V = f(x)$ when $y = 0$, $V = F(x)$ when $y = b$, $V = \phi(y)$ when $x = 0$, and $V = \chi(y)$ when $x = a$, then

$$V = \frac{2}{a} \sum_{m=1}^{m=\infty} \left[\sin \frac{m\pi x}{a} \left(\frac{\sinh \frac{m\pi}{a}(b-y)}{\sinh \frac{m\pi b}{a}} \int_0^a f(\lambda) \sin \frac{m\pi \lambda}{a} d\lambda \right. \right.$$
$$\left. \left. + \frac{\sinh \frac{m\pi y}{a}}{\sinh \frac{m\pi b}{a}} \int_0^a F(\lambda) \sin \frac{m\pi \lambda}{a} d\lambda \right) \right]$$
$$+ \frac{2}{b} \sum_{m=1}^{m=\infty} \left[\sin \frac{m\pi y}{b} \left(\frac{\sinh \frac{m\pi}{b}(a-x)}{\sinh \frac{m\pi a}{b}} \int_0^b \phi(\lambda) \sin \frac{m\pi \lambda}{b} d\lambda \right. \right.$$
$$\left. \left. + \frac{\sinh \frac{m\pi x}{b}}{\sinh \frac{m\pi a}{b}} \int_0^b \chi(\lambda) \sin \frac{m\pi \lambda}{b} d\lambda \right) \right].$$

8. If $f(x) = \phi(y) = 0$ and $F(x) = \chi(y) = 1$ the response of Ex. 7 might be decreased to

$$V = \frac{2}{\pi} \left[\frac{\pi y}{2b} - \frac{\sinh \frac{\pi}{b}\left(\frac{a}{2}-x\right)}{\sinh \frac{\pi a}{2b}} \sin \frac{\pi y}{b} + \frac{1}{2} \frac{\cosh \frac{2\pi}{b}\left(\frac{a}{2}-x\right)}{\cosh \frac{2\pi a}{2b}} \sin \frac{2\pi y}{b} \right.$$
$$\left. - \frac{1}{3} \frac{\sinh \frac{3\pi}{b}\left(\frac{a}{2}-x\right)}{\sinh \frac{3\pi a}{2b}} \sin \frac{3\pi y}{b} + \frac{1}{4} \frac{\cosh \frac{4\pi}{b}\left(\frac{a}{2}-x\right)}{\cosh \frac{4\pi a}{2b}} \sin \frac{4\pi y}{b} - \cdots \right].$$

9. Track down the temperature of the center mark of a slim square plate whose faces are impenetrable to warm; first, when three edges are kept at the temperature 0Â° and the fourth edge at the temperature 100Â°; 2d, when two inverse edges are kept at the temperature 0Â° and the other two at the temperature 100Â°; 3d, when two neighboring edges are kept at the temperature 0Â° what's more, different edges at the temperature 100Â°. See models 3, 6, and 8.

Ans., (1) 25Â°; (2) 50Â°; (3) 50Â°.

PHYSICS APPLICATIONS

60. Let us give to the thought of the progression of intensity in one aspect.

Assume that we have a limitless strong with two equal plane faces whose distance separated is c.

Take the beginning in one face and the hub of X opposite to the countenances. Allow the underlying temperature to be any given capability of x and left the two faces alone kept at the steady temperature zero; to track down the temperature anytime of the chunk whenever.

We need to address the condition
$$D_t u = a^2 D_x^2 u \qquad (1)$$

dependent upon the circumstances

$$u = 0 \quad \text{when} \quad and x and = 0 \qquad (2)$$
$$u = 0 \quad " and x and = c \qquad (3)$$
$$u = f(x) \quad " and t and = 0. \qquad (4)$$

In Art. 49 we have found
$$u = e^{-a^2 \alpha^2 t} \sin \alpha x$$

and*and*u
$$= e^{-a^2 \alpha^2 t} \cos \alpha x$$

as specific arrangements of (1).

$u = e^{-a^2 \alpha^2 t} \sin \alpha x$ fulfills (2) anything esteem is given to α. It fulfills (3) if $\alpha = \frac{m\pi}{c}$ gave m is a number. Allow us to attempt to fabricate a worth of u out of terms of the structure $A e^{-\frac{a^2 m^2 \pi^2 t}{c^2}} \sin \frac{m\pi x}{c}$ which will fulfill (4).

We have
$$f(x) = \frac{2}{c} \sum_{m=1}^{m=\infty} \left[\sin \frac{m\pi x}{c} \int_0^c f(\lambda) \sin \frac{m\pi \lambda}{c} d\lambda \right]. \qquad (5)$$

$$u = \frac{2}{c} \sum_{m=1}^{m=\infty} \left[e^{-\frac{m^2 a^2 \pi^2 t}{c^2}} \sin \frac{m\pi x}{c} \int_0^c f(\lambda) \sin \frac{m\pi \lambda}{c} d\lambda \right], \qquad (6)$$

diminishes to (5) when $t = 0$ and is our necessary arrangement.

EXAMPLES.

1. If $f(\lambda) = b$, a consistent, (6) Art. 60 diminishes to

$$u = \frac{4b}{\pi}\left[e^{-\frac{a^2 \pi^2 t}{c^2}} \sin \frac{\pi x}{c} + \frac{1}{3} e^{-\frac{9a^2 \pi^2 t}{c^2}} \sin \frac{3\pi x}{c} + \frac{1}{5} e^{-\frac{25\pi^2 a^2 t}{c^2}} \sin \frac{5\pi x}{c} + \cdots \right].$$

2. An iron chunk 10 cm. thick is put between and in touch with two other iron pieces each 10 cm. thick. The temperature of the center chunk is at first 100Â° all through, and of the external chunks 0Â° all through. The external appearances of the external chunks are kept at the temperature 0Â°. Required the temperature of a point in the center piece fifteen minutes after the sections have been put in touch. Given $a^2 = 0.185$ in C.G.S. units. *Ans.*, 10Â°.3.

3. Two iron pieces each 20 cm. thick one of which is at the temperature 0Â° what's more, the other at the temperature 100Â° all through, are set together face to face, and their external countenances are kept at the temperature 0Â°. Track down the temperature of a point in their well known face and of focuses 10 cm. from the normal face fifteen minutes after the chunks have been put together.

Ans., 22Â°.8; 15Â°.1; 17Â°.2.

PHYSICS APPLICATIONS   87

4. One face of an iron chunk 40 cm. thick is kept at the temperature 0Â° and the other face at the temperature 100Â° until the long-lasting condition of temperatures is set up. Each face is then kept at the temperature 0Â°. Required the temperature of a point in the chunk, and of focuses 10 cm. from the countenances fifteen minutes after the cooling has started. *Ans.*, 22Â°.8; 15Â°.6; 16Â°.7.

61. On the off chance that the essences of the piece treated in Art. 60 as opposed to being kept at the temperature zero are delivered impenetrable to warm, the arrangement of the issue is simple.

For this situation we need to tackle the condition

$$D_t u = a^2 D_x^2 u$$

dependent upon the circumstances

$$D_x u = 0 \quad \text{when} \quad and\, x = 0$$
$$D_x u = 0 \quad \text{"} and\, x = c$$
$$u = f(x) \quad \text{"} and\, t = 0.$$

We have just to utilize the specific arrangement

$$u = e^{-a^2 \alpha^2 t} \cos \alpha x$$

as we used

$$u = e^{-a^2 \alpha^2 t} \sin \alpha x$$

in Art. 60. We get

$$u = \frac{2}{c}\left[\frac{1}{2} \int_0^c f(\lambda) d\lambda + \sum_{m=1}^{m=\infty}\left(e^{-\frac{m^2 a^2 \pi^2 t}{c^2}} \cos \frac{m\pi x}{c} \int_0^c f(\lambda) \cos \frac{m\pi \lambda}{c} d\lambda\right)\right]. \tag{1}$$

EXAMPLES.

1. Tackle model 2 Art. 60 it are covered to assume that the external surfaces after the chunks are set together so that intensity can neither enter nor escape. Track down likewise the temperature of the external surfaces fifteen minutes later the pieces are put in touch. *Ans.*, 33Â°.3; 33Â°.3.

2. Settle model 3 Art. 60 on the speculation recently expressed, getting likewise the temperatures of focuses on the external surfaces.
$$Ans., 50Â°;\ 33Â°.9;\ 66Â°.1;\ 27Â°.2;\ 72Â°.8.$$

3. Address model 4 Art. 60 assuming that heat neither enters nor get away at the external surfaces after the extremely durable condition of temperatures has been set up. Track down additionally the temperatures of focuses in the external surfaces.
$$Ans., 50Â°;\ 39Â°.7;\ 60Â°.3;\ 35Â°.5;\ 64Â°.5.$$

4. Show that if $u = 0$ when $x = 0$, $D_x u = 0$ when $x = c$, and $u = f(x)$ when $t = 0$,

$$u = \frac{2}{c} \sum_{m=0}^{m=\infty} \left(e^{-\frac{(2m+1)^2 a^2 \pi^2 t}{4c^2}} \sin \frac{(2m+1)\pi x}{2c} \int_0^c f(\lambda) \sin \frac{(2m+1)\pi \lambda}{2c} d\lambda\right).$$

Suggestion: Accept $u = 0$ when $x = 2c$ and $f(2c - x) = f(x)$, and see (6) Art. 60.

PHYSICS APPLICATIONS

62. In the event that the temperature of the right-hand face of the chunk considered in Art. 60 is a consistent γ rather than zero we have just to add to the subsequent part of (6) Art. 60 a term u_1 which will fulfill the circumstances

$$D_t u_1 = a^2 D_x^2 u_1 \tag{1}$$

$$u_1 = 0 \quad \text{when} \quad and\, x = 0 \tag{2}$$

$$u_1 = 0 \quad \text{"} and\, t \quad = 0 \tag{3}$$

$$u_1 = \gamma \quad \text{"} and\, x \quad = c. \tag{4}$$

$u_1 = \frac{\gamma x}{c}$ clearly fulfills (1), (2), and (4); to make it fulfill (3) also we should add a term u_2 which will be equivalent to zero when $x = 0$ and when $x = c$ and to $-\frac{\gamma x}{c}$ when $t = 0$, while continuously fulfilling (1). It is given quickly by (6) Art. 60 and is

$$u_2 = -\frac{2\gamma}{c^2} \sum_{m=1}^{m=\infty} \left(e^{-\frac{m^2 a^2 \pi^2 t}{c^2}} \sin \frac{m\pi x}{c} \int_0^c \lambda \sin \frac{m\pi \lambda}{c} d\lambda \right). \tag{5}$$

$$\int_0^c \lambda \sin \frac{m\pi \lambda}{c} d\lambda = -\frac{c^2}{m\pi} \cos m\pi = (-1)^{m+1} \frac{c^2}{m\pi},$$

and

$$u_2 = \frac{2\gamma}{\pi} \sum_{m=1}^{m=\infty} \left(\frac{(-1)^m}{m} e^{-\frac{m^2 a^2 \pi^2 t}{c^2}} \sin \frac{m\pi x}{c} \right). \tag{6}$$

Hence

$$u_1 = \gamma \left[\frac{x}{c} + \frac{2}{\pi} \sum_{m=1}^{m=\infty} \left(\frac{(-1)^m}{m} e^{-\frac{m^2 a^2 \pi^2 t}{c^2}} \sin \frac{m\pi x}{c} \right) \right]. \tag{7}$$

In the event that the left-hand face of the chunk considered in Art. 60 is to be kept at a steady temperature β and the right-hand face at the temperature zero we can get the term u_3 which should be added to the second individual from (6) Art. 60 by supplanting γ by β and x by $c - x$ in (7). We then have

$$u_3 = \beta \left[\frac{c-x}{c} - \frac{2}{\pi} \sum_{m=1}^{m=\infty} \left(\frac{1}{m} e^{-\frac{m^2 a^2 \pi^2 t}{c^2}} \sin \frac{m\pi x}{c} \right) \right]. \tag{8}$$

EXAMPLES.

1. Show that if $u = \beta$ when $x = 0$, $u = \gamma$ when $x = c$, and $u = f(x)$ when $t = 0$

$$u = \beta + (\gamma - \beta) \left[\frac{x}{c} + \frac{2}{\pi} \sum_{m=1}^{m=\infty} \left(\frac{(-1)^m}{m} e^{-\frac{m^2 \pi^2 a^2 t}{c^2}} \sin \frac{m\pi x}{c} \right) \right]$$

$$+ \frac{2}{c} \sum_{m=1}^{m=\infty} \left(e^{-\frac{m^2 a^2 \pi^2 t}{c^2}} \sin \frac{m\pi x}{c} \int_0^c [f(\lambda) - \beta] \sin \frac{m\pi \lambda}{c} d\lambda \right).$$

2. Show that if $u = \beta$ when $x = 0$, $u = 0$ when $t = 0$, and $D_x u = 0$ when $x = c$

$$u = \beta \left[1 - \frac{4}{\pi} \sum_{m=0}^{m=\infty} \left(\frac{1}{2m+1} e^{-\frac{(2m+1)^2 a^2 \pi^2 t}{4c^2}} \sin \frac{(2m+1)\pi x}{2c} \right) \right]$$

$$= \beta \left[1 - \frac{4}{\pi} \left(e^{-\frac{a^2 \pi^2 t}{4c^2}} \sin \frac{\pi x}{2c} + \frac{1}{3} e^{-\frac{9 a^2 \pi^2 t}{4c^2}} \sin \frac{3\pi x}{2c} + \frac{1}{5} e^{-\frac{25 a^2 \pi^2 t}{4c^2}} \sin \frac{5\pi x}{2c} + \cdots \right) \right].$$

PHYSICS APPLICATIONS

63. In the event that the temperature of the right-hand face of the chunk just considered is a component of the time rather than a consistent and the temperature of the left-hand face is zero the issue can be settled by a strategy almost indistinguishable with that of Art. 51.

Let $\phi(x,t)$ be a component of x and t which will be zero in the event that t is under nothing furthermore, will be equivalent to

$$\frac{x}{c} + \frac{2}{\pi} \sum_{m=1}^{m=\infty} \left(\frac{(-1)^m}{m} e^{-\frac{m^2 a^2 \pi^2 t}{c^2}} \sin \frac{m\pi x}{c} \right)$$

[v. (7) Art. 62] assuming that t is equivalent to or more prominent than nothing. So that

$$\begin{aligned}
\phi(x,t) &= 0 \quad \text{if} \quad andt \quad &&< 0 \\
\phi(x,t) &= 0 \quad \text{"} andt \quad &&= 0 and \quad \text{unless} \quad x = c \\
\phi(x,t) &= 1 \quad \text{"} andt \quad &&= 0 and and \quad x = c \\
\phi(x,t) &= 1 \quad \text{"} andx \quad &&= c \\
\phi(x,t) &= 0 \quad \text{"} andx \quad &&= 0.
\end{aligned}$$

Definitively as in Art. 51 we get

$$u = \lim_{\tau \doteq 0} \sum_{k=0}^{k=n} \left[F(k\tau) \frac{[\phi(x, t - k\tau) - \phi(x, t - (k+1)\tau)]\tau}{\tau} \right] \tag{1}$$

as the expected arrangement of our concern, n being as in Art. 51 the biggest whole number in $\frac{t}{\tau}$ where t is any given worth of the time.

On our speculation the last term of (1), that is, $-F(n\tau)\phi[x, t - (n+1)\tau] = 0$; the close to the last term $F(n\tau)\phi(x, t - n\tau)$ has for its restricting worth

$$F(t)\phi(x, 0) = F(t) \left[\frac{x}{c} + \frac{2}{\pi} \sum_{m=1}^{m=\infty} \left(\frac{(-1)^m}{m} \sin \frac{m\pi x}{c} \right) \right],$$

while as in Art. 51 the restricting worth of the remainder of the aggregate is

$$-\int_0^t F(\lambda) D_\lambda \phi(x, t - \lambda) d\lambda.$$

$$D_\lambda \phi(x, t - \lambda) = \frac{2a^2 \pi}{c^2} \sum_{m=1}^{m=\infty} \left[(-1)^m m e^{-\frac{m^2 a^2 \pi^2}{c^2}(t-\lambda)} \sin \frac{m\pi x}{c} \right].$$

Consequently

$$u = F(t) \left[\frac{x}{c} + \frac{2}{\pi} \sum_{m=1}^{m=\infty} \left(\frac{(-1)^m}{m} \sin \frac{m\pi x}{c} \right) \right]$$
$$- \frac{2a^2 \pi}{c^2} \sum_{m=1}^{m=\infty} \left((-1)^m m \sin \frac{m\pi x}{c} \int_0^t F(\lambda) e^{-\frac{m^2 a^2 \pi^2}{c^2}(t-\lambda)} d\lambda \right),$$

$$u = \frac{x}{c} F(t) + \frac{2}{\pi} \sum_{m=1}^{m=\infty} \left[\frac{(-1)^m}{m} \sin \frac{m\pi x}{c} \left(F(t) \right. \right.$$
$$\left. \left. - \frac{m^2 a^2 \pi^2}{c^2} \int_0^t F(\lambda) e^{-\frac{m^2 a^2 \pi^2}{c^2}(t-\lambda)} d\lambda \right) \right]. \tag{2}$$

PHYSICS APPLICATIONS 90

Assuming we substitute $\beta = \frac{m^2 a^2 \pi^2}{c^2}(t - \lambda)$ we get

$$u = \frac{x}{c} F(t) + \frac{2}{\pi} \sum_{m=1}^{m=\infty} \left[\frac{(-1)^m}{m} \sin \frac{m\pi x}{c} \left(F(t) - \int_0^{\frac{m^2 a^2 \pi^2 t}{c^2}} e^{-\beta} F\left(t - \frac{\beta c^2}{m^2 a^2 \pi^2}\right) d\beta \right) \right]. \qquad (3)$$

EXAMPLES.

1. In the event that the temperature of the left-hand face is a component of t and the temperature of the right-hand face is zero and the underlying temperature is zero

$$u = \left(1 - \frac{x}{c}\right) F(t) - \frac{2}{\pi} \sum_{m=1}^{m=\infty} \left[\frac{1}{m} \sin \frac{m\pi x}{c} \left(F(t) - \int_0^{\frac{m^2 a^2 \pi^2 t}{c^2}} e^{-\beta} F\left(t - \frac{\beta c^2}{m^2 a^2 \pi^2}\right) d\beta \right) \right].$$

2. On the off chance that the temperature of the left-hand face is a component of t, the underlying temperature is zero, and the right-hand face is impenetrable to warm

$$u = F(t) - \frac{4}{\pi} \sum_{m=0}^{m=\infty} \left[\frac{1}{2m+1} \sin \frac{(2m+1)\pi x}{2c} \left(F(t) - \frac{(2m+1)^2 a^2 \pi^2}{4c^2} \int_0^t F(\lambda) e^{-\frac{(2m+1)^2 a^2 \pi^2}{4c^2}(t-\lambda)} d\lambda \right) \right].$$

3. If in Arts. 60- - 63 we are managing a bar of little cross-segment and of length c and heat is emanating from the outer layer of the bar into air at the temperature zero so $D_t u = a^2 D_x^2 u - b^2 u$, show that: (a) the subsequent individuals of (6) Art. 60 and (1) Art. 61 should be increased by $e^{-b^2 t}$; (b) condition (7) Art. 62 becomes

$$u_1 = \gamma \left\{ \frac{\sinh \frac{bx}{a}}{\sinh \frac{bc}{a}} + 2a^2 \pi e^{-b^2 t} \sum_{m=1}^{m=\infty} \left[(-1)^m \frac{m}{b^2 c^2 + m^2 a^2 \pi^2} e^{-\frac{m^2 a^2 \pi^2 t}{c^2}} \sin \frac{m\pi x}{c} \right] \right\};$$

(c) condition (2) Art. 63 becomes

$$u = \frac{\sinh \frac{bx}{a}}{\sinh \frac{bc}{a}} F(t) + 2a^2 \pi \sum_{m=1}^{m=\infty} \left\{ \frac{(-1)^m m}{b^2 c^2 + m^2 a^2 \pi^2} \sin \frac{m\pi x}{c} \left[F(t) - \frac{b^2 c^2 + m^2 a^2 \pi^2}{c^2} \int_0^t e^{-\frac{b^2 c^2 + m^2 a^2 \pi^2}{c^2}(t-\lambda)} F(\lambda) d\lambda \right] \right\}.$$

64. The issue of the movement of a limited extended versatile line of length l secured at the finishes and contorted at first into some given bend $y = f(x)$, and afterward permitted to swing, has been dealt with and to some degree settled in Art. 8.

The total arrangement is effectively seen to be

$$y = \frac{2}{l} \sum_{m=1}^{m=\infty} \sin \frac{m\pi x}{l} \cos \frac{m\pi a t}{l} \int_0^l f(\lambda) \sin \frac{m\pi \lambda}{l} d\lambda. \qquad (1)$$

The second individual from (1) is an intermittent capability of t having the period $\frac{2l}{a}$. The movement, then, at that point, in contrast to that on account of a boundless string (Art. 55) is a genuine vibration, an intermittent movement. The period $\frac{2l}{a}$ is the time it takes an unsettling influence to travel two times the length of the string (v. Art. 55).

PHYSICS APPLICATIONS

A cautious assessment of (1) will show that the real movement is a fair setup like that for the situation considered in Art. 55. The first aggravation breaks up into two waves one of which races to one side until it arrives at the finish of the string and is then reflected, and runs back to one side or the under side of the string, while the other wave hurries to one side and is reflected at the left-hand end of the string and runs back to one side under the string and is again reflected, runs back to one side over the string, etc endlessly.

In the event that the bend into which the string is misshaped toward the beginning is of the structure $y = b \sin \frac{m\pi x}{l}$ the arrangement is

$$y = b \sin \frac{m\pi x}{l} \cos \frac{m\pi a t}{l}. \qquad (2)$$

Regardless of what esteem t may have the bend is generally of the structure

$$y = A \sin \frac{m\pi x}{l};$$

that is, for various upsides of t we have a bunch of sine bends varying just in the adequacy and not by any stretch in that frame of mind of the bend. For this situation by the same token the entire string if $m = 1$, or each mth of the string in the event that m isn't equivalent to one, ascents and falls, and there is no evident ahead movement. At the point when this is the case we are said to have a *steady* vibration.

If $m = 1$ we get consistent movement of the string in general and if the vibration is sufficiently fast to give a melodic note the note is supposed to be the unadulterated essential note of the string. If $m = 2$ the vibration is two times as fast as when $m = 1$, the center place of the string doesn't move and is known as a hub, the two parts of the string are in inverse periods of vibration at any moment, and the note given is an octave higher than the major note and is called its unadulterated *first harmonic*.

If $m = 3$ the vibration is multiple times as fast as in the primary case, there are two hubs $x = \frac{l}{3}$ and $x = \frac{2l}{3}$, and the note is the unadulterated *second harmonic* of the key note.

For any worth of m the vibration is m times as quick as when $m = 1$, there are $m - 1$ hubs at the focuses $x = \frac{l}{m}, x = \frac{2l}{m}, \cdots x = \frac{m-1}{m} l$, and we get the $m - 1$st consonant of the principal note.

It is obvious from (1) that regardless of what the first type of the string the coming about vibration can be viewed as a mix of consistent vibrations each of which alone would give the crucial note of the string or one of its sounds, and that the intricate note coming about is actually a harmony of the crucial note and a portion of its sounds.

A finely prepared ear can frequently perceive in a mind boggling note the basic note of the string and a portion of its music and is equipped for breaking down a complex note into its part unadulterated notes definitively as Fourier's Hypothesis empowers us to investigate the complicated capability addressing the underlying type of the string into the less difficult sine-capabilities which should be consolidated to shape it.

EXAMPLES.

1. Show that in the event that a point whose separation from the finish of a harp string is $\frac{1}{n}$th the length of the string is drawn aside by the player's finger to a distance b from its place of balance and afterward delivered, the type of the vibrating string at any moment is given by the situation

$$y = \frac{2bn^2}{(n-1)\pi^2} \sum_{m=1}^{m=\infty} \left(\frac{1}{m^2} \sin \frac{m\pi}{n} \sin \frac{m\pi x}{l} \cos \frac{m\pi a t}{l} \right).$$

Show from this that every one of the music of the crucial note of the string which compare to types of vibration having hubs at the point drawn aside by the finger will be needing in the mind boggling note as a matter of fact sounded.

PHYSICS APPLICATIONS   92

2. On the off chance that an extended string begins from its place of balance, every one of its focuses having a given beginning speed, so we have

$$\begin{aligned} y &= 0 \quad \text{and when} & t &= 0 \\ D_t y &= F(x) \quad \text{and''} & t &= 0 \\ y &= 0 \quad \text{and''} & x &= 0 \\ y &= 0 \quad \text{and''} & x &= l, \end{aligned}$$

the arrangement of the issue of its vibration is simple and gives

$$y = \frac{2}{a\pi} \sum_{m=1}^{m=\infty} \left(\frac{1}{m} \sin \frac{m\pi x}{l} \sin \frac{m\pi a t}{l} \int_0^l F(\lambda) \sin \frac{m\pi \lambda}{l} d\lambda \right).$$

3. Record the answer for the situation where the string is at first misshaped furthermore, each point has a given introductory speed.

65. On the off chance that we don't disregard the opposition of the air in the issue of the vibration of an extended string the differential condition is somewhat more convoluted also, the arrangement isn't really effectively acquired. The condition is given as (IX) Art. 1.

Allow us to take care of the issue for the situation where there is no underlying speed.

Here we have and $D_t^2 y + 2k D_t y = a^2 D_x^2 y.$ and (1)

$$\begin{aligned} y &= 0 & \text{when} \quad \text{and} \, x &= 0 & (2) \\ y &= 0 & \text{''and} \, x &= l & (3) \\ y &= f(x) & \text{''and} \, t &= 0 & (4) \\ D_t y &= 0 & \text{''and} \, t &= 0. & (5) \end{aligned}$$

We get specific arrangements of (1) in the standard way. Expect $y = e^{\alpha x + \beta t}$ furthermore, substitute in (1). We have

$$\beta^2 + 2k\beta = a^2 \alpha^2$$

as the main essential connection among β and α. This gives

$$\beta = -k\hat{A} \pm \sqrt{a^2 \alpha^2 + k^2}.$$

Hence $\qquad y = e^{\alpha x - kt\hat{A} \pm t\sqrt{a^2\alpha^2 + k^2}}$ (6)

is an answer of (1) regardless of what the worth of α.

To toss it into Mathematical structure supplant α by αi, and since in genuine issues k, which is relative to the opposition, is tiny, take -1 out as a component of the extremist. We have

$$y = e^{-kt} e^{(\alpha x \hat{A} \pm t\sqrt{a^2\alpha^2 - k^2})i}.$$

Since α might be positive or negative we can get

$$y = e^{-kt} \sin(\alpha x \hat{A} \pm t\sqrt{a^2\alpha^2 - k^2})$$

and $\qquad y = e^{-kt} \cos(\alpha x \hat{A} \pm t\sqrt{a^2\alpha^2 - k^2})$

as arrangements of (1), or by joining these

$$y = e^{-kt} \sin \alpha x \cos t\sqrt{a^2\alpha^2 - k^2} \qquad (7)$$
$$y = e^{-kt} \sin \alpha x \sin t\sqrt{a^2\alpha^2 - k^2} \qquad (8)$$
$$y = e^{-kt} \cos \alpha x \cos t\sqrt{a^2\alpha^2 - k^2} \qquad (9)$$
$$y = e^{-kt} \cos \alpha x \sin t\sqrt{a^2\alpha^2 - k^2} \qquad (10)$$

PHYSICS APPLICATIONS

(7) and (8) fulfill (1) and (2) for all upsides of α. They fulfill (3) if $\alpha = \frac{m\pi}{l}$. Allow us to check whether out of them we can't develop a worth that will fulfill (4) and (5) too.

$$f(x) = \frac{2}{l} \sum_{m=1}^{m=\infty} \left(\sin \frac{m\pi x}{l} \int_0^l f(\lambda) \sin \frac{m\pi \lambda}{l} d\lambda \right). \tag{11}$$

$$y = \frac{2}{l} e^{-kt} \sum_{m=1}^{m=\infty} \left(\sin \frac{m\pi x}{l} \cos t \sqrt{\frac{m^2 \pi^2 a^2}{l^2} - k^2} \cdot \int_0^l f(\lambda) \sin \frac{m\pi \lambda}{l} d\lambda \right) \tag{12}$$

diminishes to (11) when $t = 0$ and in this way fulfills (4).

$$D_t y = -\frac{2}{l e^{kt}} \sum_{m=1}^{m=\infty} \left(\sqrt{\frac{m^2 \pi^2 a^2}{l^2} - k^2} \cdot \sin \frac{m\pi x}{l} \sin t \sqrt{\frac{m^2 \pi^2 a^2}{l^2} - k^2} \right.$$

$$\left. \cdot \int_0^l f(\lambda) \sin \frac{m\pi \lambda}{l} d\lambda \right)$$

$$- \frac{2k}{l e^{kt}} \sum_{m=1}^{m=\infty} \left(\sin \frac{m\pi x}{l} \cos t \sqrt{\frac{m^2 \pi^2 a^2}{l^2} - k^2} \cdot \int_0^l f(\lambda) \sin \frac{m\pi \lambda}{l} d\lambda \right). \tag{13}$$

When $t = 0$ the main line of the second individual from (13) evaporates however the second line lessens to

$$-\frac{2k}{l} \sum_{m=1}^{m=\infty} \left(\sin \frac{m\pi x}{l} \int_0^l f(\lambda) \sin \frac{m\pi \lambda}{l} d\lambda \right).$$

We should, then, at that point, bring into (12) an extra term which will rise to nothing when $t = 0$ and whose subordinate regarding t will drop the term above when $t = 0$.

This is effectively seen to be

$$\frac{2k}{l} e^{-kt} \sum_{m=1}^{m=\infty} \frac{1}{\sqrt{\frac{m^2 \pi^2 a^2}{l^2} - k^2}} \sin \frac{m\pi x}{l} \sin t \sqrt{\frac{m^2 \pi^2 a^2}{l^2} - k^2} \cdot \int_0^l f(\lambda) \sin \frac{m\pi \lambda}{l} d\lambda.$$

Thus our total arrangement is

$$y = \frac{2}{l} e^{-kt} \sum_{m=1}^{m=\infty} \left[\left(\cos t \sqrt{\frac{m^2 \pi^2 a^2}{l^2} - k^2} \right.\right.$$

$$\left.\left. + \frac{k}{\sqrt{\frac{m^2 \pi^2 a^2}{l^2} - k^2}} \sin t \sqrt{\frac{m^2 \pi^2 a^2}{l^2} - k^2} \right) \sin \frac{m\pi x}{l} \int_0^l f(\lambda) \sin \frac{m\pi \lambda}{l} d\lambda \right]. \tag{14}$$

Here the way that e^{-kt}, which diminishes quickly as t increments, is a component of the entire second part shows that the adequacy of the vibration quickly diminishes.

Contrasting this arrangement and that given in Art. 64 for the situation where there is no obstruction we see that the time of some random term

$$A \sin \frac{m\pi x}{l} \cos t \sqrt{\frac{m^2 \pi^2 a^2}{l^2} - k^2},$$

is more prominent than that of the comparing term $A_1 \sin \frac{m\pi x}{l} \cos \frac{m\pi a t}{l}$ in Art. 64. As such the impact of the obstruction of the air is to smooth fairly every part a piece of the note given by the string. More than this since the times of the various terms of (14) are presently not careful submultiples of the time of the initial term, the part notes are presently not in awesome agreement

PHYSICS APPLICATIONS

with the principal note of the string, and the best amazing congruity between the major note and its sounds isn't exactly acknowledged in any genuine case.

When k is tiny, as on account of a fine string, the takeoff from wonderful congruity is extremely slight; however on account of a coarse string or more terrible still of a versatile lace, where the obstruction of the air is significant, the unmusical person of the sound is entirely perceptible.

EXAMPLES.

1. Tackle Ex. 1 Art. 64 taking into account the opposition of the air.

2. Tackle Ex. 2 Art. 64 taking into account the opposition of the air;

$$y = \frac{2}{l} e^{-kt} \sum_{m=1}^{m=\infty} \left(\frac{1}{\sqrt{\frac{m^2\pi^2 a^2}{l^2} - k^2}} \sin \frac{m\pi x}{l} \sin t \sqrt{\frac{m^2\pi^2 a^2}{l^2} - k^2} \cdot \int_0^l F(\lambda) \sin \frac{m\pi\lambda}{l} d\lambda \right).$$

3. Track down a specific arrangement of (1) Art. 65 with the understanding that it is of the structure $y = T.X$, where T is a component of t alone and X an element of x alone.

66. We give now to several issues that require the adjustment furthermore, expansion of Fourier's Hypothesis, *the cooling of a circle in air*, and the *vibration of an extended rectangular membrane*, however as a prologue to the previous we will initially think about the accompanying extremely straightforward issue; to see as the temperature of any place of a circle whose underlying temperature is any given capability of r the distance of the point from the middle, and whose surface is kept at the steady temperature b.

Here we are to address

$$D_t(ru) = a^2 D_r^2(ru), \tag{1}$$

see [v] Art. 1, dependent upon the circumstances

$$u = f(r) \quad \text{when} \quad t = 0 \tag{2}$$
$$u = b \quad \text{''} \quad r = c \tag{3}$$

assuming c is the sweep.

Let $v = ru$, then our conditions become

$$D_t v = a^2 D_r^2 v \tag{4}$$
$$v = rf(r) \quad \text{when} \quad t = 0 \tag{5}$$
$$v = bc \quad \text{''} \quad r = c \tag{6}$$
$$v = 0 \quad \text{''} \quad r = 0. \tag{7}$$

Our concern is presently definitively that of Art. 62 and we have as our answer

$$ru = \frac{2}{c} \sum_{m=1}^{m=\infty} \left(e^{-\frac{m^2 a^2 \pi^2}{c^2} t} \sin \frac{m\pi r}{c} \int_0^c \lambda f(\lambda) \sin \frac{m\pi\lambda}{c} d\lambda \right)$$
$$+ b \left[r + \frac{2c}{\pi} \sum_{m=1}^{m=\infty} \left(\frac{(-1)^m}{m} e^{-\frac{m^2 a^2 \pi^2}{c^2} t} \sin \frac{m\pi r}{c} \right) \right]. \tag{8}$$

PHYSICS APPLICATIONS

EXAMPLES.

1. If $f(r) = b$ (8) Art. 66 diminishes to $u = b$ and there is no difference in temperature.

2. Assuming the underlying temperature is consistent and equivalent to β

$$u = b + \frac{2c}{\pi r}(\beta - b)\left[e^{-\frac{a^2\pi^2}{c^2}t}\sin\frac{\pi r}{c} - \frac{1}{2}e^{-\frac{4a^2\pi^2}{c^2}t}\sin\frac{2\pi r}{c}\right.$$
$$\left. + \frac{1}{3}e^{-\frac{9a^2\pi^2}{c^2}t}\sin\frac{3\pi r}{c} - \cdots\right].$$

3. An iron circle 40 cm. in breadth is warmed to the temperature 100Â° centigrade all through; its surface is then kept at the consistent temperature 0Â°. Find the temperature of a point 10 cm. from the middle, and track down the temperature of the middle, 15 minutes in the wake of cooling has started. Given $a^2 = 0.185$ in C.G.S. units.   Ans., 2Â°.1; 3Â°.3.

67. If rather than having the temperature of the outer layer of the circle steady, the circle is set in air which is kept at the consistent temperature zero, the issue is significantly more convoluted. For this situation the surface temperature can never again be basically communicated yet is given by a new differential condition

$$D_r u + hu = 0 \quad \text{when} \quad r = c, \tag{1}$$

where h is an exploratory consistent relying on what is known as the surface conductivity of the circle.

Our conditions, then, at that point, are

$$D_t(ru) = a^2 D_r^2(ru) \tag{2}$$
$$u = f(r) \quad \text{when} \quad t = 0 \tag{3}$$
$$D_r u + hu = 0 \quad \text{when} \quad r = c. \tag{4}$$

As in Art. 66 let $v = ru$; then, at that point, we have

$$D_t v = anda^2 D_r^2 v \tag{5}$$
$$v = rf(r) \quad \text{when} \quad andt = 0 \tag{6}$$
$$v = 0 \quad \text{"and } r = 0 \tag{7}$$
$$D_r v + \left(h - \frac{1}{c}\right)v = 0 \quad \text{when} \quad r = c. \tag{8}$$

$v = e^{-a^2\alpha^2 t}\cos\alpha r$ and $v = e^{-a^2\alpha^2 t}\sin\alpha r$ have previously been seen as specific arrangements of (5) (see Art. 60).

$$v = e^{-a^2\alpha^2 t}\sin\alpha r \tag{9}$$

fulfills (7) for all upsides of α.

Substitute this worth of v in (8) and we have

$$\alpha c\cos\alpha c + (hc - 1)\sin\alpha c = 0. \tag{10}$$

On the off chance that α_k is a worth of α which is a base of the supernatural condition (10)

$$v = e^{-a^2\alpha_k^2 t}\sin\alpha_k r \tag{11}$$

will fulfill (5), (7), and (8).

It stays to see whether out of terms of the structure surrendered (11) we can develop a worth of v which will fulfill (6).

When $t=0$ the second individual from (11) decreases to $\sin\alpha_k r$. In the event that, we can communicate $rf(r)$ as an amount of terms of the structure $b_k \sin\alpha_k r$ where α_k is a root of (10)

$$v = \sum b_k e^{-a^2\alpha_k^2 t} \sin\alpha_k r \tag{12}$$

will fulfill the conditions in general (5), (6), (7), and (8), and will be the required arrangement.

Here, then, at that point, we have another issue comparable to that of creating in a Fourier's Series, but instead more confounded, to be specific, to foster any capability of x in a progression of the structure $\sum a_m \sin\alpha_m x$ where α_m is a foundation of the situation (10); or on the other hand on the off chance that we call $ac = \phi$ and $hc - 1 = p$, where $a_m = \frac{\phi_m}{c}$, ϕ_m being a root of the situation

$$\phi \cos\phi + p \sin\phi = 0 \tag{13}$$

or all the more just of

$$\phi + p\tan\phi = 0; \tag{14}$$

recalling that the series and the capability should be equivalent for all upsides of x among nothing and c.

On the off chance that ϕ_m is a foundation of (14) $-\phi_m$ is likewise a root.

Since $\sin\frac{\phi_m}{c}x = -\sin\left(-\frac{\phi_m}{c}x\right)$ the conditions of the expected turn of events which relate to negative roots might be joined with those comparing to positive roots, and along these lines we really want think about just certain roots.

$\phi = 0$ is a foundation of (14) however as $\sin 0 = 0$ there will no relate term in the turn of events. In the event that we develop the bend

$$y = -\frac{1}{p}x \tag{15}$$

and the curve

$$y = \tan x \tag{16}$$

the abscissas of their places of convergence are upsides of x which fulfill $\frac{x}{p} + \tan x = 0$, or at least, are foundations of condition (14). Seeing that is simple there will continuously be an endless number of genuine positive roots, one for each of the parts of the intermittent bend $y = \tan x$ which lie to one side of the beginning. The mathematical upsides of these roots can be gotten by a simple calculation. The development proposed above shows that as m increments ϕ_m will quickly move toward the worth $(2m-1)\frac{\pi}{2}$ in the event that p is positive or on the other hand assuming p is negative what's more, mathematically not as much as solidarity, and $(2m+1)\frac{\pi}{2}$ on the off chance that p is negative and mathematically more noteworthy than solidarity.

There exist, then, an endless number of positive genuine foundations of $\phi + p\tan\phi = 0$ furthermore, thusly of

$$ac \cos ac + (hc - 1)\sin ac = 0.$$

68. The improvement called for in the last article can be acquired very effectively from a less difficult one which we will currently consider, in particular, to create $f(x)$ into a progression of the structure

$$f(x) = a_1 \sin\phi_1 x + a_2 \sin\phi_2 x + a_3 \sin\phi_3 x + \cdots \tag{1}$$

where $\phi_1, \phi_2, \phi_3 \cdots$ are foundations of the situation

$$\phi \cos\phi + p \sin\phi = 0, \tag{2}$$

PHYSICS APPLICATIONS

the improvement to hold really great for all upsides of x between $x = 0$ and $x = 1$.

Allow us to continue as in Arts. 24 and 27. Call $\frac{1}{n+1} = \Delta x$ and structure n conditions by filling in for x thus in the situation

$$f(x) = a_1 \sin \phi_1 x + a_2 \sin \phi_2 x + a_3 \sin \phi_3 x + \cdots + a_n \sin \phi_n x \tag{3}$$

the qualities $\Delta x, 2\Delta x, 3\Delta x, \cdots n\Delta x$; this being comparable to making the qualities of the aggregate and the capability concur for the n upsides of x subbed.

To decide any coefficient a_m duplicate the main condition by $\Delta x . \sin(\phi_m \, \Delta x)$, the second by $\Delta x . \sin(2\phi_m \Delta x)$, the third by $\Delta x . \sin(3\phi_m \Delta x)$, etc, the nth condition by $\Delta x . \sin(n\phi_m \Delta x)$; add the conditions and process the restricting upsides of the details of the subsequent condition as n is endlessly expanded. This as in Art. 24 apparently is identical to duplicating (3) by $\sin \phi_m x . dx$ furthermore, coordinating between the cutoff points $x = 0$ and $x = 1$.

The main individual from the subsequent condition is

$$\int_0^1 f(x) \sin \phi_m x . dx;$$

The coefficient of a_k is

$$\int_0^1 \sin \phi_k x \sin \phi_m x . dx,$$

what's more, of a_m is

$$\int_0^1 \sin^2 \phi_m x . dx.$$

$$\int_0^1 \sin \phi_k x \sin \phi_m x . dx = \frac{1}{2} \int_0^1 [\cos(\phi_k - \phi_m)x - \cos(\phi_k + \phi_m)x] dx$$
$$= \frac{1}{2} \left[\frac{\sin(\phi_k - \phi_m)}{\phi_k - \phi_m} - \frac{\sin(\phi_k + \phi_m)}{\phi_k + \phi_m} \right]$$
$$= -\frac{\phi_k \cos \phi_k \sin \phi_m - \phi_m \sin \phi_k \cos \phi_m}{\phi_k^2 - \phi_m^2} \tag{4}$$

But$and\phi_k \cos \phi_k \qquad +p \sin \phi_k = 0$
and$and\phi_m \cos \phi_m \qquad +p \sin \phi_m = 0 \quad$ by (2).

Thus the numerator of the second individual from (4) is zero, and the coefficient of a_k disappears assuming k isn't equivalent to m.

$$\int_0^1 \sin^2 \phi_m x . dx = \frac{1}{2\phi_m}[\phi_m - \sin \phi_m \cos \phi_m] = \frac{1}{2}\left[1 - \frac{\sin 2\phi_m}{2\phi_m}\right]. \tag{5}$$

Therefore
$$a_m = \frac{2}{1 - \frac{\sin 2\phi_m}{2\phi_m}} \int_0^1 f(x) \sin \phi_m x . dx. \tag{6}$$

The coefficient of the essential in (6) can be changed as follows so as not to include mathematical

PHYSICS APPLICATIONS

capabilities.

$$\phi_m \cos \phi_m + p \sin \phi_m = 0, \qquad \text{by (2)}$$

$$\phi_m \cos^2 \phi_m + \frac{p}{2} \sin 2\phi_m = 0,$$

$$\frac{\sin 2\phi_m}{2\phi_m} = -\frac{\cos^2 \phi_m}{p}. \tag{7}$$

$$\phi_m^2 \cos^2 \phi_m = p^2 \sin^2 \phi_m,$$

$$(\phi_m^2 + p^2) \cos^2 \phi_m = p^2,$$

$$\frac{\cos^2 \phi_m}{p} = \frac{p}{\phi_m^2 + p^2}. \tag{8}$$

Consequently by (7) and (8)

$$1 - \frac{\sin 2\phi_m}{2\phi_m} = \frac{\phi_m^2 + p(p+1)}{\phi_m^2 + p^2},$$

and
$$a_m = \frac{2(\phi_m^2 + p^2)}{\phi_m^2 + p(p+1)} \int_0^1 f(\alpha) \sin \phi_m \alpha . d\alpha. \tag{9}$$

In this way our expected improvement is

$$f(x) = \sum_{m=1}^{m=\infty} \left(\frac{2(\phi_m^2 + p^2)}{\phi_m^2 + p(p+1)} \sin \phi_m x \int_0^1 f(\alpha) \sin \phi_m \alpha . d\alpha \right). \tag{10}$$

From (10) it effectively follows that for upsides of x among 0 and c

$$f(x) = a_1 \sin a_1 x + a_2 \sin a_2 x + a_3 \sin a_3 x + \cdots \tag{11}$$

where
$$a_m = \frac{2}{c} \cdot \frac{\alpha_m^2 c^2 + p^2}{\alpha_m^2 c^2 + p(p+1)} \int_0^c f(\lambda) \sin \alpha_m \lambda . d\lambda, \tag{12}$$

furthermore, α_m is a foundation of the situation

$$\alpha c \cos \alpha c + p \sin \alpha c = 0. \tag{13}$$

It is to be seen that assuming p is endless (13) decreases to $\sin \alpha c = 0$, α_m becomes $\frac{m\pi}{c}$ and (11) and (12) give our guideline Fourier sine series (v. Art. 31), and consequently the standard Fourier improvement in sine series is only a exceptional instance of the issue recently tackled.

Besides since the Fourier technique for deciding the coefficients of such a series requires that

$$\int_0^c \sin \alpha_m x \sin \alpha_n x . dx = 0,$$

that is that
$$\frac{\sin(\alpha_m - \alpha_n)c}{\alpha_m - \alpha_n} - \frac{\sin(\alpha_m + \alpha_n)c}{\alpha_m + \alpha_n} = 0$$

or diminishing, that
$$\frac{\alpha_m c \cos \alpha_m c}{\sin \alpha_m c} = \frac{\alpha_n c \cos \alpha_n c}{\sin \alpha_n c},$$

or on the other hand that α_m and α_n ought to be foundations of the situation

$$\frac{\alpha c \cos \alpha c}{\sin \alpha c} = p$$

where p is some steady, it follows that we have gotten in (11) the most general sine advancement that can be acquired by Fourier's strategy.

PHYSICS APPLICATIONS

EXAMPLES.

1. Show that the arrangement of the issue of Art. 67 is

$$ru = \sum_{m=1}^{m=\infty} b_m e^{-a^2 \alpha_m^2 t} \sin \alpha_m r,$$

where
$$b_m = \frac{2}{c} \cdot \frac{\alpha_m^2 c^2 + (hc-1)^2}{\alpha_m^2 c^2 + hc(hc-1)} \int_0^c \lambda f(\lambda) \sin \alpha_m \lambda \, d\lambda$$

furthermore, α_m is a foundation of

$$\alpha c \cos \alpha c + (hc - 1) \sin \alpha c = 0.$$

2. In the event that the underlying temperature of the circle is steady and equivalent to β

$$ru = \sum_{m=1}^{m=\infty} b_m e^{-a^2 \alpha_m^2 t} \sin \alpha_m r$$

$$b_m = 2\beta h \cdot \frac{\alpha_m^2 c^2 + (hc-1)^2}{\alpha_m^2 c^2 + hc(hc-1)} \cdot \frac{\sin \alpha_m c}{\alpha_m^2}$$

$$= \frac{2\beta hc}{\alpha_m} \cdot \frac{[\alpha_m^2 c^2 + (hc-1)^2]^{\frac{1}{2}}}{\alpha_m^2 c^2 + hc(hc-1)}.$$

3. In the event that the temperature of the air is a steady γ rather than zero the surface condition of condition is

$$D_r u + h(u - \gamma) = 0 \quad \text{when} \quad r = c.$$

The replacement of $u_1 = u - \gamma$, notwithstanding, brings the issue under Ex. 1 furthermore, we get

$$r(u - \gamma) = \sum_{m=1}^{m=\infty} b_m e^{-a^2 \alpha_m^2 t} \sin \alpha_m r$$

where
$$b_m = \frac{2}{c} \cdot \frac{\alpha_m^2 c^2 + (hc-1)^2}{\alpha_m^2 c^2 + hc(hc-1)} \int_0^c \lambda [f(\lambda) - \gamma] \sin \alpha_m \lambda \, d\lambda.$$

4. An iron circle 40 cm. in breadth is warmed to the temperature 100Â° centigrade all through; it is then permitted to cool in air which is kept at the steady temperature 0Â°. Track down the temperature at the middle; at a point 10 cm. from the middle; and at the surface; 15 minutes subsequent to cooling has started. Given $a^2 = 0.185$ and $h = \frac{1}{800}$ in C.G.S. units. (v. Ex. 3, Art. 66.)

Ans., 97Â°.67; 97Â°.36; 96Â°.46.

5. Show that assuming in the piece considered in Art. 60 one face is presented to air at the temperature zero, so we have $D_t u = a^2 D_x^2 u$, $u = 0$ when $x = 0$, $u = f(x)$ when $t = 0$, and $D_x u + hu = 0$ when $x = c$, then, at that point,

$$u = \sum_{m=1}^{m=\infty} a_m e^{-a^2 \alpha_m^2 t} \sin \alpha_m x$$

where
$$a_m \text{ and} = 2 \frac{\alpha_m^2 + h^2}{\alpha_m^2 c + h(hc+1)} \int_0^c f(\lambda) \sin \alpha_m \lambda \, d\lambda,$$

α_m being a foundation of $\alpha c \cos \alpha c + hc \sin \alpha c = 0$.

PHYSICS APPLICATIONS

6. On the off chance that in the issue of Art. 57 heat escapes from one side of the plate into air at the temperature zero so we have $D_x^2 u + D_y^2 u = 0$, $u = 0$ when $x = 0$, $u = f(x)$ when $y = 0$, and $D_x u + hu = 0$ when $x = a$, then, at that point,

$$u = \sum_{m=1}^{m=\infty} a_m e^{-\alpha_m y} \sin \alpha_m x$$

where
$$a_m = 2 \frac{\alpha_m^2 + h^2}{\alpha_m^2 a + h(ha+1)} \int_0^a f(\lambda) \sin \alpha_m \lambda \, d\lambda,$$

α_m being a base of $\alpha a \cos \alpha a + ha \sin \alpha a = 0$.

7. Assuming in the issue of Art. 59 there is spillage at one side of the sheet so that we have $D_x^2 V + D_y^2 V = 0$, $V = 0$ when $x = 0$, $V = 0$ when $y = b$, $V = f(x)$ when $y = 0$, and $D_x V + hV = 0$ when $x = a$, then

$$V = \sum_{m=1}^{m=\infty} a_m \frac{\sinh \alpha_m (b-y)}{\sinh \alpha_m b} \sin \alpha_m x,$$

where a_m has the worth given in Ex. 6.

69. Assuming we have a limitless strong with one plane face which is presented to air at the temperatures $U = F(t)$ and intensity can stream just at right points to this face, we can take care of the issue promptly for the situation where the underlying temperatures are zero. We have

$$D_t u = a^2 D_x^2 u$$

dependent upon the circumstances

$$u = 0 \quad \text{when} \quad t = 0$$

and
$$D_x u + h(U - u) = 0 \quad \text{when} \quad x = 0.$$

Let
$$v = u - \frac{1}{h} D_x u. \tag{1}$$

Then v will fulfill the condition

$$D_t v = a^2 D_x^2 v,$$

furthermore, we will likewise have $v = U$ when $x = 0$.

Since $U = F(t)$
$$v = \frac{2}{\sqrt{\pi}} \int_{\frac{x}{2a\sqrt{t}}}^{\infty} e^{-\beta^2} F\left(t - \frac{x^2}{4a^2 \beta^2}\right) d\beta \tag{2}$$

by Art. 51 (10).

$$D_x u - hu = -hv \quad \text{by (1)}.$$

Hence
$$u e^{-hx} = -h \int e^{-hx} v \, dx + C;$$

v. Int. Cal. § 4, page 314.

Deciding C by the way that $ue^{-hx} = 0$ when $x = \infty$ we have

$$u = h e^{hx} \int_x^{\infty} e^{-hx} v \, dx. \tag{3}$$

Subbing the worth of v from (2) we have

$$u = \frac{2h e^{hx}}{\sqrt{\pi}} \int_x^{\infty} e^{-hx} dx \int_{\frac{x}{2a\sqrt{t}}}^{\infty} e^{-\beta^2} F\left(t - \frac{x^2}{4a^2 \beta^2}\right) d\beta, \tag{4}$$

PHYSICS APPLICATIONS

as our expected arrangement.

For an augmentation of this technique to the progression of intensity in two and three aspects also, for the understanding of the outcomes by the guide of the hypothesis of *Images*, see E. W. Hobson, Proc. Lond. Math. Soc., Vol. XIX.

EXAMPLES.

1. On the off chance that the temperature of the air is an occasional capability of the time, say $\rho_m \sin(m\alpha t + \lambda_m)$ and we care just for the restricting worth of u as t increments, show that this worth is

$$\frac{h\rho_m e^{-\frac{x}{a}\sqrt{\frac{m\alpha}{2}}}}{\left(h + \frac{1}{a}\sqrt{\frac{m\alpha}{2}}\right)^2 + \frac{m\alpha}{2a^2}} \left[\left(h + \frac{1}{a}\sqrt{\frac{m\alpha}{2}}\right)\sin\left(m\alpha t - \frac{x}{a}\sqrt{\frac{m\alpha}{2}} + \lambda_m\right) \right.$$
$$\left. - \frac{1}{a}\sqrt{\frac{m\alpha}{2}}\cos\left(m\alpha t - \frac{x}{a}\sqrt{\frac{m\alpha}{2}} + \lambda_m\right)\right].$$

v. Art. 52 and Art. 51 Ex. 4.

Note that $and \int e^{ax}\sin bx.dx \quad = \dfrac{e^{ax}(a\sin bx - b\cos bx)}{a^2 + b^2}$

and $and \int e^{ax}\cos bx.dx \quad = \dfrac{e^{ax}(a\cos bx + b\sin bx)}{a^2 + b^2}$

v. Int. Cal. Table of Int. (235) and (236).

2. If $D_x^2 V + D_y^2 V = 0$, $V = 0$ when $y = 0$ and $D_x V + h[F(y) - V] = 0$ when $x = 0$ show that

$$V = \frac{he^{hx}}{\pi}\int_x^\infty e^{-hx}dx \int_0^\infty F(\lambda)d\lambda \left[\frac{x}{x^2 + (\lambda - y)^2} - \frac{x}{x^2 + (\lambda + y)^2}\right];$$

v. Art. 47 Ex. 1.

70. The answer for a momentary intensity wellspring of solidarity Q at the point $x = \lambda$ in the event that intensity escapes at the beginning into air at the temperature zero, so that $D_x u - hu = 0$ when $x = 0$, can be acquired by the guide of Art. 53.

Let $u = u_1 + u_2$ where u_1 is the temperature that would be because of the given source assuming we had no limit at the beginning, so that

$$u_1 = \frac{Q}{2a\sqrt{\pi t}} e^{-\frac{(\lambda-x)^2}{4a^2 t}}. \qquad \text{[Art. 53 (2)]}$$

$$D_x u - hu = D_x u_1 - hu_1 + D_x u_2 - hu_2 = 0 \quad \text{when} \quad x = 0.$$

Therefore $and\ D_x u_2 - hu_2 \quad = -(D_x u_1 - hu_1) \tag{1}$

when $x = 0$.

But $and - (D_x u_1 - hu_1) = -\dfrac{Q}{2a\sqrt{\pi t}}\left(\dfrac{\lambda - x}{2a^2 t} - h\right)e^{-\frac{(\lambda-x)^2}{4a^2 t}}$

$$= -\frac{Q}{2a\sqrt{\pi t}}\left(\frac{\lambda}{2a^2 t} - h\right)e^{-\frac{\lambda^2}{4a^2 t}}$$

when $x = 0$.

This is handily seen to be the worth to which

$$-\frac{Q}{2a\sqrt{\pi t}}\left(\frac{\lambda + x}{2a^2 t} - h\right)e^{-\frac{(\lambda+x)^2}{4a^2 t}}$$

diminishes when $x = 0$, and this last articulation is

$$(D_x + h)\frac{Q}{2a\sqrt{\pi t}}e^{-\frac{(\lambda+x)^2}{4a^2 t}}$$

furthermore, hence fulfills the condition

$$D_t u = a^2 D_x^2 u; \tag{2}$$

since $\frac{Q}{2a\sqrt{\pi t}}e^{-\frac{(\lambda+x)^2}{4a^2 t}}$ is the temperature because of a source at $x = -\lambda$.

In the event that, we decide u_2 from the condition that

$$D_x u_2 - h u_2 = -\frac{Q}{2a\sqrt{\pi t}}\left(\frac{\lambda+x}{2a^2 t} - h\right)e^{-\frac{(\lambda+x)^2}{4a^2 t}} \tag{3}$$

taking consideration not to present any inconsistent consistent or erratic capability of t in our coordination, u_2 will fulfill condition (2) and condition (1).

Coordinating (3) [v. Int. Cal. § 4, page 314] and deciding the constants of coordination appropriately we get

$$u_2 = \frac{Q}{2a\sqrt{\pi t}}\left[e^{-\frac{(\lambda+x)^2}{4a^2 t}} - 2he^{hx}\int_x^\infty e^{-hx-\frac{(\lambda+x)^2}{4a^2 t}}dx\right]. \tag{4}$$

Thusly the arrangement of our concern is

$$u = \frac{Q}{2a\sqrt{\pi t}}\left[e^{-\frac{(\lambda-x)^2}{4a^2 t}} + e^{-\frac{(\lambda+x)^2}{4a^2 t}} - 2he^{hx}\int_x^\infty e^{-hx-\frac{(\lambda+x)^2}{4a^2 t}}dx\right]. \tag{5}$$

In the event that we supplant Q by $f(\lambda)d\lambda$ and coordinate from 0 to ∞ we get as the arrangement for the situation where $u = f(x)$ when $t = 0$ and $x > 0$, and $D_x u - hu = 0$ when $x = 0$

$$u = \frac{1}{2a\sqrt{\pi t}}\int_0^\infty f(\lambda)d\lambda\left[e^{-\frac{(\lambda-x)^2}{4a^2 t}} + e^{-\frac{(\lambda+x)^2}{4a^2 t}} - 2he^{hx}\int_x^\infty e^{-hx-\frac{(\lambda+x)^2}{4a^2 t}}dx\right]. \tag{6}$$

For a translation of this outcome by the hypothesis of Pictures and the expansion of the technique to the conduction of intensity in n aspects see G. H. Bryan, Proc. Lond. Math. Soc., Vol. XXII.

EXAMPLE.

Show that if $u = f(x)$ when $t = 0$ and $D_x u + h[F(t) - u] = 0$ when $x = 0$ we should take u equivalent to the amount of the second individuals from (6) Art. 70 what's more, of (4) Art. 69.

71. As one more issue requiring a slight expansion of Fourier's Hypothesis allow us to think about the vibration of a rectangular extended flexible film secured at the edges, that is of a rectangular drumhead.

On the off chance that two of the sides are taken as tomahawks and the plane of harmony of the film as the plane of XY the condition for the movement of the layer is

$$D_t^2 z = c^2(D_x^2 z + D_y^2 z) \tag{1}$$

see [x] Art. 1.

PHYSICS APPLICATIONS

Allow the film to be misshaped toward the begin some given structure $z = f(x,y)$ and afterward permitted to swing. Our conditions of conditions are then, at that point,

$$z = 0 \quad \text{when} \quad and\, x = 0 \tag{2}$$
$$z = 0 \quad "and\, x = a \tag{3}$$
$$z = 0 \quad "and\, y = 0 \tag{4}$$
$$z = 0 \quad "and\, y = b \tag{5}$$
$$z = f(x,y) \quad "and\, t = 0 \tag{6}$$
$$D_t z = 0 \quad "and\, t = 0. \tag{7}$$

We can get a specific arrangement of (1) by our standard gadget. Accept

$$z = e^{\alpha x + \beta y + \gamma t}$$

also, substitute in (1). We get $\gamma^2 = c^2(\alpha^2 + \beta^2)$ as the as it were connection that need hold between α, β, and γ, all together that $z = e^{\alpha x + \beta y + \gamma t}$ might be a arrangement. This gives

$$\gamma = \pm c\sqrt{\alpha^2 + \beta^2}.$$

Therefore
$$z = e^{\alpha x + \beta y \pm ct\sqrt{\alpha^2 + \beta^2}}$$

is an answer of (1) regardless of what values are given to α and β.

Supplant α and β by αi and βi and we have

$$z = e^{(\alpha x + \beta y \pm ct\sqrt{\alpha^2 + \beta^2})i}$$

as an answer, and from this we get

$$z = \sin(\alpha x + \beta y \pm ct\sqrt{\alpha^2 + \beta^2}) \tag{8}$$
and
$$z = \cos(\alpha x + \beta y \pm ct\sqrt{\alpha^2 + \beta^2}) \tag{9}$$

as specific arrangements of (1), α and β being unlimited.

From (8) and (9) we can get arrangements of the accompanying structures

$$\left.\begin{aligned} z &= \sin \alpha x \sin \beta y \sin ct\sqrt{\alpha^2 + \beta^2} \\ z &= \sin \alpha x \sin \beta y \cos ct\sqrt{\alpha^2 + \beta^2} \\ z &= \sin \alpha x \cos \beta y \sin ct\sqrt{\alpha^2 + \beta^2} \\ z &= \sin \alpha x \cos \beta y \cos ct\sqrt{\alpha^2 + \beta^2} \\ z &= \cos \alpha x \sin \beta y \sin ct\sqrt{\alpha^2 + \beta^2} \\ z &= \cos \alpha x \sin \beta y \cos ct\sqrt{\alpha^2 + \beta^2} \\ z &= \cos \alpha x \cos \beta y \sin ct\sqrt{\alpha^2 + \beta^2} \\ z &= \cos \alpha x \cos \beta y \cos ct\sqrt{\alpha^2 + \beta^2}, \end{aligned}\right\} \tag{10}$$

every one of which will fulfill condition (1). The second of these will fulfill too (2), (4) and (7) anything values be taken for α and β. It will fulfill (3) and (5) if α and β are equivalent $\frac{m\pi}{a}$ and $\frac{n\pi}{b}$ separately.

On the off chance that, we can so consolidate terms of the structure

$$\sin \frac{m\pi x}{a} \sin \frac{n\pi y}{b} \cos c\pi t \sqrt{\frac{m^2}{a^2} + \frac{n^2}{b^2}}$$

as to fulfill (6) our concern will be totally addressed.

PHYSICS APPLICATIONS

This should be possible in the event that we can communicate $f(x,y)$ as an amount of terms of the structure $A \sin \frac{m\pi x}{a} \sin \frac{n\pi y}{b}$, the total and the capability being equivalent when x lies among 0 and a and y among 0 and b.

$f(x,y)$ can be communicated as far as $\sin \frac{m\pi x}{a}$ by Fourier's Hypothesis if we view y as consistent. We have

$$f(x,y) = \sum_{m=1}^{m=\infty} a_m \sin \frac{m\pi x}{a} \tag{11}$$

where

$$a_m = \frac{2}{a} \int_0^a f(\lambda, y) \sin \frac{m\pi \lambda}{a} d\lambda. \tag{12}$$

$f(\lambda, y)$ in (12) is a component of y and might be created by Fourier's Hypothesis.

We have and $\quad f(\lambda, y) = \sum_{n=1}^{n=\infty} b_n \sin \frac{n\pi y}{b} \tag{13}$

where

$$b_n = \frac{2}{b} \int_0^b f(\lambda, \mu) \sin \frac{n\pi \mu}{b} d\mu. \tag{14}$$

Filling in for $f(\lambda, y)$ in (12) the worth just acquired we have

$$a_m = \frac{2}{a}\frac{2}{b} \sum_{n=1}^{n=\infty} \left(\int_0^a d\lambda \int_0^b f(\lambda, \mu) \sin \frac{m\pi\lambda}{a} \sin \frac{n\pi\mu}{b} d\mu \right) \sin \frac{n\pi y}{b}$$

also,

$$f(x,y) = \frac{4}{ab} \sum_{m=1}^{m=\infty} \sum_{n=1}^{n=\infty} \left(\sin \frac{m\pi x}{a} \sin \frac{n\pi y}{b} \int_0^a d\lambda \int_0^b f(\lambda,\mu) \sin \frac{m\pi\lambda}{a} \sin \frac{n\pi\mu}{b} d\mu \right). \tag{15}$$

Hence

$$z = \sum_{m=1}^{m=\infty} \sum_{n=1}^{n=\infty} \left(A_{m,n} \sin \frac{m\pi x}{a} \sin \frac{n\pi y}{b} \cos c\pi t \sqrt{\frac{m^2}{a^2} + \frac{n^2}{b^2}} \right), \tag{16}$$

where

$$A_{m,n} = \frac{4}{ab} \int_0^a d\lambda \int_0^b f(\lambda, \mu) \sin \frac{m\pi \lambda}{a} \sin \frac{n\pi \mu}{b} d\mu. \tag{17}$$

is our necessary arrangement.

EXAMPLES.

1. Show that assuming the layer begins from its place of harmony yet with a given beginning speed put forth for each point so that $z = 0$ when $t = 0$ and $D_t z = F(x,y)$ when $t = 0$ the arrangement is

$$z = \frac{1}{c\pi} \sum_{m=1}^{m=\infty} \sum_{n=1}^{n=\infty} \left(A_{m,n} \frac{1}{\sqrt{\frac{m^2}{a^2} + \frac{n^2}{b^2}}} \sin \frac{m\pi x}{a} \sin \frac{n\pi y}{b} \sin c\pi t \sqrt{\frac{m^2}{a^2} + \frac{n^2}{b^2}} \right)$$

where

$$A_{m,n} = \frac{4}{ab} \int_0^a d\lambda \int_0^b F(\lambda, \mu) \sin \frac{m\pi \lambda}{a} \sin \frac{n\pi \mu}{b} d\mu.$$

2. In the event that there is both starting contortion and beginning speed

$$z = \frac{4}{ab} \sum_{m=1}^{m=\infty} \sum_{n=1}^{n=\infty} \sin \frac{m\pi x}{a} \sin \frac{n\pi y}{b} \left[A_{m,n} \cos c\pi t \sqrt{\frac{m^2}{a^2} + \frac{n^2}{b^2}} \right.$$

$$\left. + B_{m,n} \sin c\pi t \sqrt{\frac{m^2}{a^2} + \frac{n^2}{b^2}} \right]$$

PHYSICS APPLICATIONS 105

where
$$A_{m,n} = \int_0^a d\lambda \int_0^b f(\lambda,\mu) \sin\frac{m\pi\lambda}{a} \sin\frac{n\pi\mu}{b} d\mu,$$

and
$$B_{m,n} = \frac{1}{c\pi\sqrt{\frac{m^2}{a^2}+\frac{n^2}{b^2}}} \int_0^a d\lambda \int_0^b F(\lambda,\mu) \sin\frac{m\pi\lambda}{a} \sin\frac{n\pi\mu}{b} d\mu.$$

3. Get a specific arrangement of (1) Art. 71 by expecting $z = T.X.Y$ where T is an element of t alone, X of x alone, and Y of y alone.

72. various fascinating ends can be drawn from the consequences of Art. 71 and Exs. 1 and 2.

(*a*) Nobody of the three upsides of z is in everyday an occasional capability of t, what's more, therefore a vibrating rectangular layer won't overall give a melodic note.

(*b*) An extended rectangular layer can be made to give a melodic note by beginning the vibration appropriately. For assuming that the underlying conditions are such that the arrangement decreases to a solitary term, as will be the situation if the underlying mutilation in the issue of Art. 71 be to such an extent that $f(x,y) = A_{m,n} \sin\frac{m\pi x}{a} \sin\frac{n\pi y}{b}$, or on the other hand the underlying speed in Ex. 1 be to such an extent that $F(x,y) = B_{m,n} \sin\frac{m\pi x}{a} \sin\frac{n\pi y}{b}$, or on the other hand the underlying contortion and starting speed in Ex. 2 be the qualities recently given, then the vibration will be occasional and will have the period

$$T = \frac{2}{c\sqrt{\frac{m^2}{a^2}+\frac{n^2}{b^2}}}. \tag{1}$$

Since T is a component of m and n and m and n are any entire numbers, the same film is fit for giving an extraordinary assortment of melodic notes of various pitches. On the off chance that m and n are both solidarity we get the most reduced note the layer can give, which is called its crucial note. Its period

$$T_1 = \frac{2}{c\sqrt{\frac{1}{a^2}+\frac{1}{b^2}}} = \frac{2ab}{c\sqrt{a^2+b^2}} \tag{2}$$

In the event that m and n are both equivalent to k we get

$$T_k = \frac{2ab}{kc\sqrt{a^2+b^2}}; \tag{3}$$

accordingly the layer can be made to give any consonant of its central note.

More than this, since as we have seen

$$T_{m,n} = \frac{2}{c\sqrt{\frac{m^2}{a^2}+\frac{n^2}{b^2}}}$$

is the time of any note the film can give, and since if m and n are supplanted by mk and nk we get

$$T_{mk,nk} = \frac{2}{ck\sqrt{\frac{m^2}{a^2}+\frac{n^2}{b^2}}}$$

the layer can sound every one of the music of any note which it can give.

(*c*) For the situation thought about above, where the arrangement diminishes to the single term

$$z = \sin\frac{m\pi x}{a} \sin\frac{n\pi y}{b}\left[A_{m,n} \cos c\pi t\sqrt{\frac{m^2}{a^2}+\frac{n^2}{b^2}} + B_{m,n} \sin c\pi t\sqrt{\frac{m^2}{a^2}+\frac{n^2}{b^2}}\right],$$

PHYSICS APPLICATIONS   106

in the event that $x = \frac{a}{m}$, or $\frac{2a}{m}$, or $\frac{3a}{m}$ $\cdots$ or $\frac{(m-1)a}{m}$, $z = 0$ for all upsides of t, and the lines $x = \frac{a}{m}$, $x = \frac{2a}{m}$, $\cdots$ $x = \frac{(m-1)a}{m}$ stay very still during the entirety vibration and are hubs. Exactly the same thing is valid for the lines

$$y = \frac{b}{n},\ y = \frac{2b}{n},\ y = \frac{3b}{n},\ \cdots\ y = \frac{(n-1)b}{n}.$$

73. Assuming that the film is square it might have substantially more convoluted hubs than if the length and expansiveness are inconsistent, as for this situation the time of any term of the overall arrangement diminishes to

$$T = \frac{2a}{c\sqrt{m^2 + n^2}} \tag{1}$$

furthermore, there will overall be two terms having a similar period, and a melodic note of the pitch relating to that period might be created by introductory conditions that acquire the two terms. Consequently

$$z = \sin\frac{m\pi x}{a} \sin\frac{n\pi y}{a} \left[A_{m,n} \cos\frac{c\pi t}{a}\sqrt{m^2+n^2} + B_{m,n} \sin\frac{c\pi t}{a}\sqrt{m^2+n^2} \right]$$
$$+ \sin\frac{n\pi x}{a} \sin\frac{m\pi y}{a} \left[A_{n,m} \cos\frac{c\pi t}{a}\sqrt{m^2+n^2} + B_{n,m} \sin\frac{c\pi t}{a}\sqrt{m^2+n^2} \right]$$

is a type of vibration that will give a melodic note. Allow us to compose this

$$z = \cos\frac{c\pi t}{a}\sqrt{m^2+n^2} \left[A\sin\frac{m\pi x}{a} \sin\frac{n\pi y}{a} + B\sin\frac{n\pi x}{a} \sin\frac{m\pi y}{a} \right]$$
$$+ \sin\frac{c\pi t}{a}\sqrt{m^2+n^2} \left[C\sin\frac{m\pi x}{a} \sin\frac{n\pi y}{a} + D\sin\frac{n\pi x}{a} \sin\frac{m\pi y}{a} \right] \tag{2}$$

also, in concentrating on the types of melodic vibration of which the film is proficient we might take A, B, C, and D at joy. Think about the basic case where $A = C$ and $B = D$; then, at that point (2) decreases to

$$z = \left(A\sin\frac{m\pi x}{a} \sin\frac{n\pi y}{a} + B\sin\frac{n\pi x}{a} \sin\frac{m\pi y}{a} \right) \left(\cos\frac{c\pi t}{a}\sqrt{m^2+n^2} \right.$$
$$\left. + \sin\frac{c\pi t}{a}\sqrt{m^2+n^2} \right). \tag{3}$$

Upsides of x and y that will diminish the principal enclosure in (3) to zero will compare to points of the layer staying still during the vibration.

Allow us to think about a couple of cases finally.

(a) If $m = 1$ and $n = 1$, the main bracket in (3) becomes

$$(A+B)\sin\frac{\pi x}{a} \sin\frac{\pi y}{a},$$

which is equivalent to zero just when $x = 0$ or $y = 0$, or $x = a$ or $y = a$, that is, for the four edges of the layer. On the off chance that, the film is sounding its essential note it has no hubs.

(b) If $m = 1$ and $n = 2$, we have

$$A\sin\frac{\pi x}{a} \sin\frac{2\pi y}{a} + B\sin\frac{2\pi x}{a} \sin\frac{\pi y}{a} = 0$$

to give the hubs.

Let $B = 0$, then $\sin\frac{\pi x}{a} \sin\frac{2\pi y}{a} = 0$, which is fulfilled by $y = \frac{a}{2}$; and notwithstanding the edges the line $y = \frac{a}{2}$ is very still and is a hub.

PHYSICS APPLICATIONS

If $A = 0$ $\quad x = \frac{a}{2}$ is a hub.
On the off chance that $A = B$

$$\sin\frac{\pi x}{a}\sin\frac{2\pi y}{a} + \sin\frac{2\pi x}{a}\sin\frac{\pi y}{a} = 0$$

$$2\sin\frac{\pi x}{a}\sin\frac{\pi y}{a}\cos\frac{\pi y}{a} + 2\sin\frac{\pi x}{a}\cos\frac{\pi x}{a}\sin\frac{\pi y}{a} = 0$$

$$\sin\frac{\pi x}{a}\sin\frac{\pi y}{a}\left(\cos\frac{\pi y}{a} + \cos\frac{\pi x}{a}\right) = 0.$$

The primary component gives the four edges of the layer. The second composed equivalent to zero gives

$$\cos\frac{\pi y}{a} = -\cos\frac{\pi x}{a} = \cos\left(\pi - \frac{\pi x}{a}\right)$$

$$\frac{\pi y}{a} = \pi - \frac{\pi x}{a}$$

$$x + y = a,$$

which is a corner to corner of the square.

If $B = -A$

$$\sin\frac{\pi x}{a}\sin\frac{2\pi y}{a} - \sin\frac{2\pi x}{a}\sin\frac{\pi y}{a} = 0$$

$$\cos\frac{\pi y}{a} = \cos\frac{\pi x}{a}$$

$$x - y = 0,$$

which is the other corner to corner of the square.

Different relations between A and B will give Geometrical bends of the structure

$$\cos\frac{\pi y}{a} = -\frac{B}{A}\cos\frac{\pi x}{a}$$

which are handily built and which clearly all concur in going through the center place of the square.

We give the figures for a couple of the cases

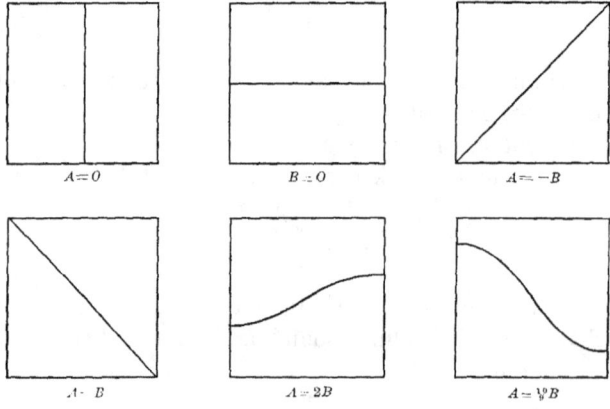

(c) If $m = n = 2$ we have

$$(A + B)\sin\frac{2\pi x}{a}\sin\frac{2\pi y}{a} = 0$$

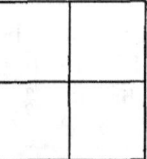

PHYSICS APPLICATIONS

to give the hubs, which are simply the lines

$$x = \frac{a}{2}, \quad \text{and} \quad y = \frac{a}{2}.$$

This structure gives the octave of the central note.

(d) If $m = 1$ and $n = 3$ we have

$$A \sin \frac{\pi x}{a} \sin \frac{3\pi y}{a} + B \sin \frac{3\pi x}{a} \sin \frac{\pi y}{a} = 0$$

to give the hubs.

If $A = 0$ we get $\quad x = \dfrac{a}{3} \quad$ and $\quad x = \dfrac{2a}{3}$ \hfill (1)

If $B = 0$ we get $\quad y = \dfrac{a}{3} \quad$ and $\quad y = \dfrac{2a}{3}$ \hfill (2)

If $A = -B$ we get

$$\sin \frac{\pi x}{a} \sin \frac{3\pi y}{a} - \sin \frac{3\pi x}{a} \sin \frac{\pi y}{a} = 0$$

$$\sin \frac{\pi x}{a} \sin \frac{\pi y}{a} \left[4 \cos^2 \frac{\pi y}{a} - 1 - 4 \cos^2 \frac{\pi x}{a} + 1 \right] = 0$$

$$\cos^2 \frac{\pi y}{a} - \cos^2 \frac{\pi x}{a} = 0$$

$$\left(\cos \frac{\pi y}{a} - \cos \frac{\pi x}{a} \right) \left(\cos \frac{\pi y}{a} + \cos \frac{\pi x}{a} \right) = 0$$

or $\quad and\ x - y = 0 \qquad\qquad$ and $\quad x + y = a.$ \hfill (3)

If $A = B$ we get $and \cos^2 \dfrac{\pi y}{a} \qquad + \cos^2 \dfrac{\pi x}{a} = \dfrac{1}{2}$

or $\quad and \cos \dfrac{2\pi y}{a} + \qquad\qquad \cos \dfrac{2\pi x}{a} - -1,$ \hfill (4)

a Geometrical bend effortlessly built.

For different relations among A and B we get more confounded Geometrical bends going under the general structure

$$A \cos \frac{2\pi y}{a} + B \cos \frac{2\pi x}{a} = -\frac{A+B}{2} \tag{5}$$

which all concur in containing the focuses

$$\left(\frac{a}{3}, \frac{a}{3} \right), \left(\frac{a}{3}, \frac{2a}{3} \right), \left(\frac{2a}{3}, \frac{a}{3} \right), \text{ and } \left(\frac{2a}{3}, \frac{2a}{3} \right).$$

MISCELLANEOUS PROBLEMS.

I. *Logarithmic Potential. Polar Coördinates.*

1. Show that $D_x^2 V + D_y^2 V = 0$ becomes

$$D_r^2 V + \frac{1}{r} D_r V + \frac{1}{r^2} D_\phi^2 V = 0$$

assuming we change to Polar Coördinates.

2. If in
$$D_r^2 V + \frac{1}{r} D_r V + \frac{1}{r^2} D_\phi^2 V = 0 \qquad (1)$$

we let $V = R.\Phi$ we get

$$\left.\begin{array}{l}\Phi = A\cos\alpha\phi + B\sin\alpha\phi \\ R = A_1 r^\alpha + B_1 r^{-\alpha}\end{array}\right\} \text{ or } \left.\begin{array}{l}\Phi = Ane^{\alpha\phi} + Be^{-\alpha\phi} \\ R = A_1 \cos(\alpha\log r) + B_1 \sin(\alpha\log r);\end{array}\right\}$$

whence

$$\begin{array}{l}V = r^\alpha \cos\alpha\phi \text{ and } V = e^{\alpha\phi}\cos(\alpha\log r) \text{ and } V = \cosh\alpha\phi \cos(\alpha\log r) \\ V = r^\alpha \sin\alpha\phi \text{ and } V = e^{\alpha\phi}\sin(\alpha\log r) \text{ and } V = \cosh\alpha\phi \sin(\alpha\log r) \\ V = \frac{1}{r^\alpha} \cos\alpha\phi \text{ and } V = e^{-\alpha\phi}\cos(\alpha\log r) \text{ and } V = \sinh\alpha\phi \cos(\alpha\log r) \\ V = \frac{1}{r^\alpha} \sin\alpha\phi \text{ and } V = e^{-\alpha\phi}\sin(\alpha\log r) \text{ and } V = \sinh\alpha\phi \sin(\alpha\log r)\end{array}$$

are specific arrangements of (1).

3. Show that assuming V fulfills (1) Ex. 2 and $V = f(\phi)$ when $r = a$

$$V = \frac{1}{2}b_0 + \sum_{m=1}^{m=\infty} \left(\frac{r}{a}\right)^m (b_m \cos m\phi + a_m \sin m\phi) \quad \text{for} \quad r < a$$

and

$$V = \frac{1}{2}b_0 + \sum_{m=1}^{m=\infty} \left(\frac{a}{r}\right)^m (b_m \cos m\phi + a_m \sin m\phi) \quad \text{for} \quad r > a,$$

where

$$b_m = \frac{1}{\pi}\int_{-\pi}^{\pi} f(\phi)\cos m\phi.d\phi \quad \text{and} \quad a_m = \frac{1}{\pi}\int_{-\pi}^{\pi} f(\phi)\sin m\phi.d\phi$$

4. Show that assuming V fulfills (1) Ex. 2 and $V = f(r)$ when $\phi = 0$ and $r > 0$

$$V = \frac{1}{\pi}\int_{-\infty}^{\infty} f(e^\lambda)d\lambda \int_0^\infty \frac{\cosh\alpha(\pi-\phi)}{\cosh\alpha\pi}\cos\alpha(\lambda - \log r).d\alpha$$

$$= \frac{1}{\pi}\sin\frac{\phi}{2}\int_{-\infty}^{\infty} f(e^\lambda)\frac{\cosh\frac{1}{2}(\lambda - \log r)}{\cosh(\lambda - \log r) - \cos\phi}d\lambda.$$

5. If $V = 1$ when $\phi = 0$ and $0 < r < 1$, and $V = 0$ when $\phi = 0$ and $r > 1$

$$V = \frac{1}{\pi}\left\{\frac{\pi}{2} - \tan^{-1}\left[\frac{\sinh\frac{\log r}{2}}{\sin\frac{\phi}{2}}\right]\right\} = \frac{1}{\pi}\left[\frac{\pi}{2} - \tan^{-1}\left(\frac{r-1}{2\sqrt{r}.\sin\frac{\phi}{2}}\right)\right].$$

MISCELLANEOUS PROBLEMS.

6. If $V = f(r)$ when $\phi = 0$ and $V = 0$ when $\phi = \beta$

$$V = \frac{1}{\pi} \int_{-\infty}^{\infty} f(e^\lambda) d\lambda \int_0^\infty \frac{\sinh(\beta-\phi)\alpha}{\sinh \beta \alpha} \cos \alpha(\lambda - \log r).d\alpha$$

$$= \frac{1}{2\beta} \sin \frac{\pi\phi}{\beta} \int_{-\infty}^{\infty} \frac{f(e^\lambda) d\lambda}{\cosh \frac{\pi}{\beta}(\lambda - \log r) - \cos \frac{\pi}{\beta}\phi},$$

if $0 < \phi < \beta$.

7. If $V = 0$ when $\phi = 0$ and $V = F(r)$ when $\phi = \beta$

$$V = \frac{1}{\pi} \int_{-\infty}^{\infty} F(e^\lambda) d\lambda \int_0^\infty \frac{\sinh \phi \alpha}{\sinh \beta \alpha} \cos \alpha(\lambda - \log r).d\alpha$$

$$= \frac{1}{2\beta} \sin \frac{\pi\phi}{\beta} \int_{-\infty}^{\infty} \frac{F(e^\lambda) d\lambda}{\cosh \frac{\pi}{\beta}(\lambda - \log r) + \cos \frac{\pi}{\beta}\phi}.$$

8. If $V = \chi(r)$ when $\phi = 0$ and $r < a$, $V = 0$ when $\phi = \beta$, and $V = 0$ when $r = a$

$$V = \frac{1}{2\beta} \sin \frac{\pi\phi}{\beta} \int_{-\infty}^0 \chi(ae^\lambda) \left[\frac{d\lambda}{\cosh \frac{\pi}{\beta}\left(\lambda - \log \frac{r}{a}\right) - \cos \frac{\pi\phi}{\beta}} \right.$$

$$\left. - \frac{d\lambda}{\cosh \frac{\pi}{\beta}\left(\lambda + \log \frac{r}{a}\right) - \cos \frac{\pi\phi}{\beta}} \right].$$

9. If $V = 0$ when $r = 1$, $V = 1$ when $\phi = 0$, $V = 0$ when $\phi = \frac{\pi}{2}$

$$V = \frac{2}{\pi} \tan^{-1} \left[\frac{1-r^2}{1+r^2} \operatorname{ctn} \phi \right].$$

10. If $V = 0$ when $r = 1$, $V = 1$ when $\phi = 0$, $V = 1$ when $\phi = \frac{\pi}{2}$

$$V = \frac{2}{\pi} \tan^{-1} \left[\frac{1-r^4}{2r^2 \sin 2\phi} \right].$$

11. If $V = f(\phi)$ when $r = a$, $V = 0$ when $\phi = 0$, and $V = 0$ when $\phi = \beta$

$$V = \sum_{m=1}^{m=\infty} a_m \left(\frac{r}{a}\right)^{\frac{m\pi}{\beta}} \sin \frac{m\pi\phi}{\beta} \quad \text{if} \quad r < a$$

$$V = \sum_{m=1}^{m=\infty} a_m \left(\frac{a}{r}\right)^{\frac{m\pi}{\beta}} \sin \frac{m\pi\phi}{\beta} \quad \text{if} \quad r > a$$

where $\quad a_m = \frac{2}{\beta} \int_0^\beta f(\phi) \sin \frac{m\pi\phi}{\beta} d\phi \quad \text{and} \quad 0 < \phi < \beta.$

12. If $V = f(\phi)$ when $r = a$, $V = 0$ when $r = b$, $V = 0$ when $\phi = 0$, what's more, $V = 0$ when $\phi = \beta$, then, at that point, if $a < r < b$ and $0 < \phi < \beta$

$$V = \sum_{m=1}^{m=\infty} \left\{ \frac{a^{\frac{m\pi}{\beta}} b^{\frac{m\pi}{\beta}}}{a^{\frac{2m\pi}{\beta}} - b^{\frac{2m\pi}{\beta}}} \left[\left(\frac{r}{b}\right)^{\frac{m\pi}{\beta}} - \left(\frac{b}{r}\right)^{\frac{m\pi}{\beta}} \right] a_m \sin \frac{m\pi\phi}{\beta} \right\}$$

$$a_m = \frac{2}{\beta} \int_0^\beta f(\phi) \sin \frac{m\pi\phi}{\beta} d\phi.$$

MISCELLANEOUS PROBLEMS.   111

13. If $V = F(\phi)$ when $r = b$, $V = 0$ when $r = a$, $V = 0$ when $\phi = 0$, what's more, $V = 0$ when $\phi = \beta$, then, at that point, if $a < r < b$ and $0 < \phi < \beta$

$$V = \sum_{m=1}^{m=\infty} \left\{ \frac{a^{\frac{m\pi}{\beta}} b^{\frac{m\pi}{\beta}}}{b^{\frac{2m\pi}{\beta}} - a^{\frac{2m\pi}{\beta}}} \left[\left(\frac{r}{a}\right)^{\frac{m\pi}{\beta}} - \left(\frac{a}{r}\right)^{\frac{m\pi}{\beta}} \right] a_m \sin\frac{m\pi\phi}{\beta} \right\}$$

where
$$a_m = \frac{2}{\beta} \int_0^\beta F(\phi) \sin\frac{m\pi\phi}{\beta} d\phi.$$

14. If $V = \chi(r)$ when $\phi = 0$, $V = 0$ when $\phi = \beta$, $V = 0$ when $r = a$, what's more, $V = 0$ when $r = b$, then, at that point, if $a < r < b$ and $0 < \phi < \beta$

$$V = \sum_{m=1}^{m=\infty} \left\{ a_m \frac{\sinh\frac{m\pi(\beta-\phi)}{\log b - \log a}}{\sinh\frac{m\pi\beta}{\log b - \log a}} \sin\frac{m\pi(\log r - \log a)}{\log b - \log a} \right\}$$

where
$$a_m = \frac{2}{\log b - \log a} \int_0^{\log\frac{b}{a}} \chi(ae^x) \sin\frac{m\pi x}{\log b - \log a} dx.$$

15. If $V = \psi(r)$ when $\phi = \beta$, $V = 0$ when $\phi = 0$, $V = 0$ when $r = a$, furthermore, $V = 0$ when $r = b$, then, at that point, if $a < r < b$ and $0 < \phi < \beta$

$$V = \sum_{m=1}^{m=\infty} \left\{ a_m \frac{\sinh\frac{m\pi\phi}{\log b - \log a}}{\sinh\frac{m\pi\beta}{\log b - \log a}} \sin\frac{m\pi(\log r - \log a)}{\log b - \log a} \right\}$$

where
$$a_m = \frac{2}{\log b - \log a} \int_0^{\log\frac{b}{a}} \psi(ae^x) \sin\frac{m\pi x}{\log b - \log a} dx.$$

II. *Potential Capability in Space.*

1. Show that

$$f(x,y) = \frac{1}{\pi^2} \int_0^\infty d\alpha \int_0^\infty d\beta \int_0^\infty d\lambda \int_0^\infty f(\lambda,\mu) \cos\alpha(\lambda - x) \cos\beta(\mu - y).d\mu,$$

for all upsides of x and y.

2. Track down specific arrangements of $D_x^2 V + D_y^2 V + D_z^2 V = 0$ in the structures

$$V = e^{\hat{A} \pm z\sqrt{\alpha^2 + \beta^2}} \cos(\alpha x \hat{A} \pm \beta y)$$
$$V = e^{\hat{A} \pm z\sqrt{\alpha^2 + \beta^2}} \sin(\alpha x \hat{A} \pm \beta y)$$
$$V = \sinh z\sqrt{\alpha^2 + \beta^2}.\sin(\alpha x \hat{A} \pm \beta y)$$
$$V = \cosh z\sqrt{\alpha^2 + \beta^2}.\sin(\alpha x \hat{A} \pm \beta y)$$
&c.

3. Given $D_x^2 V + D_y^2 V + D_z^2 V = 0$, and $V = f(x,y)$ when $z = 0$, tackle for positive upsides of z.

Result: and $V = \frac{1}{2\pi} \int_{-\infty}^\infty d\lambda \int_{-\infty}^\infty \frac{zf(\lambda,\mu)d\mu}{[z^2 + (\lambda - x)^2 + (\mu - y)^2]^{\frac{3}{2}}}.$

MISCELLANEOUS PROBLEMS. 112

4. Affirm the consequence of the last model by showing that assuming $f(x, y)$ is autonomous of y

$$V = \frac{1}{\pi} \int_{-\infty}^{\infty} \frac{zf(\lambda, \mu)d\lambda}{z^2 + (\lambda - x)^2} \qquad \text{(v. Ex. 3 Art. 45)}.$$

5. If $D_x^2 V + D_y^2 V + D_z^2 V = 0$, and $V = 1$ when $z = 0$ for all places inside the square shape limited by the lines $x = a$, $x = -a$, $y = b$, and $y = -b$; and $V = 0$ when $z = 0$ for all focuses beyond this square shape, then

$$2\pi V = \frac{b-y}{\sqrt{(b-y)^2}} \left\{ \frac{\pi}{2} + \frac{1}{2} \sin^{-1} \frac{(a-x)^2(b-y)^2 - z^2[(a-x)^2 + (b-y)^2 + z^2]}{(a-x)^2(b-y)^2 + z^2[(a-x)^2 + (b-y)^2 + z^2]} \right.$$

$$\left. + \frac{1}{2} \sin^{-1} \frac{(a+x)^2(b-y)^2 - z^2[(a+x)^2 + (b-y)^2 + z^2]}{(a+x)^2(b-y)^2 + z^2[(a+x)^2 + (b-y)^2 + z^2]} \right\}$$

$$+ \frac{b+y}{\sqrt{(b+y)^2}} \left\{ \frac{\pi}{2} + \frac{1}{2} \sin^{-1} \frac{(a-x)^2(b+y)^2 - z^2[(a-x)^2 + (b+y)^2 + z^2]}{(a-x)^2(b+y)^2 + z^2[(a-x)^2 + (b+y)^2 + z^2]} \right.$$

$$\left. + \frac{1}{2} \sin^{-1} \frac{(a+x)^2(b+y)^2 - z^2[(a+x)^2 + (b+y)^2 + z^2]}{(a+x)^2(b+y)^2 + z^2[(a+x)^2 + (b+y)^2 + z^2]} \right\}$$

if $-a < x < a$, and

$$4\pi V = \frac{b-y}{\sqrt{(b-y)^2}} \left\{ \sin^{-1} \frac{(a-x)^2(b-y)^2 - z^2[(a-x)^2 + (b-y)^2 + z^2]}{(a-x)^2(b-y)^2 + z^2[(a-x)^2 + (b-y)^2 + z^2]} \right.$$

$$\left. - \sin^{-1} \frac{(a+x)^2(b-y)^2 - z^2[(a+x)^2 + (b-y)^2 + z^2]}{(a+x)^2(b-y)^2 + z^2[(a+x)^2 + (b-y)^2 + z^2]} \right\}$$

$$+ \frac{b+y}{\sqrt{(b+y)^2}} \left\{ \sin^{-1} \frac{(a-x)^2(b+y)^2 - z^2[(a-x)^2 + (b+y)^2 + z^2]}{(a-x)^2(b+y)^2 + z^2[(a-x)^2 + (b+y)^2 + z^2]} \right.$$

$$\left. - \sin^{-1} \frac{(a+x)^2(b+y)^2 - z^2[(a+x)^2 + (b+y)^2 + z^2]}{(a+x)^2(b+y)^2 + z^2[(a+x)^2 + (b+y)^2 + z^2]} \right\};$$

if $x < -a$ or $x > a$.

6. Assuming the worth of the potential capability V is given at each place of the base of a limitless rectangular crystal and in the event that the sides of the crystal are at potential no the worth of V anytime inside the crystal is

$$V = \frac{4}{ab} \sum_{m=1}^{m=\infty} \sum_{n=1}^{n=\infty} e^{-\pi z \sqrt{\frac{m^2}{a^2} + \frac{n^2}{b^2}}} \sin \frac{m\pi x}{a} \sin \frac{n\pi y}{b}$$

$$\int_0^a d\lambda \int_0^b f(\lambda, \mu) \sin \frac{m\pi \lambda}{a} \sin \frac{n\pi \mu}{b} d\mu.$$

On the off chance that $V = 1$ on the foundation of the crystal this lessens to

$$V = \frac{16}{\pi^2} \sum_{m=0}^{m=\infty} \sum_{n=0}^{n=\infty} e^{-\pi z \sqrt{\frac{(2m+1)^2}{a^2} + \frac{(2n+1)^2}{b^2}}} \frac{\sin \frac{(2m+1)\pi x}{a} \sin \frac{(2n+1)\pi y}{b}}{(2m+1)(2n+1)}.$$

7. If the worth of the expected capability on five countenances of a rectangular parallelopiped, whose length, broadness, and level are a, b, and c, is zero, and assuming the worth of V is given for each place of the 6th face, then for any point inside the parallelopiped

$$V = \sum_{m=1}^{m=\infty} \sum_{n=1}^{n=\infty} A_{m,n} \frac{\sinh \pi(c-z)\sqrt{\frac{m^2}{a^2} + \frac{n^2}{b^2}}}{\sinh \pi c \sqrt{\frac{m^2}{a^2} + \frac{n^2}{b^2}}} \sin \frac{m\pi x}{a} \sin \frac{n\pi y}{b}$$

MISCELLANEOUS PROBLEMS.

where
$$A_{m,n} = \frac{4}{ab}\int_0^a d\lambda \int_0^b f(\lambda, \mu) \sin\frac{m\pi\lambda}{a} \sin\frac{n\pi\mu}{b} d\mu.$$

8. In the event that the worth of the potential capability is given on two inverse countenances of a rectangular parallelopiped and is zero on the four leftover countenances, then, at that point, inside the parallelopiped

$$V = \sum_{m=1}^{m=\infty}\sum_{n=1}^{n=\infty} A_{m,n} \frac{\sinh \pi(c-z)\sqrt{\frac{m^2}{a^2}+\frac{n^2}{b^2}}}{\sinh \pi c\sqrt{\frac{m^2}{a^2}+\frac{n^2}{b^2}}} \sin\frac{m\pi x}{a} \sin\frac{n\pi y}{b}$$

$$+ \sum_{m=1}^{m=\infty}\sum_{n=1}^{n=\infty} B_{m,n} \frac{\sinh \pi z\sqrt{\frac{m^2}{a^2}+\frac{n^2}{b^2}}}{\sinh \pi c\sqrt{\frac{m^2}{a^2}+\frac{n^2}{b^2}}} \sin\frac{m\pi x}{a} \sin\frac{n\pi y}{b}$$

where
$$A_{m,n} = \frac{4}{ab}\int_0^a d\lambda \int_0^b f(\lambda, \mu) \sin\frac{m\pi\lambda}{a} \sin\frac{n\pi\mu}{b} d\mu$$

and
$$B_{m,n} = \frac{4}{ab}\int_0^a d\lambda \int_0^b F(\lambda, \mu) \sin\frac{m\pi\lambda}{a} \sin\frac{n\pi\mu}{b} d\mu.$$

9. On the off chance that the worth of the potential capability is given at each point on a superficial level of a rectangular parallelopiped, what is its worth anytime inside the parallelopiped?

III. Conduction of Intensity in a Plane.

1. Track down specific arrangements of $D_t u = a^2(D_x^2 u + D_y^2 u)$ of the structures

$$u = e^{-a^2(\alpha^2+\beta^2)t} \sin(\alpha x \pm \beta y)$$
$$u = e^{-a^2(\alpha^2+\beta^2)t} \cos(\alpha x \pm \beta y).$$

2. Given the underlying temperature of each and every point in a dainty plane plate, find the temperature of any point whenever,

$$u = \frac{1}{4a^2\pi t}\int_{-\infty}^{\infty} d\lambda \int_{-\infty}^{\infty} e^{-\frac{(\lambda-x)^2+(\mu-y)^2}{4a^2 t}} f(\lambda, \mu) d\mu$$

$$= \frac{1}{\pi}\int_{-\infty}^{\infty} e^{-\beta^2} d\beta \int_{-\infty}^{\infty} e^{-\gamma^2} f(x+2a\sqrt{t}.\beta, y+2a\sqrt{t}.\gamma) d\gamma.$$

3. For a quick *source* of solidarity Q at (λ, μ)

$$u = \frac{Q}{4\pi a^2 t} e^{-\frac{(\lambda-x)^2+(\mu-y)^2}{4a^2 t}} \quad \text{andv. Art. 53.}$$

For a prompt *doublet* of solidarity P at $(0, \mu)$ with its hub opposite to the pivot of Y

$$u = \frac{Px}{8\pi a^4 t^2} e^{-\frac{x^2+(\mu-y)^2}{4a^2 t}} \quad \text{andv. Art. 54.}$$

For a super durable doublet of solidarity P at $(0, \mu)$ with its pivot opposite to the hub of Y

$$u = \frac{P}{2\pi a^2} \frac{x}{x^2 + (\mu-y)^2} e^{-\frac{x^2+(\mu-y)^2}{4a^2 t}}.$$

MISCELLANEOUS PROBLEMS.

Assuming the strength of the doublet were $Pd\mu$ and the intensity were consistently created and consumed along the component $d\mu$ of the pivot of Y starting at $(0, \mu)$ we ought to have

$$u = \frac{P}{2\pi a^2} e^{-\frac{x^2+(\mu-y)^2}{4a^2 t}} \frac{x d\mu}{x^2+(\mu-y)^2} = \frac{P}{2\pi a^2} e^{-\frac{x^2+(\mu-y)^2}{4a^2 t}} d\tan^{-1}\frac{\mu-y}{x},$$

also, since $d\tan^{-1}\frac{\mu-y}{x}$ is the point ARA', where A and A' are the focuses $(0, \mu)$ furthermore $(0, \mu + d\mu)$ and R is the point (x,y), $u = 0$ when $x = 0$ except if $\mu < y < \mu + d\mu$, in which case $u = \frac{P}{2a^2}$ if x approaches zero from the positive side; and $u = 0$ when $t = 0$ besides in the component $d\mu$. In the event that, $u = 0$ when $t = 0$ and $u = f(y)$ when $x = 0$ we have just to assume a doublet of solidarity $2a^2 f(x)dx$ put in every component of the pivot of Y and then, at that point, to incorporate; we get

$$u = \frac{1}{\pi} \int_{-\infty}^{\infty} e^{-\frac{x^2+(\mu-y)^2}{4a^2 t}} \frac{xf(\mu)}{x^2+(\mu-y)^2} d\mu.$$

For a long-lasting doublet of solidarity $F(t)$ at $(0, \mu)$ we have

$$u = \frac{x}{8\pi a^4} \int_0^t e^{-\frac{x^2+(\mu-y)^2}{4a^2 (t-\tau)}} (t-\tau)^{-2} F(\tau) d\tau.$$

$$= \frac{1}{2\pi a^2} \left[\frac{xF(0)}{x^2+(\mu-y)^2} e^{-\frac{x^2+(\mu-y)^2}{4a^2 t}} + \int_0^t \frac{xF'(\tau)}{x^2+(\mu-y)^2} e^{-\frac{x^2+(\mu-y)^2}{4a^2 (t-\tau)}} d\tau \right].$$

From the thinking over this should be zero when $t = 0$ besides at the point $(0, \mu)$, should be $2a^2 F(t)$ at the point $(0, \mu)$, and 0 at each and every mark of the hub of Y when t isn't zero.

Consequently if $u = 0$ when $t = 0$ and $u = F(y,t)$ when $x = 0$

$$u = \frac{1}{\pi} \int_{\infty}^{\infty} \frac{xF(\mu,0)}{x^2+(\mu-y)^2} e^{-\frac{x^2+(\mu-y)^2}{4a^2 t}} d\mu + \frac{1}{\pi} \int_{-\infty}^{\infty} d\mu \int_0^t \frac{xD_\tau F(\mu,\tau)}{x^2+(\mu-y)^2} e^{-\frac{x^2+(\mu-y)^2}{4a^2 (t-\tau)}} d\tau.$$

For an expansion of this arrangement by the technique for pictures to the situation where there are other rectilinear limits and for its application to the comparing issues in the progression of intensity in three aspects see E. W. Hobson in Vol. XIX. Proc. Lond. Math. Soc.

4. In the event that the edge of a flimsy plane rectangular plate is kept at the temperature zero and the underlying temperatures of all places of the plate are given, then for any place of the plate

$$u = \frac{4}{bc} \sum_{m=1}^{m=\infty} \sum_{n=1}^{n=\infty} e^{-a^2 \pi^2 \left(\frac{m^2}{b^2}+\frac{n^2}{c^2}\right)t} \sin\frac{m\pi x}{b} \sin\frac{n\pi y}{c} \int_0^b d\lambda \int_0^c f(\lambda,\mu) \sin\frac{m\pi\lambda}{b} \sin\frac{n\pi\mu}{c} d\mu.$$

on the off chance that b is the length and c the expansiveness of the plate.

5. An enormous mass of iron at the temperature $0\text{Å}°$ contains an iron center in the state of a long crystal 40 cm. square. The center is eliminated and warmed to the temperature of $100\text{Å}°$ all through and afterward supplanted. Track down the temperature of a point in the pivot of the center fifteen minutes subsequently. Given $a^2 = .185$ in C.G.S. units. *Ans.*, $52\text{Å}°.9$.

6. In the event that the crystal portrayed in Ex. 5 subsequent to being warmed to $100\text{Å}°$ has its parallel faces saved for 15 minutes at the temperature $0\text{Å}°$ track down the temperature of a point in its pivot. *Ans.*, $20\text{Å}°.8$.

IV. Conduction of Intensity in Space.

1. Show that

$$\frac{1}{\pi^3}\int_0^\infty d\alpha \int_0^\infty d\beta \int_0^\infty d\gamma \int_{-\infty}^\infty d\lambda \int_{-\infty}^\infty d\mu \int_{-\infty}^\infty f(\lambda,\mu,\nu) \cos\alpha(\lambda-x)\cos\beta(\mu-y)\cos\gamma(\nu-z).d\nu = f(x,y,z)$$

for all upsides of x, y, and z.

2. Show that

$$f(x,y,z) = \sum_{m=1}^{m=\infty}\sum_{n=1}^{n=\infty}\sum_{p=1}^{p=\infty} A_{m,n,p}\sin\frac{m\pi x}{a}\sin\frac{n\pi y}{b}\sin\frac{p\pi z}{c}$$

where and $A_{m,n,p} = \dfrac{8}{abc}\int_0^a d\lambda \int_0^b d\mu \int_0^c f(\lambda,\mu,\nu) \sin\dfrac{m\pi\lambda}{a}\sin\dfrac{n\pi\mu}{b}\sin\dfrac{p\pi\nu}{c}d\nu,$

for $0 < x < a$, $0 < y < b$, $0 < z < c$.

3. Acquire specific arrangements of $D_t u = a^2(D_x^2 u + D_y^2 u + D_z^2 u)$ of the structures

$$u = e^{-a^2(\alpha^2+\beta^2+\gamma^2)t}\sin(\alpha x \pm \beta y \pm \gamma z).$$
$$u = e^{-a^2(\alpha^2+\beta^2+\gamma^2)t}\cos(\alpha x \pm \beta y \pm \gamma z).$$

4. Given the underlying temperature of each and every point in an endless homogeneous strong track down the temperature of any point whenever.

$$u = \frac{1}{8a^3(\pi t)^{\frac{3}{2}}}\int_{-\infty}^\infty d\lambda \int_{-\infty}^\infty d\mu \int_{-\infty}^\infty e^{-\frac{(\lambda-x)^2+(\mu-y)^2+(\nu-z)}{4a^2 t}} f(\lambda,\mu,\nu)d\nu$$

$$= \frac{1}{\pi^{\frac{3}{2}}}\int_{-\infty}^\infty e^{-\beta^2}d\beta \int_{-\infty}^\infty e^{-\gamma^2}d\gamma \int_{-\infty}^\infty e^{-\delta^2} f(x+2a\sqrt{t}.\beta, y+2a\sqrt{t}.\gamma, z+2a\sqrt{t}.\delta)d\delta.$$

5. In the event that the outer layer of a rectangular parallelopiped is kept at the temperature zero and the underlying temperatures of all places of the parallelopiped are given, then, at that point, for any mark of the parallelopiped

$$u = \sum_{m=1}^{m=\infty}\sum_{n=1}^{n=\infty}\sum_{p=1}^{p=\infty} A_{m,n,p} e^{-a^2\pi^2\left(\frac{m^2}{b^2}+\frac{n^2}{c^2}+\frac{p^2}{d^2}\right)t}\sin\frac{m\pi x}{b}\sin\frac{n\pi y}{c}\sin\frac{p\pi z}{d}$$

where $A_{m,n,p} = \dfrac{8}{bcd}\int_0^b d\lambda \int_0^c d\mu \int_0^d f(\lambda,\mu,\nu)\sin\dfrac{m\pi\lambda}{b}\sin\dfrac{n\pi\mu}{c}\sin\dfrac{p\pi\nu}{d}d\nu.$

6. An iron block 40 cm. on an edge is warmed to the uniform temperature of 100Â° Centigrade and afterward firmly encased in a huge iron mass which is at the uniform temperature of 0Â°. Track down the temperature of the focal point of the 3D shape fifteen minutes subsequently. *Ans.*, 38Â°.4.

7. An iron block 40 cm. on an edge is warmed to the uniform temperature of 100Â° and afterward its surface is saved for fifteen minutes at the temperature 0Â°. Required the temperature of its middle. *Ans.*, 9Â°.5.

Chapter 5

ZONAL HARMONICS

74. In Art. 16 we acquired
$$z = Ap_m(x) + Bq_m(x) \tag{1}$$

[v. (6) Art. 16] as the overall arrangement of Legendre's Situation

$$(1-x^2)\frac{d^2z}{dx^2} - 2x\frac{dz}{dx} + m(m+1)z = 0, \tag{2}$$

m being completely unhindered in esteem and x lying between -1 and 1; where

$$p_m(x) = 1 - \frac{m(m+1)}{2!}x^2 + \frac{m(m-2)(m+1)(m+3)}{4!}x^4 \\ - \frac{m(m-2)(m-4)(m+1)(m+3)(m+5)}{6!}x^6 + \cdots \tag{3}$$

what's more,

$$q_m(x) = x - \frac{(m-1)(m+2)}{3!}x^3 + \frac{(m-1)(m-3)(m+2)(m+4)}{5!}x^5 \\ - \frac{(m-1)(m-3)(m-5)(m+2)(m+4)(m+6)}{7!}x^7 + \cdots; \tag{4}$$

and we found

$$\left.\begin{array}{l} V = r^m p_m(\cos\theta) \\ V = \dfrac{1}{r^{m+1}} p_m(\cos\theta) \\ V = r^m q_m(\cos\theta) \\ V = \dfrac{1}{r^{m+1}} q_m(\cos\theta), \end{array}\right\} \tag{5}$$

m being unhindered in esteem, as specific arrangements of the extraordinary structure accepted by Laplace's Situation in round coÃ¶rdinates when V is autonomous of ϕ; that is, of the situation

$$rD_r^2(rV) + \frac{1}{\sin\theta}D_\theta(\sin\theta D_\theta V) = 0. \tag{6}$$

For the significant situation where m is a positive whole number we found

$$z = AP_m(x) + BQ_m(x) \tag{7}$$

---

[0] Prior to perusing this part the understudy is encouraged to re-read cautiously articles 9, 10, 13(c), 15, 16, and 18(c).

ZONAL HARMONICS.

[v. (10) Art. 16] as the overall arrangement of Legendre's Situation (2), whence

$$\left.\begin{array}{l} V = r^m P_m(\cos\theta) \\ V = \dfrac{1}{r^{m+1}} P_m(\cos\theta) \\ V = r^m Q_m(\cos\theta) \\ V = \dfrac{1}{r^{m+1}} Q_m(\cos\theta) \end{array}\right\} \quad (8)$$

are specific arrangements of (6) in the event that m is a positive number.

$$P_m(x) = \frac{(2m-1)(2m-3)\cdots 1}{m!}\left[x^m - \frac{m(m-1)}{2(2m-1)}x^{m-2} + \frac{m(m-1)(m-2)(m-3)}{2.4.(2m-1)(2m-3)}x^{m-4} - \cdots\right] \quad (9)$$

[v. (8) Art. 16] and is a limited total ending with the term which includes x assuming that m is odd and with the term including x^0 assuming that m is even.

It is known as a *Surface Zonal Harmonic*, or a *Legendre's Coefficient*, or more momentarily a *Legendrian*.

$$Q_m(x) = \frac{m!}{(2m+1)(2m-1)\cdots 1}\left[\frac{1}{x^{m+1}} + \frac{(m+1)(m+2)}{2.(2m+3)}\frac{1}{x^{m+3}} + \frac{(m+1)(m+2)(m+3)(m+4)}{2.4.(2m+3)(2m+5)}\frac{1}{x^{m+5}} + \cdots\right] \quad (10)$$

in the event that $x < -1$ or $x > 1$. [v. (9) Art. 16.]

It is known as a *Surface Zonal Harmonic* of the *second kind*.

$$\begin{aligned} Q_m(x) &= (-1)^{\frac{m+1}{2}} \frac{2^{m-1}\left[\Gamma\left(\frac{m+1}{2}\right)\right]^2}{\Gamma(m+1)} p_m(x) \\ &= (-1)^{\frac{m+1}{2}} \frac{2.4.6.\cdots(m-1)}{3.5.7.\cdots m} p_m(x) \end{aligned} \quad (11)$$

[v. (13) Art. 16] assuming m is odd and $-1 < x < 1$.

$$\begin{aligned} Q_m(x) &= (-1)^{\frac{m}{2}} \frac{2^m\left[\Gamma\left(\frac{m+1}{2}\right)\right]^2}{\Gamma(m+1)} q_m(x) \\ &= (-1)^{\frac{m}{2}} \frac{2.4.6.\cdots m}{1.3.5.\cdots(m-1)} q_m(x) \end{aligned} \quad (12)$$

[v. (14) Art. 16] assuming that m is even and $-1 < x < 1$.

In a large portion of the work that quickly follows we will respect x in $P_m(x)$ as equivalent to $\cos\theta$ and along these lines as lying between -1 and 1.[1]

75. In Article 9 the dubious coefficient a_m of x^m in $P_m(x)$ was with no obvious end goal in mind written in the structure $\frac{(2m-1)(2m-3)\cdots 1}{m!}$ because of reasons which will presently be given.

In Articles 9 and 16 $z = P_m(x)$ was gotten as a specific arrangement of Legendre's Condition

$$(1-x^2)\frac{d^2 z}{dx^2} - 2x\frac{dz}{dx} + m(m+1)z = 0 \quad (1)$$

---
[1] English scholars on Circular Sounds by and large use μ instead of x for $\cos\theta$. We will follow them, in any case, just when we ought to consequently stay away from confusion.

ZONAL HARMONICS.

by the gadget of expecting that z could be communicated as a total or a progression of terms of the structure $a_n x^n$ and afterward deciding the coefficients. We can, in any case, get a specific arrangement of Legendre's Situation by a totally unique strategy.

The likely capability because of a unit of mass gathered at a given point (x_1, y_1, z_1) is

$$V = \frac{1}{\sqrt{(x-x_1)^2 + (y-y_1)^2 + (z-z_1)^2}} \qquad (2)$$

furthermore, this should be a specific arrangement of Laplace's Situation

$$D_x^2 V + D_y^2 V + D_z^2 V = 0, \qquad (3)$$

as is effectively confirmed by direct replacement.

Assuming we change (2) to round coördinates utilizing the equations of change

$$x = r \cos \theta \text{ and}$$
$$y = r \sin \theta \cos \phi$$
$$z = r \sin \theta \sin \phi \qquad \text{we get}$$

$$V = \frac{1}{\sqrt{r^2 - 2rr_1[\cos\theta\cos\theta_1 + \sin\theta\sin\theta_1\cos(\phi-\phi_1)] + r_1^2}} \qquad (4)$$

as an answer of Laplace's Situation in Circular Coördinates

$$rD_r^2(rV) + \frac{1}{\sin\theta} D_\theta(\sin\theta D_\theta V) + \frac{1}{\sin^2\theta} D_\phi^2 V = 0 \qquad \text{[XIII] Art. 1.}$$

In the event that the given point (x_1, y_1, z_1) is assumed the pivot of X, as it should be that (4) might be autonomous of ϕ, $\theta_1 = 0$, and

$$V = \frac{1}{\sqrt{r^2 - 2rr_1 \cos\theta + r_1^2}} \qquad (5)$$

is an answer of

$$rD_r^2(rV) + \frac{1}{\sin\theta} D_\theta(\sin\theta D_\theta V) = 0. \qquad (6)$$

Condition (5) might be composed

$$V = \frac{1}{r} \frac{1}{\sqrt{1 - 2\frac{r_1}{r}\cos\theta + \frac{r_1^2}{r^2}}} \qquad (7)$$

or

$$V = \frac{1}{r_1} \frac{1}{\sqrt{1 - 2\frac{r}{r_1}\cos\theta + \frac{r^2}{r_1^2}}}. \qquad (8)$$

$\sqrt{1 - 2z\cos\theta + z^2}$ is limited and consistent for all values genuine or complex of z. It is twofold esteemed however the two parts of the capability are particular aside from for the upsides of z which make $1 - 2z\cos\theta + z^2 = 0$ to be specific $z = \cos\theta + I\sin\theta$ what's more, $z = \cos\theta - i\sin\theta$, the two of which have the modulus solidarity and which are *critical* values.

$\frac{1}{\sqrt{1-2z\cos\theta+z^2}}$ is limited and ceaseless aside from the upsides of $z = \cos\theta - i\sin\theta$ and $z = \cos\theta + i\sin\theta$ for which it becomes endless; it is twofold esteemed however has as basic qualities just these upsides of z. It is then *holomorphic* inside a circle depicted with the beginning as focus and the sweep solidarity, and can be formed into a power series which will be concurrent for all upsides of z having moduli short of what one. (Int. Cal. Arts. 207, 212, 214, 220.)

ZONAL HARMONICS. 119

On the off chance that, $r > r_1$ $\dfrac{1}{\sqrt{1 - \frac{2r_1}{r}\cos\theta + \frac{r_1^2}{r^2}}}$ can be formed into a merged series including entire powers of $\frac{r_1}{r}$.

Let $\sum p_m \dfrac{r_1^m}{r^m}$ be this series, p_m, obviously, being an element of $\cos\theta$. Then, at that point,

$$V = \frac{1}{r}\sum p_m \frac{r_1^m}{r^m}$$

[v. (7)] is an answer of (6). Substitute this worth of V in (6) and we get

$$\left[\frac{r_1^m}{r^{m+1}} m(m+1)p_m + \frac{r_1^m}{r^{m+1}} \frac{1}{\sin\theta} \frac{d}{d\theta}\left(\sin\theta \frac{dp_m}{d\theta}\right)\right] = 0.$$

As this should hold anything that the worth of r gave $r > r_1$ the coefficient of each force of r should be zero, and thus the condition

$$\frac{1}{\sin\theta}\frac{d}{d\theta}\left(\sin\theta \frac{dp_m}{d\theta}\right) + m(m+1)p_m = 0 \tag{9}$$

should be valid.

Be that as it may, as we have seen in Art. 9 the replacement of $x = \cos\theta$ in (9) decreases it to

$$(1-x^2)\frac{d^2 p_m}{dx^2} - 2x\frac{dp_m}{dx} + m(m+1)p_m = 0,$$

and therefore $z = p_m$ and

is an answer of Legendre's Situation (1).

If $r < r_1$ $\dfrac{1}{\sqrt{1 - \frac{2r}{r_1}\cos\theta + \frac{r^2}{r_1^2}}}$ can be formed into a concurrent series including entire powers of $\frac{r^2}{r_1^2}$.

Let $\sum p_m \dfrac{r^m}{r_1^m}$ be this series. Then

$$V = \frac{1}{r_1}\sum p_m \frac{r^m}{r_1^m}$$

(v. 8) is an answer of (6); subbing in (6) we get

$$\sum\left[\frac{r^m}{r_1^{m+1}} m(m+1)p_m + \frac{r^m}{r_1^{m+1}} \frac{1}{\sin\theta}\frac{d}{d\theta}\left(\sin\theta \frac{dp_m}{d\theta}\right)\right] = 0,$$

whence it follows as before that

$$z = p_m$$

is an answer of Legendre's Situation.

In any case, p_m is the coefficient of the mth force of $\frac{r}{r_1}$ in the improvement of $\left(1 - 2\frac{r}{r_1}\cos\theta + \frac{r^2}{r_1^2}\right)^{-\frac{1}{2}}$ as per powers of $\frac{r}{r_1}$, or of the mth force of $\frac{r_1}{r}$ in the advancement of $\left(1 - 2\frac{r_1}{r}\cos\theta + \frac{r_1^2}{r^2}\right)^{-\frac{1}{2}}$ as indicated by powers of $\frac{r_1}{r}$, or all the more momentarily it is the coefficient of the mth force of z in the advancement of $(1 - 2xz + z^2)^{-\frac{1}{2}}$ as per powers of z, x representing $\cos\theta$.

$$(1 - 2xz + z^2)^{-\frac{1}{2}} = [1 - z(2x - z)]^{-\frac{1}{2}}$$

ZONAL HARMONICS.

also, can be created by the Binomial Hypothesis; the coefficient of z^m is without any problem chosen and is

$$\frac{(2m-1)(2m-3)\cdots 1}{m!}\left[x^m - \frac{m(m-1)}{2(2m-1)}x^{m-2} + \frac{m(m-1)(m-2)(m-3)}{2.4.(2m-1)(2m-3)}x^{m-4} - \cdots\right].$$

Be that as it may, this is definitively $P_m(x)$. [v. Art. 74 (9)]

Consequently $P_m(x)$ is equivalent to the coefficient of the mth force of z in the improvement of $[1-2xz+z^2]^{-\frac{1}{2}}$ into a power series, the modulus of z being not as much as solidarity.

76. If $x=1$ $P_m(x)=1$. For if $x=1$ $(1-2xz+z^2)^{-\frac{1}{2}}$ diminishes to $(1-2z+z^2)^{-\frac{1}{2}}$ that is to $(1-z)^{-1}$, which forms into

$$1 + z + z^2 + z^3 + z^4 + \cdots,$$

furthermore, the coefficient of each force of z is solidarity. Subsequently

$$P_m(1) = 1. \tag{1}$$

We have seen that assuming m is even $P_m(x)$ contains just even powers of x and ends with the term including x^0, that is with the steady term.

The worth of this steady term can be selected from the equation for $P_m(x)$ [v. Art. 74 (9)]. It is $(-1)^{\frac{m}{2}}\frac{1.3.5.\cdots(m-1)}{2.4.6.\cdots m}$; or it very well may be found as follows:- - - It is plainly the worth $P_m(x)$ accepts when $x=0$; it is, then, the coefficient of z^m in the advancement of $(1+z^2)^{-\frac{1}{2}}$; yet

$$(1+z^2)^{-\frac{1}{2}} = 1 - \frac{1}{2}z^2 + \frac{1.3}{2.4}z^4 - \frac{1.3.5}{2.4.6}z^6 + \frac{1.3.5.7}{2.4.6.8}z^8 - \cdots$$

also, the coefficient of z^m, m being a considerably number, is $(-1)^{\frac{m}{2}}\frac{1.3.5\cdots(m-1)}{2.4.6\cdots m}$.

Assuming m is odd $P_m(x)$ contains just odd powers of x and ends with the term including x to the main power. The coefficient of this term can be chosen from (9) Art. 74 and is $(-1)^{\frac{m-1}{2}}\frac{3.5.7.\cdots m}{2.4.6.\cdots(m-1)}$; or it tends to be found as follows:- - - It is plainly the worth accepted by $\frac{dP_m(x)}{dx}$ when $x=0$.

It is, then, the coefficient of z^m in the advancement of $\frac{z}{(1+z^2)^{\frac{3}{2}}}$.

$$\frac{z}{(1+z^2)^{\frac{3}{2}}} = z - \frac{3}{2}z^3 + \frac{3.5}{2.4}z^5 - \frac{3.5.7}{2.4.6}z^7 + \cdots$$

what's more, the coefficient of z^m in this advancement is $(-1)^{\frac{m-1}{2}}\frac{3.5.7\cdots m}{2.4.6\cdots(m-1)}$, m being an odd number.

77. To restate:

$$P_m(x) = \frac{1.3.5\cdots(2m-1)}{m!}\left[x^m - \frac{m(m-1)}{2(2m-1)}x^{m-2}\right.$$
$$+ \frac{m(m-1)(m-2)(m-3)}{2.4.(2m-1)(2m-3)}x^{m-4}$$
$$\left. - \frac{m(m-1)(m-2)(m-3)(m-4)(m-5)}{2.4.6.(2m-1)(2m-3)(2m-5)}x^{m-6} + \cdots \right], \tag{1}$$

m being a positive whole number, is a *Surface Zonal Harmonic* or *Legendrian* of the mth request. It is a limited total ending with the principal force of x in the event that m is odd, and with the zeroth force of x on the off chance that m is even.

ZONAL HARMONICS.

$P_m(x)$ is the coefficient of the mth force of z in the advancement of $(1 - 2xz + z^2)^{-\frac{1}{2}}$ into a power series. Thus if $z < 1$

$$(1 - 2xz + z^2)^{-\frac{1}{2}} = P_0(x) + P_1(x).z + P_2(x).z^2 + P_3(x).z^3$$
$$+ P_4(x).z^4 + P_5(x).z^5 + \cdots + P_m(x).z^m + \cdots. \quad (2)$$

Whence

$$\left.\begin{aligned}\frac{1}{\sqrt{r^2 - 2rr_1\cos\theta + r_1^2}} &= \frac{1}{r}\left[P_0(\cos\theta) + \frac{r_1}{r}P_1(\cos\theta) + \frac{r_1^2}{r^2}P_2(\cos\theta) + \cdots \right. \\ &\qquad\left. + \frac{r_1^m}{r^m}P_m(\cos\theta) + \cdots\right] \quad \text{if} \quad r > r_1 \\ &= \frac{1}{r_1}\left[P_0(\cos\theta) + \frac{r}{r_1}P_1(\cos\theta) + \frac{r^2}{r_1^2}P_2(\cos\theta) + \cdots \right. \\ &\qquad\left. + \frac{r^m}{r_1^m}P_m(\cos\theta) + \cdots\right] \quad \text{if} \quad r < r_1.\end{aligned}\right\} \quad (3)$$

$$z = P_m(x)$$

is an answer of Legendre's Situation

$$(1 - x^2)\frac{d^2z}{dx^2} - 2x\frac{dz}{dx} + m(m+1)z = 0$$

when m is a positive whole number.

$$V = r^m P_m(\cos\theta)$$

and $V \qquad\qquad = \dfrac{1}{r^{m+1}} P_m(\cos\theta)$

are arrangements of the type of Laplace's Situation in Round Coördinates which is autonomous of ϕ, to be specific

$$rD_r^2(rV) + \frac{1}{\sin(\theta)}D_\theta(\sin\theta D_\theta V) = 0. \quad (4)$$

$$P_m(1) = 1. \quad (5)$$

$$P_{2m}(-x) = P_{2m}(x). \quad (6)$$

$$P_{2m+1}(-x) = -P_{2m+1}(x). \quad (7)$$

$$P_{2m+1}(0) = 0. \quad (8)$$

$$P_{2m}(0) = (-1)^m \frac{1.3.5.\cdots(2m-1)}{2.4.6.\cdots 2m}. \quad (9)$$

$$\left[\frac{dP_{2m+1}(x)}{dx}\right]_{x=0} = (-1)^m \frac{3.5.7.\cdots(2m+1)}{2.4.6.\cdots 2m}. \quad (10)$$

For comfort of reference we work out a couple of Zonal Sounds. They are gotten by subbing

ZONAL HARMONICS.

progressive numbers for m in recipe (1).

$$\left.\begin{aligned}
P_0(x) &= 1 \\
P_1(x) &= x \\
P_2(x) &= \frac{1}{2}(3x^2 - 1) \\
P_3(x) &= \frac{1}{2}(5x^3 - 3x) \\
P_4(x) &= \frac{1}{8}(35x^4 - 30x^2 + 3) \\
P_5(x) &= \frac{1}{8}(63x^5 - 70x^3 + 15x) \\
P_6(x) &= \frac{1}{16}(231x^6 - 315x^4 + 105x^2 - 5) \\
P_7(x) &= \frac{1}{16}(429x^7 - 693x^5 + 315x^3 - 35x) \\
P_8(x) &= \frac{1}{128}(6435x^8 - 12012x^6 + 6930x^4 - 1260x^2 + 35).
\end{aligned}\right\} \quad (11)$$

Any Surface Zonal Consonant might be acquired from the two of next lower orders by the guide of the recipe

$$(n+1)P_{n+1}(x) - (2n+1)xP_n(x) + nP_{n-1}(x) = 0 \quad (12)$$

which is effectively acquired and is helpful when the mathematical worth of x is given.

Separate (2) as for z and we get

$$\frac{-(z-x)}{(1 - 2xz + z^2)^{\frac{3}{2}}} = P_1(x) + 2P_2(x).z + 3P_3(x).z^2 + \cdots$$

whence

$$\frac{-(z-x)}{(1 - 2xz + z^2)^{\frac{1}{2}}} = (1 \quad 2xz + z^2)(P_1(r) + 2P_2(x).z + 3P_3(x).z^2 + \cdots).$$

Subsequently by (2)

$$(1 - 2xz + z^2)(P_1(x) and + 2P_2(x).z + 3P_3(x).z^2 + \cdots)$$
$$+ (z - x)(P_0(x) + P_1(x).z + P_2(x).z^2 + \cdots) = 0 \quad (13)$$

(13) is indistinguishably valid, subsequently the coefficient of each force of z should disappear. Choosing the coefficient of z^n and composing it equivalent to zero we have recipe (12) above.[2]

78. We are currently ready to take care of totally the issue considered in Art. 9.

We were to track down an answer of the differential condition

$$rD_r^2(rV) + \frac{1}{\sin\theta}D_\theta(\sin\theta D_\theta V) = 0 \quad (1)$$

dependent upon the condition

$$V = \frac{M}{(c^2 + r^2)^{\frac{1}{2}}} \quad \text{when} \quad \theta = 0. \quad (2)$$

We know (v. Art. 77) that

$$V = r^m P_m(\cos\theta)$$

and and $V = \dfrac{1}{r^{m+1}} P_m(\cos\theta)$

---

[2] For tables of Surface Zonal Sounds v. Supplement Tables I and II.

ZONAL HARMONICS.

are arrangements of (1).

For upsides of $r < c$

$$\frac{M}{(c^2+r^2)^{\frac{1}{2}}} = \frac{M}{c}\left[1 - \frac{1}{2}\frac{r^2}{c^2} + \frac{1.3}{2.4}\frac{r^4}{c^4} - \frac{1.3.5}{2.4.6}\frac{r^6}{c^6} + \cdots\right]. \tag{3}$$

Subsequently for upsides of $r < c$

$$V = \frac{M}{c}\left[P_0(\cos\theta) - \frac{1}{2}\frac{r^2}{c^2}P_2(\cos\theta) \right.$$
$$\left. + \frac{1.3}{2.4}\frac{r^4}{c^4}P_4(\cos\theta) - \frac{1.3.5}{2.4.6}\frac{r^6}{c^6}P_6(\cos\theta) + \cdots\right] \tag{4}$$

is our expected arrangement; in light of the fact that each term fulfills condition (1), and in this manner the entire worth fulfills (1), and when $\theta = 0$

$$P_m(\cos\theta) = P_m(1) = 1$$

[v. (5) Art. 77], and thus (4) diminishes to (3) and (2) is fulfilled.

For upsides of $r > c$

$$\frac{M}{(c^2+r^2)^{\frac{1}{2}}} = \frac{M}{r}\left[1 - \frac{1}{2}\frac{c^2}{r^2} + \frac{1.3}{2.4}\frac{c^4}{r^4} - \frac{1.3.5}{2.4.6}\frac{c^6}{r^6} + \cdots\right] \tag{5}$$

$$= M\left[\frac{1}{r} - \frac{1}{2}\frac{c^2}{r^3} + \frac{1.3}{2.4}\frac{c^4}{r^5} - \frac{1.3.5}{2.4.6}\frac{c^6}{r^7} + \cdots\right].$$

Hence for upsides of $r > c$

$$V = \frac{M}{c}\left[\frac{c}{r}P_0(\cos\theta) - \frac{1}{2}\frac{c^3}{r^3}P_2(\cos\theta) \right.$$
$$\left. + \frac{1.3}{2.4}\frac{c^5}{r^5}P_4(\cos\theta) - \frac{1.3.5}{2.4.6}\frac{c^7}{r^7}P_6(\cos\theta) + \cdots\right] \tag{6}$$

is our expected arrangement. For it fulfills (1) and lessens to (2) when $\theta = 0$.

79. As another model let us guess a guide as a meager roundabout circle accused of power, and allow it to be expected to track down the worth of the likely capability anytime in space.

Assuming that the extent of the charge is M and the sweep of the plate is a the surface thickness at a mark of the plate a good ways off r from the middle is

$$\sigma = \frac{M}{4a\pi\sqrt{a^2 - r^2}}$$

and all marks of the guide are at the potential $\frac{\pi M}{2a}$. (v. Peirce's Newtonian Expected Capability, § 61.)

The worth of the expected capability at a point in the pivot of the plate at the distance x from the plate is effortlessly seen to be

$$V = \frac{M}{a}\int_0^a \frac{r\,dr}{\sqrt{(a^2-r^2)(x^2+r^2)}}$$
$$= \frac{M}{2a}\cos^{-1}\frac{x^2 - a^2}{x^2 + a^2}.$$

ZONAL HARMONICS.

$$\frac{d}{dx}\left(\frac{M}{2a}\cos^{-1}\frac{x^2-a^2}{x^2+a^2}\right) = -\frac{M}{a^2+x^2}$$

$$= -\frac{M}{a^2}\left[1 - \frac{x^2}{a^2} + \frac{x^4}{a^4} - \frac{x^6}{a^6} + \cdots\right]$$

if $x < a$,

$$= -\frac{M}{x^2}\left[1 - \frac{a^2}{x^2} + \frac{a^4}{x^4} - \frac{a^6}{x^6} + \cdots\right]$$

if $x > a$.

Coordinating and afterward deciding the inconsistent consistent we have

$$\frac{M}{2a}\cos^{-1}\frac{x^2-a^2}{x^2+a^2} = \frac{M}{a}\left[\frac{\pi}{2} - \frac{x}{a} + \frac{x^3}{3a^3} - \frac{x^5}{5a^5} + \frac{x^7}{7a^7} - \cdots\right]$$

if $x < a$,

$$= \frac{M}{a}\left[\frac{a}{x} - \frac{a^3}{3x^3} + \frac{a^5}{5x^5} - \frac{a^7}{7x^7} + \cdots\right]$$

if $x > a$.

We have, then, to address the condition

$$rD_r^2(rV) + \frac{1}{\sin\theta}D_\theta(\sin\theta D_\theta V) = 0$$

dependent upon the circumstances

$$V = \frac{M}{a}\left[\frac{\pi}{2} - \frac{r}{a} + \frac{r^3}{3a^3} - \frac{r^5}{5a^5} + \frac{r^7}{7a^7} - \cdots\right]$$

when $\theta = 0$ and $r < a$

and $V = \frac{M}{a}\left[\frac{a}{r} - \frac{a^3}{3r^3} + \frac{a^5}{5r^5} - \frac{a^7}{7r^7} + \cdots\right]$

when $\theta = 0$ and $r > a$.

The necessary arrangement is handily seen to be

$$V = \frac{M}{a}\left[\frac{\pi}{2} - \frac{r}{a}P_1(\cos\theta) + \frac{1}{3}\frac{r^3}{a^3}P_3(\cos\theta) - \frac{1}{5}\frac{r^5}{a^5}P_5(\cos\theta) + \cdots\right]$$

if $r < a$ and $\theta < \frac{\pi}{2}$,

and

$$V = \frac{M}{a}\left[\frac{a}{r} - \frac{1}{3}\frac{a^3}{r^3}P_2(\cos\theta) + \frac{1}{5}\frac{a^5}{r^5}P_4(\cos\theta) - \frac{1}{7}\frac{a^7}{r^7}P_6(\cos\theta) + \cdots\right]$$

if $r > a$.

EXAMPLES.

1. Given that in the event that a charge M of power is put on an ellipsoidal guide the surface thickness anytime P of the guide is equivalent to $\frac{Mp}{4\pi abc}$, where p is the separation from the focal point of the guide to the digression plane at P (v. Peirce, New. Pot. Func. Â§ 61); track down the worth of the likely capability at any outside moment that the guide is the oblate spheroid produced by the revolution of the oval $\frac{x^2}{a^2} + \frac{y^2}{b^2} = 1$ about its minor pivot.

Ans. (1) On the off chance that the fact of the matter is on the pivot of transformation

$$V = \frac{M}{2\sqrt{a^2-b^2}}\left[\sin^{-1}\left(\frac{bx+a^2-b^2}{a\sqrt{x^2+a^2-b^2}}\right) - \sin^{-1}\left(\frac{bx-a^2+b^2}{a\sqrt{x^2+a^2-b^2}}\right)\right]$$

x being the separation from the middle.

(2) Assuming the fact is on the outer layer of the spheroid

$$V = \frac{M}{2\sqrt{a^2-b^2}}\left[\frac{\pi}{2} - \sin^{-1}\left(\frac{2b^2-a^2}{a^2}\right)\right] = \frac{M}{\sqrt{a^2-b^2}}\left[\frac{\pi}{2} - \tan^{-1}\left(\frac{b}{\sqrt{a^2-b^2}}\right)\right].$$

(3) If the distance r of the point from the middle is not exactly $\sqrt{a^2-b^2}$ and $\theta < \frac{\pi}{2}$

$$V = \frac{M}{\sqrt{a^2-b^2}}\left[\frac{\pi}{2} - \frac{r}{(a^2-b^2)^{\frac{1}{2}}}P_1(\cos\theta)\right.$$

$$\left. + \frac{r^3}{3(a^2-b^2)^{\frac{3}{2}}}P_3(\cos\theta) - \frac{r^5}{5(a^2-b^2)^{\frac{5}{2}}}P_5(\cos\theta) + \cdots\right].$$

(4) If the distance r of the point from the middle is more prominent than $\sqrt{a^2-b^2}$

$$V = \frac{M}{\sqrt{a^2-b^2}}\left[\frac{(a^2-b^2)^{\frac{1}{2}}}{r} - \frac{(a^2-b^2)^{\frac{3}{2}}}{3r^3}P_2(\cos\theta)\right.$$

$$\left. + \frac{(a^2-b^2)^{\frac{5}{2}}}{5r^5}P_4(\cos\theta) - \frac{(a^2-b^2)^{\frac{7}{2}}}{7r^7}P_6(\cos\theta) + \cdots\right].$$

2. In the event that the guide is the prolate spheroid created by the revolution of the oval $\frac{x^2}{a^2} + \frac{y^2}{b^2} = 1$ about its significant hub, show that assuming the fact of the matter is an outer point and is on the hub a good ways off x from the middle,

$$V = \frac{M}{2\sqrt{a^2-b^2}}\log\frac{x+\sqrt{a^2-b^2}}{x-\sqrt{a^2-b^2}}.$$

In the event that the fact of the matter isn't on the pivot and $r > \sqrt{a^2-b^2}$

$$V = \frac{M}{\sqrt{a^2-b^2}}\left[\frac{(a^2-b^2)^{\frac{1}{2}}}{r} + \frac{(a^2-b^2)^{\frac{3}{2}}}{3r^3}P_2(\cos\theta)\right.$$

$$\left. + \frac{(a^2-b^2)^{\frac{5}{2}}}{5r^5}P_4(\cos\theta) + \frac{(a^2-b^2)^{\frac{7}{2}}}{7r^7}P_6(\cos\theta) + \cdots\right].$$

80. As a third model we will track down the worth of the expected capability due to a slight homogeneous round circle, of thickness ρ, thickness k, and sweep a.

The worth of V at a point in the pivot of the circle a ways off x from its focus is promptly observed and ends up being

$$V_0 = 2\pi\rho k(\sqrt{x^2+a^2} - x) = \frac{2M}{a^2}[\sqrt{x^2+a^2} - x].$$

If $x > a$

$$\sqrt{x^2+a^2} = x\left(1 + \frac{a^2}{x^2}\right)^{\frac{1}{2}}$$

$$= x\left[1 + \frac{1}{2}\frac{a^2}{x^2} - \frac{1.1}{2.4}\frac{a^4}{x^4} + \frac{1.1.3}{2.4.6}\frac{a^6}{x^6} - \frac{1.1.3.5}{2.4.6.8}\frac{a^8}{x^8} + \cdots\right]$$

and

$$V_0 = \frac{2M}{a}\left[\frac{1}{2}\frac{a}{x} - \frac{1.1}{2.4}\frac{a^3}{x^3} + \frac{1.1.3}{2.4.6}\frac{a^5}{x^5} - \frac{1.1.3.5}{2.4.6.8}\frac{a^7}{x^7} + \cdots\right].$$

ZONAL HARMONICS.

On the off chance that $x < a$

$$\sqrt{x^2 + a^2} = a\left(1 + \frac{x^2}{a^2}\right)^{\frac{1}{2}}$$

$$= a\left[1 + \frac{1}{2}\frac{x^2}{a^2} - \frac{1.1}{2.4}\frac{x^4}{a^4} + \frac{1.1.3}{2.4.6}\frac{x^6}{a^6} + \cdots\right]$$

and $\quad V_0 = \frac{2M}{a}\left[1 - \frac{x}{a} + \frac{1}{2}\frac{x^2}{a^2} - \frac{1.1}{2.4}\frac{x^4}{a^4} + \frac{1.1.3}{2.4.6}\frac{x^6}{a^6} - \frac{1.1.3.5}{2.4.6.8}\frac{x^8}{a^8} + \cdots\right].$

Consequently the answer for any outside point is

$$V = \frac{2M}{a}\left[\frac{1}{2}\frac{a}{r} - \frac{1.1}{2.4}\frac{a^3}{r^3}P_2(\cos\theta)\right.$$
$$\left. + \frac{1.1.3}{2.4.6}\frac{a^5}{r^5}P_4(\cos\theta) - \frac{1.1.3.5}{2.4.6.8}\frac{a^7}{r^7}P_6(\cos\theta) + \cdots\right]$$

if $r > a$, and

$$V = \frac{2M}{a}\left[1 - \frac{r}{a}P_1(\cos\theta)\right.$$
$$\left. + \frac{1}{2}\frac{r^2}{a^2}P_2(\cos\theta) - \frac{1.1}{2.4}\frac{r^4}{a^4}P_4(\cos\theta) + \frac{1.1.3}{2.4.6}\frac{r^6}{a^6}P_6(\cos\theta) - \cdots\right]$$

if $r < a$ and $\theta < \frac{\pi}{2}$.

EXAMPLES.

1. The likely capability because of a homogeneous half of the globe whose pivot is taken as the polar pivot, is

$$V = \frac{M}{a}\left[\frac{a}{r} + \frac{3.1}{2.4}\frac{a^2}{r^2}P_1(\cos\theta)\right.$$
$$\left. - \frac{3.1.1}{2.4.6}\frac{a^4}{r^4}P_3(\cos\theta) + \frac{3.1.1.3}{2.4.6.8}\frac{a^6}{r^6}P_5(\cos\theta) - \cdots\right]$$

if $r > a$, and is

$$V = \frac{M}{a}\left[\frac{3}{2} + \frac{3}{2}\frac{r}{a}P_1(\cos\theta) + \frac{r^2}{a^2}P_2(\cos\theta)\right.$$
$$\left. + \frac{3.1}{2.4}\frac{r^3}{a^3}P_3(\cos\theta) - \frac{3.1.1}{2.4.6}\frac{r^5}{a^5}P_5(\cos\theta) + \cdots\right]$$

if $r < a$ and $\theta > \frac{\pi}{2}$.

2. The expected capability because of a strong circle whose thickness is relative to the separation from a diametral plane is, at an outside point,

$$V = \frac{8}{15}\frac{M}{a}\left[\frac{5.3}{2.4}\frac{a}{r} + \frac{5.3.1}{2.4.6}\frac{a^3}{r^3}P_2(\cos\theta)\right.$$
$$\left. - \frac{5.3.1.1}{2.4.6.8}\frac{a^5}{r^5}P_4(\cos\theta) + \frac{5.3.1.1.3}{2.4.6.8.10}\frac{a^7}{r^7}P_6(\cos\theta) - \cdots\right].$$

3. The expected capability because of the homogeneous oblate spheroid produced by the pivot of $\frac{x^2}{a^2} + \frac{y^2}{b^2} = 1$ about its minor hub is, at an outside point,

$$V = \frac{3}{2}\frac{M}{(a^2-b^2)}\left[\frac{x^2+a^2-b^2}{2(a^2-b^2)^{\frac{1}{2}}}\left(\sin^{-1}\frac{(a^2-b^2+bx)}{a\sqrt{x^2+a^2-b^2}} + \sin^{-1}\frac{(a^2-b^2-bx)}{a\sqrt{x^2+a^2-b^2}}\right) - x\right]$$

on the off chance that the fact of the matter is on the hub of the spheroid a ways off x from its middle.

$$V = \frac{3M}{(a^2-b^2)^{\frac{1}{2}}}\left[\frac{1}{1.3}\frac{(a^2-b^2)^{\frac{1}{2}}}{r} - \frac{1}{3.5}\frac{(a^2-b^2)^{\frac{3}{2}}}{r^3}P_2(\cos\theta) + \frac{1}{5.7}\frac{(a^2-b^2)^{\frac{5}{2}}}{r^5}P_4(\cos\theta) - \cdots\right]$$

on the off chance that $r > (a^2-b^2)^{\frac{1}{2}}$, and

$$V = \frac{3M}{(a^2-b^2)^{\frac{1}{2}}}\left[\frac{\pi}{4} - \frac{r}{(a^2-b^2)^{\frac{1}{2}}}P_1(\cos\theta) + \frac{\pi}{4}\frac{r^2}{(a^2-b^2)}P_2(\cos\theta) - \frac{1}{1.3}\frac{r^3}{(a^2-b^2)^{\frac{3}{2}}}P_3(\cos\theta) + \frac{1}{3.5}\frac{r^5}{(a^2-b^2)^{\frac{5}{2}}}P_5(\cos\theta) - \cdots\right]$$

if $r < (a^2-b^2)^{\frac{1}{2}}$ and $\theta < \frac{\pi}{2}$.

4. The expected capability because of the homogeneous prolate spheroid created by the pivot of $\frac{x^2}{a^2} + \frac{y^2}{b^2} = 1$ about its significant hub is, at an outside point,

$$V = \frac{3M}{(a^2-b^2)^{\frac{1}{2}}}\left[\frac{1}{1.3}\frac{(a^2-b^2)^{\frac{1}{2}}}{r} + \frac{1}{3.5}\frac{(a^2-b^2)^{\frac{3}{2}}}{r^3}P_2(\cos\theta) + \frac{1}{5.7}\frac{(a^2-b^2)^{\frac{5}{2}}}{r^5}P_4(\cos\theta) + \cdots\right]$$

if $r > (a^2-b^2)^{\frac{1}{2}}$.

81. The strategy utilized in the last three articles might be expressed in general as follows:--- At whatever point in an issue including the settling of the extraordinary type of Laplace's Situation

$$rD_r^2(rV) + \frac{1}{\sin\theta}D_\theta(\sin\theta D_\theta V) = 0,$$

the worth of V is given or can be found for all focuses on the pivot of X and this worth can be communicated as a total or a series including just entire powers positive or negative of the range vector of the point, the answer for a point not on the pivot can be gotten by increasing each term by the proper Zonal Consonant, subject just to the condition that the outcome assuming a series must be merged.

It will be displayed in the following article that $P_m(\cos\theta)$ is never more prominent than one nor not exactly short one. Subsequently the series being referred to will be concurrent for all upsides of r for which the first series was *absolutely convergent*.

ZONAL HARMONICS.

82. notwithstanding the structure yielded (1) Art. 77 for $P_m(x)$ different structures are frequently valuable.

It should be feasible to foster $P_m(\cos\theta)$, which might be viewed as a capability of θ, into a Fourier's Series, and such an improvement might be gotten, however with much work, by the techniques for Section II.

The improvement as far as cosines of products of θ might be acquired significantly more effectively by the accompanying gadget. We have seen in Art. 75 that $P_m(\cos\theta)$ is the coefficient of the mth power of z in the improvement of $(1 - 2z\cos\theta + z^2)^{-\frac{1}{2}}$ in a power series, and that in the event that mod $z < 1$ $(1 - 2z\cos\theta + z^2)^{-\frac{1}{2}}$ can be formed into such a series. We know by the Hypothesis of Capabilities that only one such series exists, so that the strategy by which we might decide to get the advancement won't influence the result.

$$(1 - 2z\cos\theta + z^2)^{-\frac{1}{2}} = (1 - z(e^{\theta i} + e^{-\theta i}) + z^2)^{-\frac{1}{2}}$$
$$= (1 - ze^{\theta i})^{-\frac{1}{2}}(1 - ze^{-\theta i})^{-\frac{1}{2}}.$$

$(1 - ze^{\theta i})^{-\frac{1}{2}}$ might be formed into a totally united series if mod $z < 1$, by the Binomial Hypothesis. We have

$$(1 - ze^{\theta i})^{-\frac{1}{2}} = 1 + \frac{1}{2}ze^{\theta i} + \frac{1.3}{2.4}z^2 e^{2\theta I} + \frac{1.3.5}{2.4.6}z^3 e^{3\theta i} + \frac{1.3.5.7}{2.4.6.8}z^4 e^{4\theta I} + \cdots$$

$$(1 - ze^{-\theta i})^{-\frac{1}{2}} = 1 + \frac{1}{2}ze^{-\theta i} + \frac{1.3}{2.4}z^2 e^{-2\theta I}$$
$$+ \frac{1.3.5}{2.4.6}z^3 e^{-3\theta i} + \frac{1.3.5.7}{2.4.6.8}z^4 e^{-4\theta I} + \cdots$$

The result of these series will give an improvement for $(1 - 2z\cos\theta + z^2)^{-\frac{1}{2}}$ in power series. The coefficient of z^m is effortlessly selected, and should be equivalent to $P_m(\cos\theta)$. We in this manner get

$$P_m(\cos\theta) = \frac{1.3.5.\cdots(2m-1)}{2.4.6.\cdots 2m}\left[e^{m\theta i} + e^{-m\theta i}\right.$$
$$+ \frac{1}{2}\cdot\frac{2m}{2m-1}(e^{(m-2)\theta i} + e^{-(m-2)\theta i})$$
$$\left.+ \frac{1.3}{2.4}\cdot\frac{2m(2m-2)}{(2m-1)(2m-3)}(e^{(m-4)\theta i} + e^{-(m-4)\theta i}) + \cdots\right]$$

$$P_m(\cos\theta) = \frac{1.3.5\cdots(2m-1)}{2.4.6.\cdots 2m}\left[2\cos m\theta + 2\frac{1.m}{1.(2m-1)}\cos(m-2)\theta\right.$$
$$+ 2\frac{1.3\,m(m-1)}{1.2(2m-1)(2m-3)}\cos(m-4)\theta$$
$$\left.+ 2\frac{1.3.5}{1.2.3}\frac{m(m-1)(m-2)}{(2m-1)(2m-3)(2m-5)}\cos(m-6)\theta + \cdots\right]. \quad (1)$$

Assuming that m is odd the advancement gets down to $\cos\theta$; on the off chance that m is even to $\cos(0)$, however all things considered the coefficient of $\cos(0)$, or at least, the consistent term, won't contain the component 2 which is normal to the wide range of various terms, however will be basically $\left[\frac{1.3.5\cdots(m-1)}{2.4.6.\cdots m}\right]^2$.

ZONAL HARMONICS.

We work out the upsides of $P_m(\cos\theta)$ for a couple of upsides of m

$$\left.\begin{aligned}
P_0(\cos\theta) &= 1 \\
P_1(\cos\theta) &= \cos\theta \\
P_2(\cos\theta) &= \frac{1}{4}(3\cos 2\theta + 1) \\
P_3(\cos\theta) &= \frac{1}{8}(5\cos 3\theta + 3\cos\theta) \\
P_4(\cos\theta) &= \frac{1}{64}(35\cos 4\theta + 20\cos 2\theta + 9) \\
P_5(\cos\theta) &= \frac{1}{128}[63\cos 5\theta + 35\cos 3\theta + 30\cos\theta] \\
P_6(\cos\theta) &= \frac{1}{512}[231\cos 6\theta + 126\cos 4\theta + 105\cos 2\theta + 50] \\
P_7(\cos\theta) &= \frac{1}{1024}[429\cos 7\theta + 231\cos 5\theta + 189\cos 3\theta + 175\cos\theta] \\
P_8(\cos\theta) &= \frac{1}{16384}[6435\cos 8\theta + 3432\cos 6\theta + 2772\cos 4\theta \\
&\qquad + 2520\cos 2\theta + 1225].
\end{aligned}\right\} \quad (2)$$

Since every one of the coefficients in the second individual from (1) are positive, and since every cosine has solidarity for its most extreme worth obviously $P_m(\cos\theta)$ has its most extreme worth when $\theta = 0$; yet we have displayed in Art. 76 that $P_m(1) = 1$. Hence $P_m(\cos\theta)$ is never more prominent than solidarity assuming that θ is genuine. It is likewise without any problem seen from (1) that $P_m(\cos\theta)$ can never be not exactly -1.

ZONAL HARMONICS.

83. $P_m(x)$ can be essentially communicated as a subsidiary. We have

$$P_m(x) = \frac{(2m-1)(2m-3)\cdots 1}{m!}\left[x^m - \frac{m(m-1)}{2.(2m-1)}x^{m-2}\right.$$
$$\left. + \frac{m(m-1)(m-2)(m-3)}{2.4.(2m-1)(2m-3)}x^{m-4} - \cdots\right]$$

$$\int_0^x P_m(x)dx = \frac{(2m-1)(2m-3)\cdots 1}{(m+1)!}\left[x^{m+1} - \frac{(m+1)m}{2.(2m-1)}x^{m-1}\right.$$
$$\left. + \frac{(m+1)m(m-1)(m-2)}{2.4.(2m-1)(2m-3)}x^{m-3} - \cdots\right]$$

$$\int_0^x {}^2 P_m(x)dx^2 = \int_0^x dx \int_0^x P_m(x)dx$$
$$= \frac{(2m-1)(2m-3)\cdots 1}{(m+2)!}\left[x^{m+2} - \frac{(m+2)(m+1)}{2.(2m-1)}x^m\right.$$
$$\left. + \frac{(m+2)(m+1)m(m-1)}{2.4.(2m-1)(2m-3)}x^{m-2} - \cdots\right]$$

$$\int_0^x {}^m P_m(x)dx^m = \frac{(2m-1)(2m-3)\cdots 1}{(2m)!}\left[x^{2m} - \frac{2m(2m-1)}{2(2m-1)}x^{2m-2}\right.$$
$$\left. + \frac{2m(2m-1)(2m-2)(2m-3)}{2.4.(2m-1)(2m-3)}x^{2m-4} - \cdots\right]$$

$$= \frac{(2m-1)(2m-3)\cdots 1}{(2m)!}\left[x^{2m} - mx^{2m-2} + \frac{m(m-1)}{2!}x^{2m-4}\right.$$
$$\left. - \frac{m(m-1)(m-2)}{3!}x^{2m-6} + \cdots\right].$$

The amount in sections clearly varies from $(x^2-1)^m$ by terms including lower powers of x than the mth.

Hence
$$P_m(x) = \frac{1.3.5\cdots(2m-1)}{(2m)!}\frac{d^m}{dx^m}(x^2-1)^m,$$

or
$$P_m(x) = \frac{1}{2^m m!}\frac{d^m}{dx^m}(x^2-1)^m. \text{and} \qquad (1)$$

This significant recipe is completely broad and holds not simply when $x = \cos\theta$, yet for all upsides of x.

84. The last outcome is vital to such an extent that it is worth while to affirm it by acquiring it straightforwardly from Legendre's Situation

$$(1-x^2)\frac{d^2z}{dx^2} - 2x\frac{dz}{dx} + m(m+1)z = 0 \qquad (1)$$

v. (1) Art. 75.

Allow us to separate (1) regarding x a couple of times addressing $\frac{dz}{dx}$ by z', $\frac{d^2z}{dx^2}$ by z'', $\frac{d^3z}{dz^3}$ by z''', &c. We get

$$(1-x^2)\frac{d^2z'}{dx^2} - 2.2x\frac{dz'}{dx} + [m(m+1) - 2]z' = 0,$$
$$(1-x^2)\frac{d^2z''}{dx^2} - 2.3x\frac{dz''}{dx} + [m(m+1) - 2(1+2)]z'' = 0,$$
$$(1-x^2)\frac{d^2z'''}{dx^2} - 2.4x\frac{dz'''}{dx} + [m(m+1) - 2(1+2+3)]z''' = 0,$$

ZONAL HARMONICS.

also, overall

$$(1-x^2)\frac{d^2 z^{(n)}}{dx^2} - 2(n+1)x\frac{dz^{(n)}}{dx} + [m(m+1) - 2(1+2+3+\cdots+n)]z^{(n)} = 0$$

or and
$$(1-x^2)\frac{d^2 z^{(n)}}{dx^2} - 2(n+1)x\frac{dz^{(n)}}{dx} + m(m+1) - n(n+1)]z^{(n)} = 0. \quad (2)$$

Following the similarity of these means composing conditions that will is simple separate into (1).

Let $\frac{dz_1}{dx} = z$, $\frac{d^2 z_2}{dx^2} = z$, $\frac{d^3 z_3}{dx^3} = z$, &c. Then

$$(1-x^2)\frac{d^2 z_1}{dx^2} + m(m+1)z_1 = 0$$

will separate into (1),

$$(1-x^2)\frac{d^2 z_2}{dx^2} + 2.1x\frac{dz_2}{dx} + [m(m+1) - 2.1]z_2 = 0$$

in the event that separated two times will give (1),

$$(1-x^2)\frac{d^2 z_3}{dx^2} + 2.2x\frac{dz_3}{dx} + [m(m+1) - 2(1+2)]z_3 = 0$$

on the off chance that separated multiple times will give (1), and overall

$$(1-x^2)\frac{d^2 z_n}{dx^2} + 2(n-1)x\frac{dz_n}{dx} + [m(m+1) - n(n-1)]z_n = 0 \quad (3)$$

whenever separated n times as for x will give (1).

On the off chance that $n = m+1$ (3) lessens to

$$(1-x^2)\frac{d^2 z_{m+1}}{dx^2} + 2mx\frac{dz_{m+1}}{dx} = 0, \quad (4)$$

also, the $(m+1)$st subsidiary as for x of any capability of x which fulfills (4) will be an answer of (1). (4) can be composed

$$(1-x^2)\frac{dz_m}{dx} + 2mxz_m = 0$$

also, can be promptly settled by isolating the factors and coordinating. v. Int. Cal. (1) page 314. It gives

$$z_m = C(x^2 - 1)^m.$$

Hence and $z = \dfrac{d^m z_m}{dx^m} = C\dfrac{d^m (x^2-1)^m}{dx^m}$ \quad (5)

is an answer of Legendre's Situation (1) and concurs with the worth of $P_m(x)$ acquired in Art. 83.

85. The conditions acquired in Art. 84 are so inquisitive thus basically related that it is worth while to completely think of them as somewhat more.

We have seen that

$$(1-x^2)\frac{d^2 z}{dx^2} + 2mx\frac{dz}{dx} = 0 \quad (1)$$

separates into

$$(1-x^2)\frac{d^2 z}{dx^2} + 2(m-1)x\frac{dz}{dx} + 2mz = 0; \quad (2)$$

ZONAL HARMONICS.

that assuming we separate (2) m times we get Legendre's Condition

$$(1-x^2)\frac{d^2z}{dx^2} - 2x\frac{dz}{dx} + m(m+1)z = 0; \tag{3}$$

that assuming we separate (2) $2m$ times we get

$$(1-x^2)\frac{d^2z}{dx^2} - 2(m+1)x\frac{dz}{dx} = 0; \tag{4}$$

that assuming we separate (2) $m-n$ times we have

$$(1-x^2)\frac{d^2z}{dx^2} + 2(n-1)x\frac{dz}{dx} + [m(m+1) - n(n-1)]z = 0; \tag{5}$$

also, that assuming we separate (2) $m+n$ times we have

$$(1-x^2)\frac{d^2z}{dx^2} - 2(n+1)x\frac{dz}{dx} + [m(m+1) - n(n+1)]z = 0. \tag{6}$$

By the guide of (1) we tracked down in the last article a specific arrangement of (2), in particular

$$z = (x^2-1)^m.$$

Assuming we substitute in (2) $z = u(x^2-1)^m$ following the strategy represented completely in Art. 18, we get as the overall arrangement of (2)

$$z = A(x^2-1)^m + B(x^2-1)^m \int \frac{dx}{(x^2-1)^{m+1}}, \tag{7}$$

A and B being inconsistent constants.

$\int \frac{dx}{(x^2-1)^{m+1}}$ is handily worked out [v. equation (42) page 6. Table of Integrals. Int. Cal. Appendix]. If $x < 1$ it evaporates when $x = 0$. If $x > 1$ it evaporates when $x = \infty$. On the off chance that, $x < 1$ (7) can be composed

$$z = A(x^2-1)^m + B(x^2-1)^m \int_0^x \frac{dx}{(x^2-1)^{m+1}} \tag{8}$$

furthermore, if $x > 1$

$$z = A(x^2-1)^m + B(x^2-1)^m \int_x^\infty \frac{dx}{(x^2-1)^{m+1}} \tag{9}$$

what's more, in these structures superfluous erratic constants are kept away from.

From (7) we can get the overall arrangements of (3), (4), (5), and (6).

$$z = A\frac{d^m(x^2-1)^m}{dx^m} + B\frac{d^m}{dx^m}\left[(x^2-1)^m \int \frac{dx}{(x^2-1)^{m+1}}\right] \tag{10}$$

is the overall arrangement of (3).

$$z = A\frac{d^{2m}(x^2-1)^m}{dx^{2m}} + B\frac{d^{2m}}{dx^{2m}}\left[(x^2-1)^m \int \frac{dx}{(x^2-1)^{m+1}}\right] \tag{11}$$

is the overall arrangement of (4).

$$z = A\frac{d^{m-n}(x^2-1)^m}{dx^{m-n}} + B\frac{d^{m-n}}{dx^{m-n}}\left[(x^2-1)^m \int \frac{dx}{(x^2-1)^{m+1}}\right] \tag{12}$$

ZONAL HARMONICS.

is the overall arrangement of (5).

$$z = A\frac{d^{m+n}(x^2-1)^m}{dx^{m+n}} + B\frac{d^{m+n}}{dx^{m+n}}\left[(x^2-1)^m \int \frac{dx}{(x^2-1)^{m+1}}\right] \quad (13)$$

is the overall arrangement of (6).

In every one of these structures A and B are erratic constants and the basic is to be taken from 0 to x if $x < 1$ and from x to ∞ if $x > 1$.

Obviously (10) should be indistinguishable with the structures previously acquired in Arts. 16 furthermore, 18 as broad arrangements of Legendre's Situation.

Condition (4) is easy to such an extent that it tends to be settled straightforwardly, and we get its arrangement in the structure

$$z = A_1 + B_1 \int \frac{dx}{(x^2-1)^{m+1}} \quad (14)$$

which should be identical to (11).

Looking at (14) with (7), the arrangement of (2), we see that each arrangement of (4) can be gotten from an answer of (2) by separating the last option by $(x^2-1)^m$, or at the end of the day that assuming we compose (2)

$$(1-x^2)\frac{d^2z}{dx^2} + 2(m-1)x\frac{dz}{dx} + 2mz = 0, \quad (2)$$

and (4) as

$$(1-x^2)\frac{d^2z_1}{dx^2} - 2(m+1)x\frac{dz_1}{dx} = 0 \quad (4)$$

$z = z_1(x^2-1)^m$; and the replacement of this worth in (2) will give (4), and the replacement of $z_1 = \frac{z}{(x^2-1)^m}$ in (4) will give (2).

We have, then, at that point, two different ways of getting (4) from (2); we might separate (2) $2m$ times as for x, or we might supplant z in (2) by $z_1(x^2-1)^m$.

Assuming we utilize the primary strategy we have seen that Legendre's Condition (3) is halfway somewhere in the range of (2) and (4). That is assuming we separate (2) m times we get (3) and in the event that we, separate (3) m times we get (4). Allow us to check whether the midway condition in our subsequent cycle is Legendre's Condition.

If
$$z = y(x^2-1)^{\frac{m}{2}}$$
and
$$y = z_1(x^2-1)^{\frac{m}{2}}$$
$$z = z_1(x^2-1)^m.$$

So that on the off chance that in (2) we supplant z by $y(x^2-1)^{\frac{m}{2}}$ and rehash the activity on the subsequent condition we will get (4). Making the main replacement we find,

$$(1-x^2)\frac{d^2y}{dx^2} - 2x\frac{dy}{dx} + \left[m(m+1) - \frac{m^2}{1-x^2}\right]y = 0, \quad (15)$$

not Legendre's Condition but rather a to some degree more broad structure. Obviously its answer is

$$y = A(x^2-1)^{\frac{m}{2}} + B(x^2-1)^{\frac{m}{2}} \int \frac{dx}{(x^2-1)^{m+1}}. \quad (16)$$

(2) and (4) are extraordinary types of (5) and (6). Allow us to attempt the investigation of subbing in (5) $z = y(1-x^2)^{\frac{n}{2}}$ and in (6) $z = \frac{y}{(1-x^2)^{\frac{n}{2}}}$. We find that the two replacements give a similar condition

$$(1-x^2)\frac{d^2y}{dx^2} - 2x\frac{dy}{dx} + \left[m(m+1) - \frac{n^2}{1-x^2}\right]y = 0. \quad (17)$$

ZONAL HARMONICS. 134

The arrangement of (17) can be acquired from either (12) or (13) and is

$$y = \frac{1}{(1-x^2)^{\frac{n}{2}}}\left\{A\frac{d^{m-n}(x^2-1)^m}{dx^{m-n}} + B\frac{d^{m-n}}{dx^{m-n}}\left[(x^2-1)^m \int \frac{dx}{(x^2-1)^{m+1}}\right]\right\} \quad (18)$$

or then again

$$y = (1-x^2)^{\frac{n}{2}}\left\{A_1\frac{d^{m+n}(x^2-1)^m}{dx^{m+n}} + B_1\frac{d^{m+n}}{dx^{m+n}}\left[(x^2-1)^m \int \frac{dx}{(x^2-1)^{m+1}}\right]\right\} \quad (19)$$

which obviously should be same.

86. notwithstanding the worth of $P_m(x)$ yielded (1) Art. 83 there is another significant subsidiary structure which we will continue to acquire. It is

$$P_m(\cos\theta) = \frac{(-1)^m}{m!}r^{m+1}D_x^m\left(\frac{1}{r}\right). \quad (1)$$

We have seen in Art. 75 that $\frac{1}{r}\frac{1}{\sqrt{1-2\frac{r_1}{r}\cos\theta+\frac{r_1^2}{r^2}}}$ can be formed into a joined series if $r_1 < r$ and that the $(m+1)$st term of that series is $\frac{P_m(\cos\theta)r_1^m}{r^{m+1}}$. Allow us to get this term by Taylor's Hypothesis.

$$\frac{1}{r}\frac{1}{\sqrt{1-2\frac{r_1}{r}\cos\theta+\frac{r_1^2}{r^2}}} = \frac{1}{\sqrt{r^2-2r_1r\cos\theta+r_1^2}} = \frac{1}{\sqrt{x^2+y^2+z^2-2xr_1+r_1^2}}$$

$$= \frac{1}{\sqrt{(x-r_1)^2+y^2+z^2}}$$

Concerning as an element of $(x-r_1)$ and creating as indicated by powers of r_1 by Taylor's Hypothesis we get as the $(m+1)$st term

$$\frac{(-1)^m}{m!}r_1^m D_x^m\left[\frac{1}{\sqrt{x^2+y^2+z^2}}\right] \quad \text{or} \quad \frac{(-1)^m}{m!}r_1^m D_x^m\left(\frac{1}{r}\right).$$

Hence
$$\frac{P_m(\cos\theta)}{r^{m+1}} = \frac{(-1)^m}{m!}D_x^m\left(\frac{1}{r}\right).$$

87. We have now acquired four unique structures for our *zonal harmonic*, a polynomial in x, an articulation including cosines of products of θ, a structure including a common mth subsidiary regarding x, and a structure including a fractional mth subsidiary regarding x. We will currently get a structure due to Laplace, including a clear vital.

$$\int_0^\pi \frac{d\phi}{a-b\cos\phi} = \frac{\pi}{(a^2-b^2)^{\frac{1}{2}}} \quad (1)$$

if $a^2 > b^2$ [v. Int. Cal. page 68].

$\frac{1}{(1-2xz+z^2)^{\frac{1}{2}}}$ can be communicated in the structure $\frac{1}{(a^2-b^2)^{\frac{1}{2}}}$ by taking $a = 1-zx$ furthermore, $b = z\sqrt{x^2-1}$ and regardless of what esteem x may have z can be taken so little that a^2 will be

ZONAL HARMONICS.

more noteworthy than b^2. Then by (1)

$$\frac{1}{(1-2xz+z^2)^{\frac{1}{2}}} = \frac{1}{\pi}\int_0^\pi \frac{d\phi}{1-zx-z\sqrt{x^2-1}.\cos\phi}$$

$$= \frac{1}{\pi}\int_0^\pi \frac{d\phi}{1-z(x+\sqrt{x^2-1}.\cos\phi)}$$

$$= \frac{1}{\pi}\int_0^\pi [1+(x+\sqrt{x^2-1}.\cos\phi)z+(x+\sqrt{x^2-1}.\cos\phi)^2 z^2$$

$$+(x+\sqrt{x^2-1}.\cos\phi)^3 z^3 +\cdots]d\phi$$

assuming z is taken little to the point that the modulus of $z(x+\sqrt{x^2-1}.\cos\phi)$ is under 1. However by Art. 77 (2) $P_m(x)$ is the coefficient of z^m in the advancement of $\frac{1}{(1-2xz+z^2)^{\frac{1}{2}}}$,

hence and $P_m(x) = \frac{1}{\pi}\int_0^\pi [x+\sqrt{x^2-1}.\cos\phi]^m d\phi.$ \hfill (2)

By supplanting ϕ by $\pi-\phi$ in (2) we get

$$P_m(x) = \frac{1}{\pi}\int_0^\pi [x-\sqrt{x^2-1}.\cos\phi]^m d\phi. \tag{3}$$

$\frac{1}{(1-2xz+z^2)^{\frac{1}{2}}} = \frac{1}{z}\frac{1}{\left(1-2x\frac{1}{z}+\frac{1}{z^2}\right)^{\frac{1}{2}}}$ and if mod $\frac{1}{z} < 1$ or as such if mod $z > 1$ $\frac{1}{\left(1-2x\frac{1}{z}+\frac{1}{z^2}\right)^{\frac{1}{2}}}$ can be formed into a united series including powers of $\frac{1}{z}$, and the coefficient of $\left(\frac{1}{z}\right)^m$ will be $P_m(x)$; however this will be the coefficient of z^{-m-1} in the improvement of $\frac{1}{(1-2xz+z^2)^{\frac{1}{2}}}$ as per diving powers of z, mod z being more noteworthy than 1.

On the off chance that now we let $a = zx-1$ and $b = z\sqrt{x^2-1}$, $a^2-b^2 = 1-2xz+z^2$ and z might be taken perfect to such an extent that $a^2-b^2 > 0$. Then by (1)

$$\frac{1}{(1-2xz+z^2)^{\frac{1}{2}}} = \frac{1}{\pi}\int_0^\pi \frac{d\phi}{zx-1-z\sqrt{x^2-1}.\cos\phi}$$

$$= \frac{1}{\pi}\int_0^\pi \frac{d\phi}{z(x-\sqrt{x^2-1}.\cos\phi)\left[1-\frac{1}{z(x-\sqrt{x^2-1}.\cos\phi)}\right]}$$

$$= \frac{1}{\pi}\int_0^\pi \frac{1}{(x-\sqrt{x^2-1}.\cos\phi)}\left[z^{-1}+\frac{1}{(x-\sqrt{x^2-1}.\cos\phi)}z^{-2}\right.$$

$$\left.+\frac{1}{(x-\sqrt{x^2-1}.\cos\phi)^2}z^{-3}+\cdots\right]d\phi\text{ and}$$

furthermore, the coefficient of z^{-m-1} is $\frac{1}{\pi}\int_0^\pi \frac{d\phi}{[x-\sqrt{x^2-1}.\cos\phi]^{m+1}}.$

Hence and $P_m(x) = \frac{1}{\pi}\int_0^\pi \frac{d\phi}{[x-\sqrt{x^2-1}.\cos\phi]^{m+1}}.$ and \hfill (4)

Supplant ϕ by $\pi-\phi$ and we get

$$P_m(x) = \frac{1}{\pi}\int_0^\pi \frac{d\phi}{[x+\sqrt{x^2-1}.\cos\phi]^{m+1}}. \tag{5}$$

ZONAL HARMONICS.

88. In the issues in which we have proactively utilized *Zonal Harmonics* (v. Arts. 78- - 81) we have had the option to begin with the worth of the Potential Capability anytime on the pivot of X, and it has been important to create the articulation for V based on that hub in conditions of rising or slipping powers of x. If, be that as it may, we start with the worth of V as far as θ for some given worth of r, that is on the outer layer of some circle, we should foster the capability of θ as far as *zonal harmonics* of $\cos\theta$ (v. Art. 10), and our concern becomes the accompanying:- - - To foster a given capability of $\cos\theta$ regarding zonal music of $\cos\theta$, or to foster a given capability of x regarding the capabilities $P_m(x)$, x lying among 1 and -1.

The issue looks like intently that of creating in a Fourier's series, which we have previously considered at such length.

$$\text{Let} and f(x) = A_0 P_0(x) + A_1 P_1(x) + A_2 P_2(x) + A_3 P_3(x) + \cdots and \tag{1}$$

for all upsides of x from -1 to 1 and allow it to be expected to decide the coefficients.

Assuming $f(x)$ is single-esteemed and has just limited discontinuities between $x = -1$ also, $x = 1$ we might continue as in Art. 19.

Allow us to take $n+1$ terms of (1) and endeavor to decide the coefficients. Take $n+1$ upsides of x at equivalent spans Δx between $x = -1$ and $x = 1$ with the goal that $(n+2)\Delta x = 2$; $f(-1+\Delta x)$, $f(-1+2\Delta x)$, $f(-1+3\Delta x)$, $\cdots f[-1+(n+1)\Delta x]$ will be the comparing upsides of $f(x)$. Substitute these qualities in (1) and we have

$$\left.\begin{aligned}
f(-1+\Delta x) &= A_0 P_0(-1+\Delta x) + A_1 P_1(-1+\Delta x) \\
&\quad + A_2 P_2(-1+\Delta x) + \cdots + A_n P_n(-1+\Delta x) \\
f(-1+2\Delta x) &= A_0 P_0(-1+2\Delta x) + A_1 P_1(-1+2\Delta x) \\
&\quad + A_2 P_2(-1+2\Delta x) + \cdots + A_n P_n(-1+2\Delta x) \\
&\vdots \\
f(1-\Delta x) &= A_0 P_0(1-\Delta x) + A_1 P_1(1-\Delta x) + A_2 P_2(1-\Delta x) + \cdots \\
&\quad + A_n P_n(1-\Delta x),
\end{aligned}\right\} \tag{2}$$

that is, $n+1$ conditions from which in principle the $n+1$ coefficients $A_0, A_1, \cdots A_n$ not entirely set in stone.

Following the relationship of Art. 24 let us duplicate the principal condition by $P_m(-1+\Delta x).\Delta x$, the second by $P_m(-1+2\Delta x).\Delta x$, the third by $P_m(-1+3\Delta x).\Delta x$, &c., and add the conditions. The primary individual from the coming about condition is

$$\sum_{k=1}^{k=n+1} f(-1+k\Delta x) P_m(-1+k\Delta x).\Delta x, \tag{3}$$

furthermore, the coefficient of any A as A_l in the subsequent part is

$$\sum_{k=1}^{k=n+1} P_m(-1+k\Delta x) P_l(-1+k\Delta x).\Delta x. \tag{4}$$

In the event that now n is endlessly expanded (3) approaches as its restricting worth

$$\int_{-1}^{1} f(x) P_m(x) dx \tag{5}$$

and (4) approaches
$$\int_{-1}^{1} P_m(x) P_l(x) dx. \tag{6}$$

ZONAL HARMONICS.

We have now to track down the worth of the vital (6) or as we will compose it for more prominent comfort

$$\int_{-1}^{1} P_m(x)P_n(x)dx.$$

89. $\int_{-1}^{1} P_m(x)P_n(x)dx = \dfrac{1}{2^{m+n}m!n!} \int_{-1}^{1} \dfrac{d^m(x^2-1)^m}{dx^m} \cdot \dfrac{d^n(x^2-1)^n}{dx^n} dx$ by (1) Art. 83.

$$\int_{-1}^{1} \frac{d^m(x^2-1)^m}{dx^m} \cdot \frac{d^n(x^2-1)^n}{dx^n} dx = \left[\frac{d^m(x^2-1)^m}{dx^m} \cdot \frac{d^{n-1}(x^2-1)^n}{dx^{n-1}}\right]_{-1}^{1}$$
$$- \int_{-1}^{1} \frac{d^{m+1}(x^2-1)^m}{dx^{m+1}} \cdot \frac{d^{n-1}(x^2-1)^n}{dx^{n-1}} dx \quad (1)$$

by *integration by parts*.

Presently if $z = X(x^2-1)^n$

$$\frac{dz}{dx} = 2nxX(x^2-1)^{n-1} + (x^2-1)^n \frac{dX}{dx}$$
$$= (x^2-1)^{n-1}\left[2nxX + (x^2-1)\frac{dX}{dx}\right]. \quad (2)$$

Subsequently the pth subordinate regarding x of any capability of x containing $(x^2-1)^n$ as a variable will contain $(x^2-1)^{n-p}$ as an element if $p < n$.

$\dfrac{d^{n-1}(x^2-1)^n}{dx^{n-1}}$, then, contains (x^2-1) as an element and is zero when $x = 1$ furthermore, when $x = -1$, so that (1) lessens to

$$\int_{-1}^{1} \frac{d^m(x^2-1)^m}{dx^m} \cdot \frac{d^n(x^2-1)^n}{dx^n} dx = -\int_{-1}^{1} \frac{d^{m+1}(x^2-1)^m}{dx^{m+1}} \cdot \frac{d^{n-1}(x^2-1)^n}{dx^{n-1}} dx.$$

That's what it follows

$$\int_{-1}^{1} \frac{d^m(x^2-1)^m}{dx^m} \cdot \frac{d^n(x^2-1)^n}{dx^n} dx$$
$$= (-1)^p \int_{-1}^{1} \frac{d^{m+p}(x^2-1)^m}{dx^{m+p}} \cdot \frac{d^{n-p}(x^2-1)^n}{dx^{n-p}} dx$$
$$= (-1)^p \int_{-1}^{1} \frac{d^{m-p}(x^2-1)^m}{dx^{m-p}} \cdot \frac{d^{n+p}(x^2-1)^n}{dx^{n+p}} dx. \quad (3)$$

On the off chance that $m < n$ we get from (3)

$$\int_{-1}^{1} \frac{d^m(x^2-1)^m}{dx^m} \cdot \frac{d^n(x^2-1)^n}{dx^n} dx = (-1)^m \int_{-1}^{1} \frac{d^{2m}(x^2-1)^m}{dx^{2m}} \cdot \frac{d^{n-m}(x^2-1)^n}{dx^{n-m}} dx$$
$$= (-1)^m(2m)!\left[\frac{d^{n-m-1}(x^2-1)^n}{dx^{n-m-1}}\right]_{-1}^{1} = 0,$$

ZONAL HARMONICS.

since
$$\frac{d^{2m}(x^2-1)^m}{dx^{2m}} = (2m)!.$$

In the event that $m > n$

$$\int_{-1}^{1} \frac{d^m(x^2-1)^m}{dx^m} \cdot \frac{d^n(x^2-1)^n}{dx^n} dx = (-1)^n \int_{-1}^{1} \frac{d^{m-n}(x^2-1)^m}{dx^{m-n}} \cdot \frac{d^{2n}(x^2-1)^n}{dx^{2n}} dx$$

$$= (-1)^n (2n)! \left[\frac{d^{m-n-1}(x^2-1)^m}{dx^{m-n-1}} \right]_{-1}^{1} = 0.$$

In the event that, m isn't equivalent to n

$$\int_{-1}^{1} P_m(x) P_n(x) dx = 0. \tag{4}$$

On the off chance that $m = n$ we need to find $\int_{-1}^{1} [P_m(x)]^2 dx$.

$$\int_{-1}^{1} [P_m(x)]^2 dx = \frac{1}{2^{2m}(m!)^2} \int_{-1}^{1} \frac{d^m(x^2-1)^m}{dx^m} \cdot \frac{d^m(x^2-1)^m}{dx^m} dx.$$

$$\int_{-1}^{1} \frac{d^m(x^2-1)^m}{dx^m} \cdot \frac{d^m(x^2-1)^m}{dx^m} dx = (-1)^m \int_{-1}^{1} \frac{d^{2m}(x^2-1)^m}{dx^{2m}} \cdot (x^2-1)^m dx$$

by (3),
$$= (-1)^m (2m)! \int_{-1}^{1} (x^2-1)^m dx.$$

$$\int_{-1}^{1} (x^2-1)^m dx = \int_{-1}^{1} (x-1)^m (x+1)^m dx = -\frac{m}{m+1} \int_{-1}^{1} (x-1)^{m-1} (x+1)^{m+1} dx$$

$$= (-1)^m \frac{m!}{(m+1)(m+2)\cdots 2m} \int_{-1}^{1} (x+1)^{2m} dx$$

$$= (-1)^m \frac{2^{2m+1} m!}{(m+1)(m+2)\cdots(2m+1)}.$$

Hence
$$\int_{-1}^{1} [P_m(x)]^2 dx = \frac{1}{2^{2m}(m!)^2} \frac{(-1)^m (2m)!(-1)^m m! 2^{2m+1}}{(m+1)(m+2)\cdots(2m+1)}$$

or
$$\int_{-1}^{1} [P_m(x)]^2 dx = \frac{2}{2m+1}. \tag{5}$$

90. The arrangement of the issue in Art. 88 is currently promptly acquired, and we have

$$f(x) = A_0 P_0(x) + A_1 P_1(x) + A_2 P_2(x) + \cdots \tag{1}$$

where
$$A_m = \frac{2m+1}{2} \int_{-1}^{1} f(x) P_m(x) dx. \tag{2}$$

ZONAL HARMONICS.

The capability and the series are equivalent for all upsides of x from $x = -1$ to $x = 1$, and $f(x)$ is liable to no circumstances save those which would empower us to foster it in a Fourier's Series. [v. Section III.]

Obviously (1) can be composed

$$f(\cos\theta) = A_0 P_0(\cos\theta) + A_1 P_1(\cos\theta) + A_2 P_2(\cos\theta) + \cdots$$

where
$$A_m = \frac{2m+1}{2} \int_{-1}^{1} f(\cos\theta) P_m(\cos\theta) d(\cos\theta)$$

or if $f(\cos\theta) = F(\theta)$

$$F(\theta) = A_0 P_0(\cos\theta) + A_1 P_1(\cos\theta) + A_2 P_2(\cos\theta) + \cdots \tag{3}$$

where
$$A_m = \frac{2m+1}{2} \int_{0}^{\pi} F(\theta) P_m(\cos\theta) \sin\theta . d\theta \tag{4}$$

what's more, the advancement holds great from $\theta = 0$ to $\theta = \pi$.

Assuming $f(x)$ is an even capability, that is to say, if $f(-x) = f(x)$ (1) and (2) can be fairly rearranged. For all things considered it tends to be effortlessly shown (v. Art. 77) that

$$\int_{-1}^{1} f(x) P_{2k}(x) dx = 2 \int_{0}^{1} f(x) P_{2k}(x) dx,$$

and that
$$\int_{-1}^{1} f(x) P_{2k+1}(x) dx = 0;$$

so that if $f(-x) = f(x)$

$$f(x) = A_0 P_0(x) + A_2 P_2(x) + A_4 P_4(x) + A_6 P_6(x) + \cdots \tag{5}$$

where
$$A_{2k} = (4k+1) \int_{0}^{1} f(x) P_{2k}(x) dx. \tag{6}$$

Assuming $f(x)$ is an odd capability, that is to say, if $f(-x) = -f(x)$ it tends to be displayed in like way that

$$f(x) = A_1 P_1(x) + A_3 P_3(x) + A_5 P_5(x) + A_7 P_7(x) + \cdots \tag{7}$$

where
$$A_{2k+1} = (4k+3) \int_{0}^{1} f(x) P_{2k+1}(x) dx. \tag{8}$$

Assuming it is just fundamental that the advancement ought to hold for $0 < x < 1$ any capability might be communicated in structure (5) or (7) at delight.

91. We can lay out the way that $\int_{-1}^{1} P_m(x) P_n(x) dx = 0$ by a more broad strategy than that utilized in Art. 89.

Allow X_m to be any arrangement of Legendre's Situation

$$\frac{d}{dx}\left[(1-x^2)\frac{dz}{dx}\right] + m(m+1)z = 0 \qquad \text{[v. (1) Art. 16]}.$$

ZONAL HARMONICS.

which with its most memorable subsidiary regarding x is limited, constant, and single-esteemed for upsides of x between -1 and 1, -1 and 1 being included.

Then
$$\frac{d}{dx}\left[(1-x^2)\frac{dX_m}{dx}\right] + m(m+1)X_m = 0 \qquad (1)$$

and
$$\frac{d}{dx}\left[(1-x^2)\frac{dX_n}{dx}\right] + n(n+1)X_n = 0 \qquad (2)$$

Duplicate (1) by X_n and (2) by X_m and take away and incorporate and we get

$$[m(m+1) - n(n+1)]\int_{-1}^{1} X_m X_n dx = \int_{-1}^{1} X_m \frac{d}{dx}\left[(1-x^2)\frac{dX_n}{dx}\right] dx$$
$$- \int_{-1}^{1} X_n \frac{d}{dx}\left[(1-x^2)\frac{dX_m}{dx}\right] dx.$$

Incorporate by parts,

$$[m(m+1) - n(n+1)]\int_{-1}^{1} X_m X_n dx = \left[X_m(1-x^2)\frac{dX_n}{dx} - X_n(1-x^2)\frac{dX_m}{dx}\right]_{x=-1}^{x=1}$$
$$- \int_{-1}^{1} (1-x^2)\frac{dX_n}{dx}\frac{dX_m}{dx} dx + \int_{-1}^{1}(1-x^2)\frac{dX_m}{dx}\frac{dX_n}{dx} dx. \qquad (3)$$

Whence
$$\int_{-1}^{1} X_m X_n dx = 0 \qquad (4)$$

except if $m = n$.

(3) gives immediately the significant equation

$$\int_{x}^{1} X_m X_n dx = \frac{(1-x^2)\left[X_n \frac{dX_m}{dx} - X_m \frac{dX_n}{dx}\right]}{m(m+1) - n(n+1)} \qquad (5)$$

from which come as unique cases

$$\int_{x}^{1} P_m(x) P_n(x) dx = \frac{(1-x^2)\left[P_n(x)\frac{dP_m(x)}{dx} - P_m(x)\frac{dP_n(x)}{dx}\right]}{m(m+1) - n(n+1)} \qquad (6)$$

also, since $P_0(x) = 1$

$$\int_{x}^{1} P_m(x) dx = \frac{(1-x^2)\frac{dP_m(x)}{dx}}{m(m+1)}, \qquad (7)$$

except if $m = 0$.

EXAMPLES.

1. Show that

$$\int_{0}^{1} P_m(x) dx = 0 \quad \text{assuming } m \text{ is even and isn't zero.}$$

$$= (-1)^{\frac{m-1}{2}} \frac{1}{m(m+1)} \frac{3.5.7.\cdots m}{2.4.6.\cdots (m-1)} \quad \text{if } m \text{ is odd.}$$

ZONAL HARMONICS. 141

v. Art. 91 (7) and Art. 77 (10).

2. Show that

$$\int_0^1 P_m(x)P_n(x)dx = 0 \quad \text{if } m \text{ and } n \text{ are both even or both odd.}$$

$$= (-1)^{\frac{m+n+1}{2}} \frac{m!\, n!}{2^{m+n-1}(m-n)(m+n+1)\left(\frac{m}{2}!\right)^2 \left(\frac{n-1}{2}!\right)^2}$$

assuming m is even and n odd. v. Art. 91 (6) and Art. 77 (8), (9), and (10). cf. J. W. Strutt (Master Rayleigh) Lond. Phil. Trans. 1870, page 579.

3. Show that $\displaystyle\int_0^1 [P_m(x)]^2 dx = \frac{1}{2m+1}$ v. Art. 89 (5).

92. Recipe (4) Art. 91 can be acquired straightforwardly from Laplace's Situation by the guide of *Green's Theorem* (v. Peirce's Newt. Pot. Func. § 48).

Take the extraordinary type of Green's Hypothesis, [(148) § 48 Peirce's Newt. Pot. Func.]

$$\iiint (U\nabla^2 V - V\nabla^2 U)dxdydz = \int (UD_n V - VD_n U)ds \tag{1}$$

where ∇^2 represents $(D_x^2 + D_y^2 + D_z^2)$, D_n is the incomplete subordinate along the outer typical, and the left-hand part is the space-basic through the space limited by any shut surface, and the right-hand part is the surface essential assumed control over a similar surface. (v. Int. Cal. Part XIV.)

On the off chance that U and V are arrangements of Laplace's Situation $\nabla^2 V = \nabla^2 U = 0$ and (1) decreases to

$$\int (UD_n V - VD_n U)ds = 0. \tag{2}$$

Presently $r^m X_m$ and $r^n X_n$ are arrangements of Laplace's Situation if $x = \cos\theta$ (v. Art. 16).

On the off chance that the unit circle is taken as the jumping surface and $U = r^m X_m$ and $V = r^n X_n$ (1) and (2) will hold great.

$$D_n U = D_r(r^m X_m) = mr^{m-1} X_m,$$
$$D_n V = nr^{n-1} X_n,$$
$$ds = \sin\theta.d\theta d\phi,$$

also (2) becomes $\displaystyle\int_0^{2\pi} d\phi \int_0^{\pi} (nX_m X_n - mX_m X_n)\sin\theta.d\theta = 0$

or $$2\pi(n-m)\int_0^{\pi} X_m X_n \sin\theta.d\theta = 0. \text{and} \tag{3}$$

Since $x = \cos\theta$, $\sin\theta.d\theta = -dx$ and (3) decreases to

$$\int_{-1}^{1} X_m X_n dx = 0^3 \tag{4}$$

except if $m = n$.

---

[3] It ought to be noticed that this confirmation is not any more broad than that of the last article, for, in request that Green's Hypothesis ought to apply to $r^m X_m$ this capability and its most memorable subsidiaries must be limited ceaseless and single-esteemed inside and on the outer layer of the unit circle. (v. Peirce, Newt. Pot. Func. § 48.)

ZONAL HARMONICS.

93. We can now tackle totally the issue of Art. 10 which was in that article conveyed to where it was simply important to create a certain capability of θ in the structure

$$A_0 P_0(\cos\theta) + A_1 P_1(\cos\theta) + A_2 P_2(\cos\theta) + \cdots$$

given that $$f(\theta) = 1 \quad \text{from} \quad \theta = 0 \quad \text{to} \quad \theta = \frac{\pi}{2}$$

andand $f(\theta) = 0$ from $\theta = \dfrac{\pi}{2}$ to $\theta = \pi$.

This adds up to exactly the same thing as forming $F(x)$ into the series

$$F(x) = A_0 P_0(x) + A_1 P_1(x) + A_2 P_2(x) + A_3 P_3(x) + \cdots$$

where $\qquad F(x) = 0 \quad \text{from} \quad x = -1 \quad \text{to} \quad x = 0$
and $\qquad F(x) = 1 \quad \text{from} \quad x = 0 \quad \text{to} \quad x = 1$.

By Art. 90 (1) and (2)

$$A_0 = \frac{1}{2}\int_0^1 P_0(x)dx = \frac{1}{2}\int_0^1 dx = \frac{1}{2},$$

and any coefficient $$A_m = \frac{(2m+1)}{2}\int_0^1 P_m(x)dx.$$

By Art. 91, Ex. 1

$$\int_0^1 P_m(x)dx = 0 \quad \text{assuming that } m \text{ is even}$$

$$= (-1)^{\frac{m-1}{2}} \frac{1}{m(m+1)} \frac{3.5.7.\cdots m}{2.4.6.\cdots(m-1)} \quad \text{if } m \text{ is odd.}$$

Hence $\quad A_m = 0 \quad$ assuming that m is even

$$= (-1)^{\frac{m-1}{2}} \frac{2m+1}{2m+2} \cdot \frac{1.3.5.\cdots(m-2)}{2.4.6.\cdots(m-1)} \quad \text{assuming that } m \text{ is odd.}$$

Then $$F(x) = \frac{1}{2} + \frac{3}{4}P_1(x) - \frac{7}{8}\cdot\frac{1}{2}P_3(x) + \frac{11}{12}\cdot\frac{1.3}{2.4}P_5(x) - \cdots \qquad (1)$$

what's more, $\quad u = \frac{1}{2} + \frac{3}{4}rP_1(\cos\theta) - \frac{7}{8}\cdot\frac{1}{2}r^3 P_3(\cos\theta) + \frac{11}{12}\cdot\frac{1.3}{2.4}r^5 P_5(\cos\theta) + \cdots \qquad (2)$

for any point inside the circle.

94. In the event that in an issue on the Potential Capability the worth of V is given at each mark of a round surface and has round symmetry[4] about a width of that surface the worth of V anytime in space can be gotten.

We need to address Laplace's Condition in the structure

$$rD_r^2(rV) + \frac{1}{\sin\theta}D_\theta(\sin\theta D_\theta V) = 0 \qquad (1)$$

dependent upon the circumstances

$$V = f(\theta) \quad \text{when} \quad r = a$$
$$V = 0 \quad\quad \text{''} \quad\quad r = \infty.$$

---

[4] See note on page 10.

ZONAL HARMONICS.

We have
$$f(\theta) = A_0 P_0(\cos\theta) + A_1 P_1(\cos\theta) + A_2 P_2(\cos\theta) + \cdots$$

where
$$A_m = \frac{(2m+1)}{2} \int_0^\pi f(\theta) P_m(\cos\theta) \sin\theta \, d\theta.$$
v. Art. 90 (4).

Subsequently
$$V = A_0 + A_1 \left(\frac{r}{a}\right) P_1(\cos\theta) + A_2 \left(\frac{r}{a}\right)^2 P_2(\cos\theta) + A_3 \left(\frac{r}{a}\right)^3 P_3(\cos\theta) + \cdots \quad (2)$$

is the expected answer for a point inside the circle, and

$$V = A_0 \left(\frac{a}{r}\right) + A_1 \left(\frac{a}{r}\right)^2 P_1(\cos\theta) + A_2 \left(\frac{a}{r}\right)^3 P_2(\cos\theta) + A_3 \left(\frac{a}{r}\right)^4 P_3(\cos\theta) + \cdots \quad (3)$$

is the necessary answer for an outer point.

EXAMPLES.

1. If on the outer layer of a circle of sweep c V is consistent and equivalent to a show that $V = a$ for any point inside the circle and $V = \frac{ac}{r}$ for any outer point.

2. Two equivalent flimsy hemispherical shells of sweep c set together to frame a round surface are isolated by a dainty non-directing layer. Charges of statical power are put on the two halves of the globe one of which is then viewed as at potential a and the other at potential b. Track down the worth of the possible capability anytime.

$$V = \frac{a+b}{2} + (b-a) \left[\frac{3}{4} \frac{r}{c} P_1(\cos\theta) - \frac{7}{8} \cdot \frac{1}{2} \frac{r^3}{c^3} P_3(\cos\theta) \right.$$
$$\left. + \frac{11}{12} \cdot \frac{1.3}{2.4} \frac{r^5}{c^5} P_5(\cos\theta) - \cdots \right]$$

for an inside point

$$V = \frac{a+b}{2} \cdot \frac{c}{r} + (b-a) \left[\frac{3}{4} \frac{c^2}{r^2} P_1(\cos\theta) - \frac{7}{8} \cdot \frac{1}{2} \frac{c^4}{r^4} P_3(\cos\theta) \right.$$
$$\left. + \frac{11}{12} \cdot \frac{1.3}{2.4} \frac{c^6}{r^6} P_5(\cos\theta) - \cdots \right]$$

for an outer point.

3. If $V_1 = f(\cos\theta)$ when $r = a$ and $V_1 = 0$ when $r = b$ show that for $a < r < b$

$$V_1 = \sum_{m=0}^{m=\infty} A_m \left(\frac{b^{m+1}}{r^{m+1}} - \frac{r^m}{b^m} \right) \left(\frac{b^{m+1}}{a^{m+1}} - \frac{a^m}{b^m} \right)^{-1} P_m(\cos\theta)$$

where
$$A_m = \frac{2m+1}{2} \int_{-1}^{1} f(x) P_m(x) \, dx.$$

4. If $V_2 = F(\cos\theta)$ when $r = b$ and $V_2 = 0$ when $r = a$ show that for $a < r < b$

$$V_2 = \sum_{m=0}^{m=\infty} B_m \left(\frac{r^m}{a^m} - \frac{a^{m+1}}{r^{m+1}} \right) \left(\frac{b^m}{a^m} - \frac{a^{m+1}}{b^{m+1}} \right)^{-1} P_m(\cos\theta)$$

where
$$B_m = \frac{2m+1}{2} \int_{-1}^{1} F(x) P_m(x) \, dx.$$

ZONAL HARMONICS.

5. Assuming the worth of the potential capability is given for arbitrary reasons on the surfaces of a round shell however has roundabout symmetry[5] about a measurement $V = V_1 + V_2$ (v. Exs. 3 and 4).

6. Two concentric empty round guides are protected and charged. The inward one of range a is at potential p, and the external one of sweep b is at potential q. Track down V for any point in space.

$$V = p \quad \text{if} \quad r < a,$$
$$V = \frac{pa}{b-a}\left(\frac{b}{r} - 1\right) + \frac{qb}{b-a}\left(1 - \frac{a}{r}\right) \quad \text{if} \quad a < r < b,$$
$$V = \frac{qb}{r} \quad \text{if} \quad r > b.$$

7. If $V = 0$ on the foundation of a half of the globe and $V = f(\cos\theta)$ on the curved surface, show that for a point inside the side of the equator

$$V = \sum_{k=0}^{k=\infty} A_{2k+1}\left(\frac{r}{a}\right)^{2k+1} P_{2k+1}(\cos\theta)$$

where
$$A_{2k+1} = (4k+3)\int_0^1 f(x) P_{2k+1}(x) dx \qquad \text{[v. Art. 90 (8)]}.$$

8. If the raised surface of a strong side of the equator of span a is kept at the consistent temperature solidarity and the base at the steady temperature zero show that after the long-lasting condition of temperatures is set up the temperature of any inward point is

$$u = \frac{3}{2}\frac{r}{a}P_1(\cos\theta) - \frac{7}{4}\cdot\frac{1}{2}\frac{r^3}{a^3}P_3(\cos\theta) + \frac{11}{6}\cdot\frac{1.3}{2.4}\frac{r^5}{a^5}P_5(\cos\theta) - \cdots$$

9. A circle of range a and with darkened surface is presented to the direct beams of the sun in air at the temperature zero. Find the *stationary temperature* of any interior point.

Suggestion: $D_r u + hu - Mf(\theta) = 0$ when $r = a$.

Let
$$u = \sum A_m \frac{r^m}{a^m} P_m(\cos\theta), \quad \text{and} \quad f(\theta) = \sum B_m P_m(\cos\theta).$$

Then we have

$$m\frac{A_m}{a}P_m(\cos\theta) + h\sum A_m P_m(\cos\theta) - M\sum B_m P_m(\cos\theta) = 0,$$

whence
$$A_m = \frac{MB_m}{h + \frac{m}{a}}.$$

Here $f(\theta) = \cos\theta$ if $0 < \theta < \frac{\pi}{2}$ and $f(\theta) = 0$ if $\frac{\pi}{2} < \theta < \pi$.

$$f(\theta) = \frac{1}{4} + \frac{1}{2}P_1(\cos\theta) + \frac{5}{16}P_2(\cos\theta) - \frac{3}{32}P_4(\cos\theta) + \cdots$$
$$+ (-1)^{k+1}\frac{(4k+1)(2k)!}{(4k+4)(2k-1)2^{2k}(k!)^2}P_{2k}(\cos\theta) + \cdots$$

v. Art. 91 Exs. (2) and (3). cf. J. W. Strutt (Ruler Rayleigh), Lond. Phil. Trans. vol. 160, page 587.

---

[5] See note on page 12

ZONAL HARMONICS.

95. The recipes of Art. 90 empower us to foster a given capability of x in terms of *Zonal Surface Harmonics*, the advancement turning out as expected for upsides of x between -1 and $+1$. If, be that as it may, we can show by outside contemplations that a given capability of x can be communicated in Zonal Surface Sounds, the advancement turning out as expected for all upsides of x, the equations of Art. 90 will give us the improvement being referred to.

For instance in the event that n is a positive number x^n can be communicated as far as Zonal Surface Music regardless of what the worth of x, and no Consonant of higher request than n will enter. For the equations giving the upsides of $P_1(x)$, $P_2(x)$, $\cdots P_n(x)$ (v. Art. 77) might be viewed as n arithmetical conditions of the principal degree as far as x, x^2, x^3, $\cdots x^n$ and $P_1(x)$, $P_2(x)$, $\cdots P_n(x)$.

From these situations the $n-1$ amounts x, x^2, x^3, $\cdots x^{n-1}$, can be killed, also, there will result a condition of the primary degree in x^n and $P_1(x)$, $P_2(x)$, $\cdots P_n(x)$, which will empower us to communicate x^n in the structure

$$A_0 + A_1 P_1(x) + A_2 P_2(x) + \cdots + A_n P_n(x),$$

regardless of what the worth of x, and we will have a similar recipe when $-1 < x < 1$ as when $x > 1$ or $x < -1$.

Allow us to get this turn of events. By Art. 90 (1) and (2)

$$x^n = A_0 P_0(x) + A_1 P_1(x) + A_2 P_2(x) + \cdots \quad (1)$$

where
$$A_m = \frac{2m+1}{2} \int_{-1}^{1} x^n P_m(x) dx. \quad (2)$$

Then
$$A_m = \frac{2m+1}{2} \frac{1}{2^m m!} \int_{-1}^{1} x^n \frac{d^m (x^2-1)^m}{dx^m} dx \quad \text{and} \quad \text{by (1) Art. 83.}$$

By *integration by parts* we get

$$\int_{-1}^{1} x^n \frac{d^m(x^2-1)^m}{dx^m} dx = n(n-1)(n-2)\cdots(n-m+1) \int_{-1}^{1} x^{n-m}(1-x^2)^m dx,$$
$$\text{if } m < n+1, \quad (3)$$
$$= 0 \text{ if } m > n.$$

By *integration by parts* we promptly get the decrease equation

$$\int_{-1}^{1} x^p(1-x^2)^q dx = \frac{2q}{p+1} \int_{-1}^{1} x^{p+2}(1-x^2)^{q-1} dx \quad \text{and whence}$$

$$\int_{-1}^{1} x^{n-m}(1-x^2)^m dx = \frac{2^m m!}{(n-m+1)(n-m+3)\cdots(n+m-1)} \int_{-1}^{1} x^{n+m} dx.$$

$$\int_{-1}^{1} x^{n+m} dx = \frac{2}{(n+m+1)} \quad \text{if } n+m \text{ is even,}$$
$$= 0 \text{ if } n+m \text{ is odd.}$$

Hence
$$A_m = \frac{(2m+1)n(n-1)(n-2)\cdots(n-m+1)}{(n-m+1)(n-m+3)(n-m+5)\cdots(n+m+1)}$$

$$\text{if } m < n+1 \text{ and } m+n \text{ is even,}$$

$$= 0 \text{ if } m > n \text{ or on the other hand if } m+n \text{ is odd.}$$

ZONAL HARMONICS.

Accordingly

$$x^n = \frac{n!}{1.3.5\cdots(2n+1)}\left[(2n+1)P_n(x) + (2n-3)\frac{(2n+1)}{2}P_{n-2}(x)\right.$$
$$+ (2n-7)\frac{(2n+1)(2n-1)}{2.4}P_{n-4}(x)$$
$$\left.+ (2n-11)\frac{(2n+1)(2n-1)(2n-3)}{2.4.6}P_{n-6}(x) + \cdots\right] \quad (4)$$

the subsequent part finishing with the term $\frac{1}{n+1}P_0(x)$ assuming n is even and with the term $\frac{3}{n+2}P_1(x)$ assuming that n is odd.

For comfort of reference we work out a couple of abilities of x.

$$\left.\begin{aligned}
x^0 &= 1 = P_0(x) \\
x &= P_1(x) \\
x^2 &= \frac{2}{3}P_2(x) + \frac{1}{3}P_0(x) \\
x^3 &= \frac{2}{5}P_3(x) + \frac{3}{5}P_1(x) \\
x^4 &= \frac{8}{35}P_4(x) + \frac{4}{7}P_2(x) + \frac{1}{5}P_0(x) \\
x^5 &= \frac{8}{63}P_5(x) + \frac{4}{9}P_3(x) + \frac{3}{7}P_1(x) \\
x^6 &= \frac{16}{231}P_6(x) + \frac{24}{77}P_4(x) + \frac{10}{21}P_2(x) + \frac{1}{7}P_0(x) \\
x^7 &= \frac{16}{429}P_7(x) + \frac{8}{39}P_5(x) + \frac{14}{33}P_3(x) + \frac{1}{3}P_1(x) \\
x^8 &= \frac{128}{6435}P_8(x) + \frac{64}{495}P_6(x) + \frac{48}{143}P_4(x) + \frac{40}{99}P_2(x) + \frac{1}{9}P_0(x).
\end{aligned}\right\} \quad (5)$$

In the event that a given capability of x can be communicated as a *terminating power series* it can be formed into a Zonal Symphonious Series by the guide of (4). Considering that

$$f(x) = a_0 + a_1x + a_2x^2 + a_3x^3 + \cdots,$$

let
$$f(x) = B_0 + B_1P_1(x) + B_2P_2(x) + B_3P_3(x) + \cdots;$$

then, at that point, selecting cautiously the coefficient of $P_m(x)$ we have

$$B_m = \frac{m!}{1.3.5.\cdots(2m-1)}\left[a_m + \frac{(m+1)(m+2)}{2.(2m+3)}a_{m+2}\right.$$
$$\left.+ \frac{(m+1)(m+2)(m+3)(m+4)}{2.4.(2m+3)(2m+5)}a_{m+4} + \cdots\right]. \quad (6)$$

96. The improvement of $\frac{dP_n(x)}{dx}$ is valuable and is effectively acquired.

Let
$$\frac{dP_n(x)}{dx} = A_0P_0(x) + A_1P_1(x) + A_2P_2(x) + \cdots$$

Then
$$A_m = \frac{2m+1}{2}\int_{-1}^{1} P_m(x)\frac{dP_n(x)}{dx}dx \quad (1)$$

by Art. 90 (2);

$$\int_{-1}^{1} P_m(x)\frac{dP_n(x)}{dx}dx = \Big[P_m(x)P_n(x)\Big]_{x=-1}^{x=1} - \int_{-1}^{1} P_n(x)\frac{dP_m(x)}{dx}dx. \quad (2)$$

ZONAL HARMONICS.

$$\left[P_m(x)P_n(x)\right]_{x=-1}^{x=1} = 0 \quad \text{if } m+n \text{ is even}$$
$$= 2 \quad \text{if } m+n \text{ is odd.}$$

Since $P_n(x)$ is an arithmetical polynomial of the nth degree in x, $\frac{dP_n(x)}{dx}$ is an mathematical polynomial of the $n-1$st degree in x. Subsequently in (1) m is less than n; subsequently $\frac{dP_m(x)}{dx}$ is a mathematical polynomial in x of lower degree than n and

$$\int_{-1}^{1} P_n(x) \frac{dP_m(x)}{dx} dx = 0 \qquad \text{by Art. 95 (3).}$$

We get then $\quad A_m and = 2m+1 \quad$ if $m+n$ is odd and $m < n$,

$$= 0 \quad \text{if } m+n \text{ is even or } m > n-1;$$

and

$$\frac{dP_n(x)}{dx} = (2n-1)P_{n-1}(x) + (2n-5)P_{n-3}(x) + (2n-9)P_{n-5}(x) + \cdots \qquad (3)$$

the subsequent part finishing with the term $3P_1(x)$ assuming n is even and with the term $P_0(x)$ on the off chance that n is odd.

From (3) various basic equations are promptly gotten. For instance

$$\frac{dP_{n+1}(x)}{dx} - \frac{dP_{n-1}(x)}{dx} = (2n+1)P_n(x) \qquad (4)$$

$$\int_{x}^{1} P_n(x)dx = \frac{1}{2n+1}[P_{n-1}(x) - P_{n+1}(x)]. \qquad (5)$$

$$(2n+1)x\frac{dP_n(x)}{dx} = n\frac{dP_{n+1}(x)}{dx} + (n+1)\frac{dP_{n-1}(x)}{dx} \qquad (6)$$

[v. (4) and Article 77 (12)].

$$(x^2-1)\frac{dP_n(x)}{dx} = nxP_n(x) - nP_{n-1}(x) \qquad (7)$$

[v. (5) and Article 91 (7).]

97. By the guide of the recipes of Art. 96 various important turns of events can be gotten. Allow us to get $\cos n\theta$ and $\sin n\theta$, n being any certain genuine. $z = \cos n\theta$ and $z = \sin n\theta$ are arrangements of the situation

$$\frac{d^2z}{d\theta^2} + n^2 z = 0$$

or on the other hand assuming we let $x = \cos\theta$, of the situation

$$(1-x^2)\frac{d^2z}{dx^2} - x\frac{dz}{dx} + n^2 z = 0. \qquad (1)$$

Let $\quad a_0 P_0(x) + a_1 P_1(x) + a_2 P_2(x) + \cdots$

be the necessary improvement of $\cos n\theta$ or of $\sin n\theta$.

Then $\quad \displaystyle\sum_{m=0}^{m=\infty} a_m \left[(1-x^2)\frac{d^2 P_m(x)}{dx^2} - x\frac{dP_m(x)}{dx} + n^2 P_m(x)\right] = 0 \qquad$ by (1).

ZONAL HARMONICS.   148

$z = P_m(x)$ is an answer of Legendre's Situation (v. Art. 77). Consequently

$$(1-x^2)\frac{d^2 P_m(x)}{dx^2} - x\frac{dP_m(x)}{dx} = x\frac{dP_m(x)}{dx} - m(m+1)P_m(x),$$

also (1) becomes

$$\sum_{m=0}^{m=\infty} a_m \left[x\frac{dP_m(x)}{dx} + [n^2 - m(m+1)]P_m(x) \right] = 0. \tag{2}$$

Recipes (4) and (6) of Art. 96 empower us to toss (2) into the structure

$$\sum_{m=0}^{m=\infty} a_m \left[\frac{n^2 - m^2}{2m+1}\frac{dP_{m+1}(x)}{dx} - \frac{n^2 - (m+1)^2}{2m+1}\frac{dP_{m-1}(x)}{dx} \right] = 0. \tag{3}$$

(3) should be indistinguishably valid. Consequently the coefficient of $\frac{dP_{m+1}(x)}{dx}$ must equivalent zero, and we have

$$a_{m+2} = \frac{2m+5}{2m+1} \cdot \frac{n^2 - m^2}{n^2 - (m+3)^2} a_m. \tag{4}$$

Assuming we are creating $\cos n\theta$

$$a_0 = \frac{1}{2} \int_0^\pi \cos n\theta \sin \theta . d\theta \qquad\qquad \text{by Art. 90 (4),}$$

$$= \frac{1}{4} \int_0^\pi [\sin(n+1)\theta - \sin(n-1)\theta] d\theta,$$

$$a_0 = -\frac{1}{2} \cdot \frac{1 + \cos n\pi}{n^2 - 1}; \tag{5}$$

and
$$a_1 = \frac{3}{2} \int_0^\pi \cos n\theta \cos \theta \sin \theta . d\theta \qquad\qquad \text{by Art. 90 (4),}$$

$$a_1 = -\frac{3}{2} \cdot \frac{1 - \cos n\pi}{n^2 - 4}. \tag{6}$$

(4), (5), and (6) give us

$$\cos n\theta = -\frac{1 + \cos n\pi}{2(n^2 - 1)} \left[and P_0(\cos\theta) + 5\frac{n^2}{n^2 - 3^2} P_2(\cos\theta) \right.$$
$$\left. + 9\frac{n^2(n^2 - 2^2)}{(n^2 - 3^2)(n^2 - 5^2)} P_4(\cos\theta) + \cdots \right]$$
$$- \frac{1 - \cos n\pi}{2(n^2 - 2^2)} \left[and 3P_1(\cos\theta) + 7\frac{n^2 - 1^2}{n^2 - 4^2} P_3(\cos\theta) \right.$$
$$\left. + 11\frac{(n^2 - 1^2)(n^2 - 3^2)}{(n^2 - 4^2)(n^2 - 6^2)} P_5(\cos\theta) + \cdots \right]. \tag{7}$$

On the off chance that n is an entire number $1 + \cos n\pi$ or $1 - \cos n\pi$ will disappear and the series will end with the term including $P_n(\cos\theta)$. For this case (7) might be revamped

$$\cos n\theta = \frac{1}{2} \cdot \frac{2.4.6. \cdots 2n}{3.5.7. \cdots (2n+1)} \left[(2n+1)P_n(\cos\theta) \right.$$
$$+ (2n-3)\frac{n^2 - (n+1)^2}{n^2 - (n-2)^2} P_{n-2}(\cos\theta)$$
$$\left. + (2n-7)\frac{[n^2 - (n+1)^2][n^2 - (n-1)^2]}{[n^2 - (n-2)^2][n^2 - (n-4)^2]} P_{n-4}(\cos\theta) + \cdots \right]. \tag{8}$$

ZONAL HARMONICS.

On the off chance that we are creating $\sin n\theta$

$$a_0 = \frac{1}{2}\int_0^\pi \sin n\theta \sin\theta \, d\theta = -\frac{1}{2}\cdot\frac{\sin n\pi}{n^2-1},$$

$$a_1 = \frac{3}{2}\int_0^\pi \sin n\theta \cos\theta \sin\theta \, d\theta = \frac{3}{2}\cdot\frac{\sin n\pi}{n^2-2^2} \quad \text{and}$$

$$\sin n\theta = -\frac{1}{2}\cdot\frac{\sin n\pi}{n^2-1}\left[P_0(\cos\theta) + 5\frac{n^2}{n^2-3^2}P_2(\cos\theta)\right.$$
$$\left. + 9\frac{n^2(n^2-2^2)}{(n^2-3^2)(n^2-5^2)}P_4(\cos\theta) + \cdots\right]$$
$$+\frac{1}{2}\cdot\frac{\sin n\pi}{n^2-2^2}\left[3P_1(\cos\theta) + 7\frac{n^2-1^2}{n^2-4^2}P_3(\cos\theta)\right.$$
$$\left. + 11\frac{(n^2-1^2)(n^2-3^2)}{(n^2-4^2)(n^2-6^2)}P_5(\cos\theta) + \cdots\right]. \qquad (9)$$

In the event that n is an entire number $\sin n\pi = 0$, and every one of the details of (9) evaporate with the exception of those including $P_{n-1}(\cos\theta)$, $P_{n+1}(\cos\theta)$, $P_{n+3}(\cos\theta)$ &c., which become uncertain. For this case it is important to process a_{n-1} freely.

We have

$$a_{n-1} \, and = \frac{2n-1}{2}\int_0^\pi \sin n\theta P_{n-1}(\cos\theta)\sin\theta \, d\theta$$

$$= \frac{2n-1}{4}\int_0^\pi [\cos(n-1)\theta - \cos(n+1)\theta]P_{n-1}(\cos\theta)d\theta.$$

Hence $and \, a_{n-1} = \dfrac{2n-1}{4}\cdot\dfrac{1.3.5.\cdots(2n-3)}{2.4.6.\cdots(2n-2)}\pi$ \hfill [v. Art. 82 (1)],

furthermore,

$$\sin n\theta = and \frac{\pi}{4}\cdot\frac{1.3.\cdots(2n-3)}{2.4.\cdots(2n-2)}\left[(2n-1)P_{n-1}(\cos\theta)\right.$$
$$+(2n+3)\frac{n^2-(n-1)^2}{n^2-(n+2)^2}P_{n+1}(\cos\theta)$$
$$\left.+(2n+7)\frac{[n^2-(n-1)^2][n^2-(n+1)^2]}{[n^2-(n+2)^2][n^2-(n+4)^2]}P_{n+3}(\cos\theta)+\cdots\right]. \qquad (10)$$

EXAMPLES.

1. Show that

$$\csc\theta = \frac{\pi}{2}\left[1 + 5\left(\frac{1}{2}\right)^2 P_2(\cos\theta) + 9\left(\frac{1.3}{2.4}\right)^2 P_4(\cos\theta) + 13\left(\frac{1.3.5}{2.4.6}\right)^2 P_6(\cos\theta) + \cdots\right]$$

whence

$$\frac{1}{\sqrt{1-x^2}} = \frac{\pi}{2}\left[1 + 5\left(\frac{1}{2}\right)^2 P_2(x) + 9\left(\frac{1.3}{2.4}\right)^2 P_4(x) + 13\left(\frac{1.3.5}{2.4.6}\right)^2 P_6(x) + \cdots\right]$$

[v. Art. 90 (4) and Art. 82].

ZONAL HARMONICS.

2. Show that

$$\operatorname{ctn}\theta = \frac{\pi}{2}\left[3\left(\frac{1}{2}\right)P_1(\cos\theta) + 7\left(\frac{3}{4}\right)\left(\frac{1}{2}\right)^2 P_3(\cos\theta) + 11\left(\frac{5}{6}\right)\left(\frac{1.3}{2.4}\right)^2 P_5(\cos\theta) + \cdots\right]$$

whence

$$\frac{1}{\sqrt{1-x^2}} = \frac{\pi}{2}\left[3\left(\frac{1}{2}\right)P_1(x) + 7\left(\frac{3}{4}\right)\left(\frac{1}{2}\right)^2 P_3(x) + 11\left(\frac{5}{6}\right)\left(\frac{1.3}{2.4}\right)^2 P_5(x) + \cdots\right]$$

[v. Art. 90 (4) and Art. 82].

3. By incorporating the aftereffect of Ex. 1 and working on by the guide of Art. 96 (5), get the turn of events

$$\sin^{-1} x = \frac{\pi}{2}\left[3 and\left(\frac{1}{2}\right)^2 P_1(x) + 7\left(\frac{1}{2.4}\right)^2 P_3(x)\right.$$
$$\left. + 11\left(\frac{1.3}{2.4.6}\right)^2 P_5(x) + 15\left(\frac{1.3.5}{2.4.6.8}\right)^2 P_7(x) + \cdots\right]$$

whence $\quad \theta = \frac{\pi}{2}\left[P_0(\cos\theta) - 3\left(\frac{1}{2}\right)^2 P_1(\cos\theta) - 7\left(\frac{1}{2.4}\right)^2 P_3(\cos\theta)\right.$

$$\left. -11\left(\frac{1.3}{2.4.6}\right)^2 P_5(\cos\theta) - \cdots\right].$$

4. By coordinating the consequence of Ex. 2 and improving on by the guide of Art. 96 (5) acquire

$$\sqrt{1-x^2} = \frac{\pi}{2}\left[\frac{1}{2} - 5\left(\frac{1}{4}\right)\left(\frac{1}{2}\right)^2 P_2(x) - 9\left(\frac{3}{6}\right)\left(\frac{1}{2.4}\right)^2 P_4(x)\right.$$
$$\left. -13\left(\frac{5}{8}\right)\left(\frac{1.3}{2.4.6}\right)^2 P_6(x) + \cdots\right]$$

whence

$$\sin\theta = \frac{\pi}{2}\left[\frac{1}{2}P_0(\cos\theta) - 5\left(\frac{1}{4}\right)\left(\frac{1}{2}\right)^2 P_2(\cos\theta) - 9\left(\frac{3}{6}\right)\left(\frac{1}{2.4}\right)^2 P_4(\cos\theta) - \cdots\right].$$

To make more clear the relationship of improvement in Zonal Consonant Series with improvement in Fourier's Series we give on page 151 a cut addressing the initial seven Surface Zonal Music $P_1(\cos\theta)$, $P_2(\cos\theta)$, $\cdots P_7(\cos\theta)$, which are obviously fairly confounded Geometrical bends looking like generally $\cos\theta$, $\cos 2\theta$, $\cdots \cos 7\theta$; and on page 152, the initial four progressive approximations to the Zonal Symphonious Series

$$\frac{1}{2} + \frac{3}{4}P_1(\cos\theta) - \frac{7}{8}\cdot\frac{1}{2}P_3(\cos\theta) + \frac{11}{12}\cdot\frac{1.3}{2.4}P_5(\cos\theta) - \cdots \qquad [\text{I}]$$

[v. (1) Art. 93], and

$$\frac{\pi}{2}\left[P_0(\cos\theta) - 3\left(\frac{1}{2}\right)^2 P_1(\cos\theta) - 7\left(\frac{1}{2.4}\right)^2 P_3(\cos\theta)\right.$$
$$\left. -11\left(\frac{1.3}{2.4.6}\right)^2 P_5(\cos\theta) - \cdots\right] \qquad [\text{II}]$$

(v. Ex. 3 Art. 97).

[I] is equivalent to 1 from $\theta = 0$ to $\theta = \frac{\pi}{2}$, and to 0 from $\theta = \frac{\pi}{2}$ to $\theta = \pi$; and [II] is equivalent to θ from $\theta = 0$ to $\theta = \pi$.

The figures on page 152 are built on exactly a similar rule as those on pages 52 and 53, with which they ought to be painstakingly looked at.

ZONAL HARMONICS.

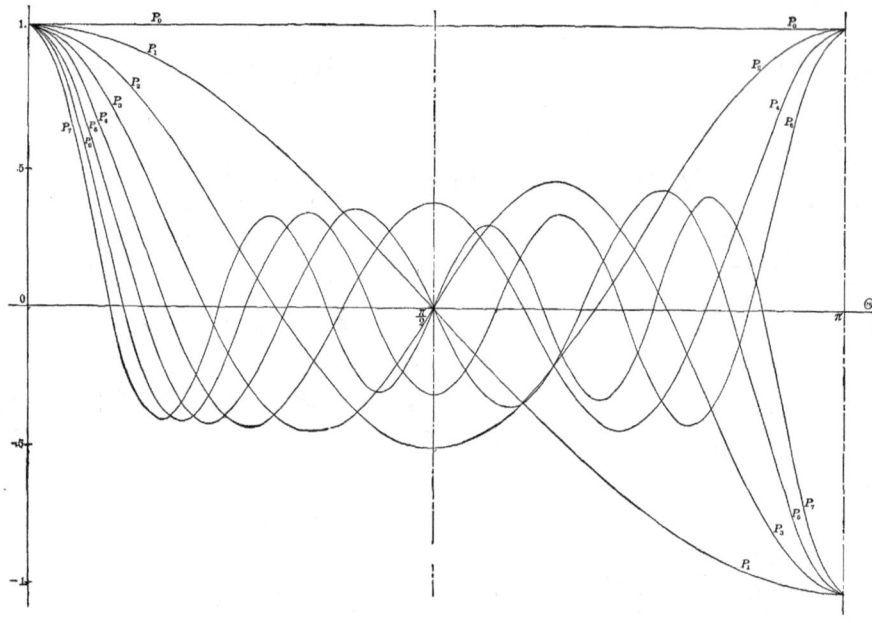

The bends $y = P_0(\cos\theta)$, $y = P_1(\cos\theta)$, ... $y = P_7(\cos\theta)$. (v. page 150.)

98. By applying *Gauss' Theorem* (B. O. Peirce, Newt. Pot. Func. § 31) or the exceptional Type of *Green's Theorem*,

$$\iiint \nabla^2 V\,dx\,dy\,dz = \int D_n V\,ds = -4\pi \iiint \rho\,dx\,dy\,dz,$$

[Peirce, N. P. F. § 49 (149)] to a crate cut from an endlessly meager shell of drawing in issue by a container of power whose end is a component of the outer layer of the shell we promptly get the significant outcome

$$4\pi\rho\kappa = D_n V_1 - D_n V_2. \tag{1}$$

where ρ is the thickness and κ the thickness of the shell, V_1 the worth of the possible capability because of the shell at an inside point and V_2 its worth at an outer point, and where D_n is the halfway subsidiary along the outside typical to the external surface of the shell.

On the off chance that we need to manage a surface circulation of issue we have just to supplant $\rho\kappa$ in (1) by σ where σ is the surface thickness, whence

$$4\pi\sigma = D_n V_1 - D_n V_2 \tag{2}$$

(v. Peirce, N. P. F. §§ 45, 46, and 47).

ZONAL HARMONICS.

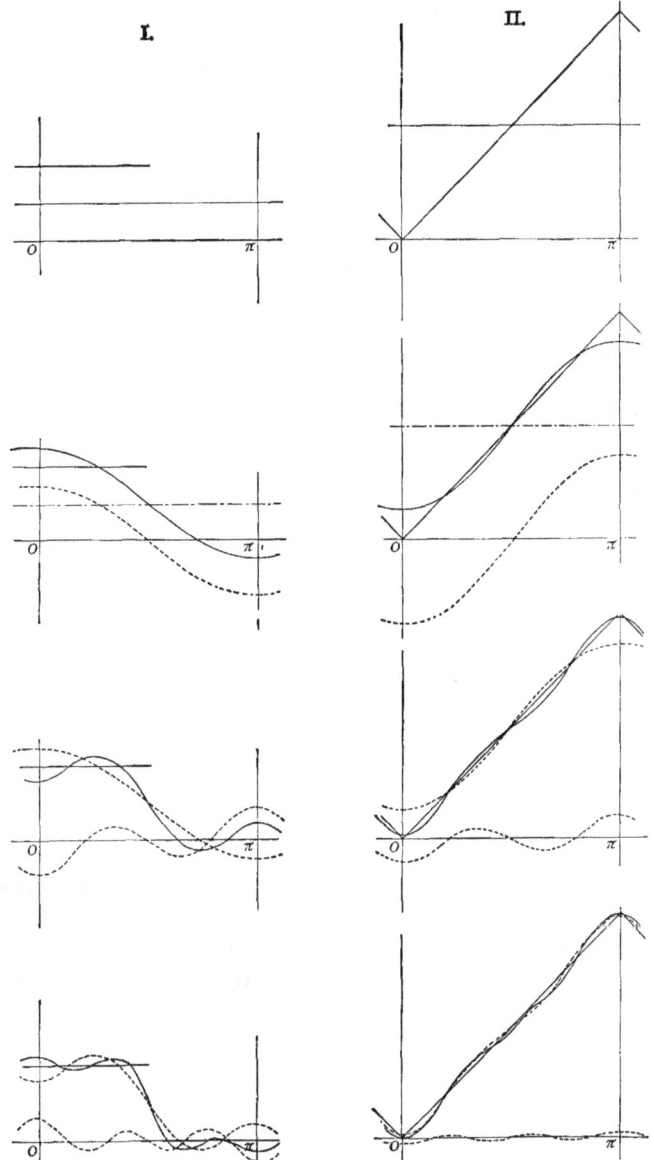

Recipes (1) and (2) empower us to take care of issues in fascination when we know the thickness of the drawing in mass, and issues in Statical Power at the point when we know the circulation of the charge, by strategies closely resembling that of Art. 94.

For instance let us track down the worth of the likely capability due to a slender material round shell of thickness ρ and span a.

Since V should be an answer of Laplace's Situation and should be limited both when $r = 0$ and $r = \infty$ we have

$$V_1 = \sum A_m r^m P_m(\cos\theta)$$
$$V_2 = \sum B_m \frac{1}{r^{m+1}} P_m(\cos\theta).$$

ZONAL HARMONICS.

V_1 and V_2 should move toward similar restricting qualities as r approaches a. Consequently

$$\frac{B_m}{a^{m+1}} = A_m a^m$$

or $B_m = A_m a^{2m+1}$.

$$D_n V_1 = D_r V_1 = \sum m r^{m-1} A_m P_m(\cos\theta),$$

$$D_n V_2 = D_r V_2 = -\sum (m+1)\frac{A_m a^{2m+1}}{r^{m+2}} P_m(\cos\theta).$$

Hence by (1)

$$4\pi\rho\kappa = \sum (2m+1) A_m a^{m-1} P_m(\cos\theta)$$

on the off chance that κ is the thickness of the shell.

Let $\rho = f(\cos\theta) = \sum C_m P_m(\cos\theta)$

where $C_m = \dfrac{2m+1}{2} \displaystyle\int_{-1}^{1} f(x) P_m(x) dx$ by Art. 90 (2).

Then $4\pi\kappa C_m = (2m+1) A_m a^{m-1}$, and

$$A_m = \frac{4\pi\kappa C_m}{(2m+1)a^{m-1}}, \quad \text{and} \quad B_m = \frac{4\pi\kappa}{2m+1} C_m a^{m+2},$$

and $V_1 = 4\pi a\kappa \sum \dfrac{C_m}{2m+1}\dfrac{r^m}{a^m} P_m(\cos\theta),$ and (3)

and $V_2 = 4\pi a\kappa \sum \dfrac{C_m}{2m+1}\dfrac{a^{m+1}}{r^{m+1}} P_m(\cos\theta).$ and (4)

99. We can now get the worth of the expected capability due to a circular shell of limited thickness, given that its thickness can be communicated as an amount of terms of the structure $Cr^k P_m(\cos\theta)$.

Let a be the sweep of the external surface and b be the span of the internal surface of the shell.

first.- - - Let $\rho = Cr^k P_m(\cos\theta)$. Then, at that point, for the shell of span s and thickness ds

$$V_1 = 4\pi s ds \frac{Cs^k}{2m+1}\frac{r^m}{s^m} P_m(\cos\theta) \quad \text{by (3) Art. 98,}$$

and $V_2 = 4\pi s ds \dfrac{Cs^k}{2m+1}\dfrac{s^{m+1}}{r^{m+1}} P_m(\cos\theta)$ by (4) Art. 98.

Then, at that point, if $r < b$

$$V = \int_b^a V_1 = \frac{4\pi C}{(2m+1)}\frac{(a^{k-m+2} - b^{k-m+2})}{(k-m+2)} r^m P_m(\cos\theta), \tag{1}$$

on the off chance that $r > a$

$$V = \int_b^a V_2 = \frac{4\pi C}{(2m+1)}\frac{(a^{k+m+3} - b^{k+m+3})}{(k+m+3)}\frac{P_m(\cos\theta)}{r^{m+1}}, \tag{2}$$

what's more, if $b < r < a$

$$V = \int_b^r V_2 + \int_r^a V_1 = \frac{4\pi C}{2m+1}\left[\frac{r^{k+m+3} - b^{k+m+3}}{(k+m+3)r^{m+1}} + \frac{a^{k-m+2} - r^{k-m+2}}{(k-m+2)} r^m\right] P_m(\cos\theta). \tag{3}$$

2d.- - - If $\rho = \sum C_m r^k P_m(\cos\theta)$ the arrangements will comprise of amounts of terms of the structures surrendered (1), (2), and (3).

ZONAL HARMONICS.

EXAMPLES.

1. In the event that the shell is homogeneous

$$V = 2\pi\rho(a^2 - b^2) \quad \text{if} \quad r < b,$$
$$V = \frac{4}{3}\pi\rho(a^3 - b^3)\frac{1}{r} = \frac{M}{r} \quad \text{if} \quad r > a,$$
$$V = 2\pi\rho\left[a^2 - \frac{2b^3}{3r} - \frac{r^2}{3}\right] \quad \text{if} \quad b < r < a.$$

2. In the event that the thickness is any given capability of the separation from the middle $V = \frac{M}{r}$ if $r > a$, and $V = $ a consistent if $r < b$.

3. On the off chance that the thickness anytime of a strong circle is corresponding to the square of the separation from a diametral plane

$$V = \frac{M}{a}\left[\frac{a}{r} + \frac{2}{7}\frac{a^3}{r^3}P_2(\cos\theta)\right] \quad \text{if} \quad r > a.$$

4. In the event that the thickness anytime of a strong circle is corresponding to its distance from a diametral plane

$$V = \frac{M}{a}\left[\frac{a}{r} + \frac{1}{6}\frac{a^3}{r^3}P_2(\cos\theta) - \frac{1.1}{6.8}\frac{a^5}{r^5}P_4(\cos\theta) + \frac{1.1.3}{6.8.10}\frac{a^7}{r^7}P_6(\cos\theta) - \cdots\right]$$

if $r > a$. Think about Ex. 2 Art. 80.

100. We have seen in Art. 18 (c) (3) that

$$Q_m(x) = CP_m(x)\int \frac{dx}{(1-x^2)[P_m(x)]^2}, \tag{1}$$

no consistent term being perceived with $\int \frac{dx}{(1-x^2)[P_m(x)]^2}$. $\frac{1}{(1-x^2)[P_m(x)]^2}$ is a reasonable part and becomes limitless just for $x = 1$, $x = -1$, and for the underlying foundations of $P_m(x) = 0$, which are all genuine and lie between -1 and 1, as can be demonstrated by the guide of the connection $P_m(x) = \frac{1}{2^m m!}\frac{d^m(x^2-1)^m}{dx^m}$.

In the event that $x^2 > 1$ $\int_x^\infty \frac{dx}{(1-x^2)[P_m(x)]^2}$ is limited and determinate and contains no steady term. Subsequently if $x^2 > 1$

$$Q_m(x) = -P_m(x)\int_x^\infty \frac{dx}{(1-x^2)[P_m(x)]^2} = P_m(x)\int_x^\infty \frac{dx}{(x^2-1)[P_m(x)]^2} \tag{2}$$

for the consistent variable of $Q_m(x)$ has been picked with the goal that $C = -1$.

If $x^2 < 1$ the second individual from (2) isn't limited and determinate, and we are tossed back to the structure (1), and C ends up being solidarity.

(1) gives us promptly

$$Q_0(x) = \frac{1}{2}\log\frac{1+x}{1-x} \tag{3}$$

$$Q_1(x) = -1 + \frac{x}{2}\log\frac{1+x}{1-x} \tag{4}$$

ZONAL HARMONICS.

if $x^2 < 1$.

(2) gives us
$$Q_0(x) = \frac{1}{2} \log \frac{x+1}{x-1} \tag{5}$$
$$Q_1(x) = -1 + \frac{x}{2} \log \frac{x+1}{x-1} \tag{6}$$

in the event that $x^2 > 1$.

From Art. 85 (10) that's what it follows

$$Q_m(x) = C \frac{d^m}{dx^m}\left[(x^2-1)^m \int_0^x \frac{dx}{(x^2-1)^{m+1}}\right] \quad \text{if} \quad x^2 < 1,$$

$$= C \frac{d^m}{dx^m}\left[(x^2-1)^m \int_x^\infty \frac{dx}{(x^2-1)^{m+1}}\right] \quad \text{if} \quad x^2 > 1.$$

C still up in the air and is equivalent to $\frac{(-1)^{m+1}2^m m!}{(2m)!}$ if $x^2 < 1$, and is equivalent to $\frac{(-1)^m 2^m m!}{(2m)!}$ if $x^2 > 1$.

Hence
$$Q_m(x) = \frac{(-1)^{m+1}2^m m!}{(2m)!} \frac{d^m}{dx^m}\left[(x^2-1)^m \int_0^x \frac{dx}{(x^2-1)^{m+1}}\right] \tag{7}$$

if $x^2 < 1$,

and
$$Q_m(x) = \frac{(-1)^m 2^m m!}{(2m)!} \frac{d^m}{dx^m}\left[(x^2-1)^m \int_x^\infty \frac{dx}{(x^2-1)^{m+1}}\right] \tag{8}$$

in the event that $x^2 > 1$.

(7) and (8) give us for $Q_0(x)$ and $Q_1(x)$ the qualities previously written in (3), (4), (5), and (6).

By the rehashed use of the recipe

$$(m+1)Q_{m+1}(x) - (2m+1)xQ_m(x) + mQ_{m-1}(x) = 0, \tag{9}$$

which might be gotten for the situation where $x^2 < 1$ from Art. 16 (13) and (14), also, for the situation where $x^2 > 1$ from Art. 16 (9), any Surface Zonal Symphonious of the Subsequent Kind can be gotten from $Q_0(x)$ and $Q_1(x)$ as yielded (3), (4), (5), and (6).

Comparable to recipes for $p_m(x)$ and $q_m(x)$ can be gotten easily from Art. 16 (4) and (5). They are

$$(m+1)^2 q_{m+1}(x) and (2m+1)xp_m(x) - m^2 q_{m-1}(x) = 0 \tag{10}$$
and $\quad and p_{m+1}(x) + \qquad (2m+1)xq_m(x) - p_{m-1}(x) = 0 \tag{11}$

furthermore, they hold really great for any worth of m.

EXAMPLES.

1. Affirm the upsides of $Q_0(x)$ and $Q_1(x)$ given in Art. 100 (3), (4), (5), and (6) by extending them and contrasting them and Art. 16 (13), (14), and (9).

2. On the off chance that the worth of V on the outer layer of a cone of upheaval can be communicated as far as entire powers positive or negative of r, V can be found for any point in space, cf. Art. 81.

In the event that $V = \sum \left(A_m r^m + \frac{B_m}{r^{m+1}}\right)$ when $\theta = \alpha$,

$$V = \sum \left(A_m r^m + \frac{B_m}{r^{m+1}}\right) \frac{P_m(\cos\theta)}{P_m(\cos\alpha)}.$$

ZONAL HARMONICS. 156

3. If $V = \sum \left(A_m r^m + \dfrac{B_m}{r^{m+1}} \right)$ when $\theta = \alpha$, and $V = 0$ when $\theta = \beta$,

$$V = \sum \left(A_m r^m + \dfrac{B_m}{r^{m+1}} \right) \left[\dfrac{Q_m(\cos\beta)P_m(\cos\theta) - P_m(\cos\beta)Q_m(\cos\theta)}{P_m(\cos\alpha)Q_m(\cos\beta) - P_m(\cos\beta)Q_m(\cos\alpha)} \right].$$

4. Find V for focuses relating to upsides of θ among α and β when V can be given as far as entire powers of r for $\theta = \alpha$ and for $\theta = \beta$.

5. Track down by the strategy for Art. 16 arrangements of Legendre's Situation of the structure

$$z = {}_1P_m(x) = 1 + \dfrac{m(m+1)}{2}(x-1) + \dfrac{(m-1)m(m+1)(m+2)}{2^2(2!)^2}(x-1)^2$$
$$+ \dfrac{(m-2)(m-1)m(m+1)(m+2)(m+3)}{2^3(3!)^2}(x-1)^3 + \cdots,$$

$$z = {}_{-1}P_m(x) = 1 - \dfrac{m(m+1)}{2}(x+1) + \dfrac{(m-1)m(m+1)(m+2)}{2^2(2!)^2}(x+1)^2$$
$$+ \dfrac{(m-2)(m-1)m(m+1)(m+2)(m+3)}{2^3(3!)^2}(x+1)^3 + \cdots.$$

In the event that m is an entire number, ${}_1P_m(x) = P_m(x)$ and ${}_{-1}P_m(x) = (-1)^m P_m(x)$. No matter what the worth of m, ${}_1P_m(x)$ is totally joined for $-1 < x < 3$, furthermore, ${}_{-1}P_m(x)$ is totally focalized for $-3 < x < 1$.

6. By the guide of (7) that's what art. 16 show

$$V = \tfrac{1}{\sqrt{r}} \sin(n \log r) k_n(\cos\theta),\, and\, V = \tfrac{1}{\sqrt{r}} \sin(n \log r) l_n(\cos\theta),$$
$$V = \tfrac{1}{\sqrt{r}} \cos(n \log r) k_n(\cos\theta),\, and\, V = \tfrac{1}{\sqrt{r}} \cos(n \log r) l_n(\cos\theta),$$

are arrangements of Laplace's Situation

$$rD_r^2(rV) + \dfrac{1}{\sin\theta} D_\theta(\sin\theta D_\theta V) = 0, \qquad \text{if}$$

$$k_n(x) = p_{-\frac{1}{2}+ni}(x) = 1 + \dfrac{n^2 + \left(\tfrac{1}{2}\right)^2}{2!} x^2 + \dfrac{\left[n^2 + \left(\tfrac{1}{2}\right)^2\right]\left[n^2 + \left(\tfrac{5}{2}\right)^2\right]}{4!} x^4$$
$$+ \dfrac{\left[n^2 + \left(\tfrac{1}{2}\right)^2\right]\left[n^2 + \left(\tfrac{5}{2}\right)^2\right]\left[n^2 + \left(\tfrac{9}{2}\right)^2\right]}{6!} x^6 + \cdots$$

and

$$l_n(x) = -q_{-\frac{1}{2}+ni}(x) = x + \dfrac{n^2 + \left(\tfrac{3}{2}\right)^2}{3!} x^3 + \dfrac{\left[n^2 + \left(\tfrac{3}{2}\right)^2\right]\left[n^2 + \left(\tfrac{7}{2}\right)^2\right]}{5!} x^5$$
$$+ \dfrac{\left[n^2 + \left(\tfrac{3}{2}\right)^2\right]\left[n^2 + \left(\tfrac{7}{2}\right)^2\right]\left[n^2 + \left(\tfrac{11}{2}\right)^2\right]}{7!} x^7 + \cdots$$

$k_n(x)$ and $l_n(x)$ are concurrent if $x^2 < 1$, yet are different if $x^2 = 1$.

7. Show by the guide of Model 5 that

$$V = \tfrac{1}{\sqrt{r}} \sin(n \log r) K_n(\cos\theta),\, and\, V = \tfrac{1}{\sqrt{r}} \sin(n \log r) K_n(-\cos\theta),$$
$$V = \tfrac{1}{\sqrt{r}} \cos(n \log r) K_n(\cos\theta),\, and\, V = \tfrac{1}{\sqrt{r}} \cos(n \log r) K_n(-\cos\theta),$$

ZONAL HARMONICS.

are arrangements of
$$rD_r^2(rV) + \frac{1}{\sin\theta}D_\theta(\sin\theta D_\theta V) = 0$$

if
$$K_n(x) =\, _1P_{-\frac{1}{2}+ni}(x) = 1 - \frac{n^2 + \left(\frac{1}{2}\right)^2}{2}(x-1)$$
$$+ \frac{\left[n^2+\left(\frac{1}{2}\right)^2\right]\left[n^2+\left(\frac{3}{2}\right)^2\right]}{2^2(2!)^2}(x-1)^2$$
$$- \frac{\left[n^2+\left(\frac{1}{2}\right)^2\right]\left[n^2+\left(\frac{3}{2}\right)^2\right]\left[n^2+\left(\frac{5}{2}\right)^2\right]}{2^2(3!)^2}(x-1)^3 + \cdots$$

and
$$K_n(-x) =\, _{-1}P_{-\frac{1}{2}+ni}(x) = 1 + \frac{n^2 + \left(\frac{1}{2}\right)^2}{2}(x+1)$$
$$+ \frac{\left[n^2+\left(\frac{1}{2}\right)^2\right]\left[n^2+\left(\frac{3}{2}\right)^2\right]}{2^2(2!)^2}(x+1)^2$$
$$+ \frac{\left[n^2+\left(\frac{1}{2}\right)^2\right]\left[n^2+\left(\frac{3}{2}\right)^2\right]\left[n^2+\left(\frac{5}{2}\right)^2\right]}{2^3(3!)^2}(x+1)^3 + \cdots.$$

$K_n(\cos\theta)$ is joined with the exception of $\theta = \pi$, and $K_n(-\cos\theta)$ is merged aside from $\theta = 0$.

$k_n(x), l_n(x), K_n(x),$ and $K_n(-x)$ are here and there called *Conal Harmonics*. They are specific upsides of z which fulfill Legendre's Condition written in the structure

$$(1-x^2)\frac{d^2z}{dx^2} - 2x\frac{dz}{dx} - \left(n^2 + \frac{1}{4}\right)z = 0.$$

For an intricate treatment of them see E. W. Hobson on "A Class of Circular Music of Mind boggling Degree." Trans. Camb. Phil. Soc., Vol. XIV.

8. If $V = f(r)$ when $\theta = \beta$,

$$V = \frac{1}{\pi\sqrt{r}} \int_{-\infty}^{\infty} d\lambda \int_0^\infty e^{\frac{\lambda}{2}} f(e^\lambda) \frac{K_\alpha(\cos\theta)}{K_\alpha(\cos\beta)} \cos[\alpha(\lambda - \log r)]d\alpha; \quad \text{if} \quad \theta < \beta.$$

9. If $V = f(r)$ when $\theta = \beta$ and $r < a$, and $V = 0$ when $r = a$,

$$V = \frac{2}{\pi}\sqrt{\frac{a}{r}} \int_{-\infty}^0 d\lambda \int_0^\infty e^{\frac{\lambda}{2}} f(ae^\lambda) \frac{K_\alpha(\cos\theta)}{K_\alpha(\cos\beta)} \sin\alpha\lambda \sin\left(\alpha\log\frac{r}{a}\right) d\alpha; \quad \text{if} \quad \theta < \beta.$$

10. If $V = f(r)$ when $\theta = \beta$ and $a < r < b$, and $V = 0$ when $r = a$ also, when $r = b$,

$$V = \sum_{m=1}^{m=\infty} A_m \frac{K_{m'}(\cos\theta)}{K_{m'}(\cos\beta)} \sin\left[\frac{m\pi(\log r - \log a)}{\log b - \log a}\right]$$

where and $m' = \dfrac{m\pi}{\log b - \log a}$ and

$$A_m = \frac{2}{\log b - \log a}\sqrt{\frac{a}{r}} \int_0^{\log\frac{b}{a}} e^{\frac{x}{2}} f(ae^x) \sin\frac{m\pi x}{\log b - \log a} dx; \quad \text{if} \quad \theta < \beta.$$

ZONAL HARMONICS.

11. If $\theta > \beta$ $\cos\theta$ should be supplanted by $(-\cos\theta)$ in models 8, 9, and 10.

12. If $V = f(r)$ when $\theta = \beta$, and $V = 0$ when $\theta = \gamma$,

$$V = \frac{1}{\pi\sqrt{r}} \int_{-\infty}^{\infty} d\lambda \int_0^{\infty} e^{\frac{\lambda}{2}} f(e^\lambda) \frac{k_\alpha(\cos\theta)l_\alpha(\cos\gamma) - k_\alpha(\cos\gamma)l_\alpha(\cos\theta)}{k_\alpha(\cos\beta)l_\alpha(\cos\gamma) - k_\alpha(\cos\gamma)l_\alpha(\cos\beta)} \cos[\alpha(\lambda - \log r)] d\alpha;$$

in the event that $\beta < \theta < \gamma$.

13. If $V = f(r)$ when $\theta = \beta$ and $a < r < b$, $V = 0$ when $\theta = \gamma$ and $a < r < b$, and $V = 0$ when $r = a$ and when $r = b$,

$$V = \sum_{m=1}^{m=\infty} A_m \frac{k_{m'}(\cos\theta)l_{m'}(\cos\gamma) - k_{m'}(\cos\gamma)l_{m'}(\cos\theta)}{k_{m'}(\cos\beta)l_{m'}(\cos\gamma) - k_{m'}(\cos\gamma)l_{m'}(\cos\beta)} \sin\frac{m\pi(\log r - \log a)}{\log b - \log a},$$

where
$$m' = \frac{m\pi}{\log b - \log a}$$
and

$$A_m = \frac{2}{\log b - \log a} \sqrt{\frac{a}{r}} \int_0^{\log\frac{b}{a}} e^{\frac{x}{2}} f(ae^x) \sin\frac{m\pi x}{\log b - \log a} dx;$$

if $\beta < \theta < \gamma$ and $a < r < b$.

14. If $V = f(r)$ when $\theta = \beta$ and $a < r < b$, and $V = 0$ when $r = a$ also, $D_r V + hV = 0$ when $r = b$,

$$V = \sum_{m=1}^{m=\infty} A_m \frac{K_{\alpha_m}(\cos\theta)}{K_{\alpha_m}(\cos\beta)} \sin\left(\alpha_m \log\frac{r}{a}\right),$$
where

$$A_m = \frac{2(\alpha_m^2 + h^2 b^2)}{\alpha_m^2(\log b - \log a) + hb[hb(\log b - \log a) + 1]} \int_0^{\log\frac{b}{a}} e^{\frac{x}{2}} f(ae^x) \sin\alpha_m x . dx$$

what's more, α_m is a base of the situation

$$\alpha\cos\left(\alpha\log\frac{b}{a}\right) + hb\sin\left(\alpha\log\frac{b}{a}\right) = 0 \qquad\qquad \text{v. Art. 68 Ex. 5.}$$

Chapter 6

SPHERICAL HARMONICS

101. When we are managing issues in finding the *potential function* because of powers which have not roundabout symmetry[1] about a hub and are utilizing Round CoÃ¶rdinates, we need to address Laplace's Condition in the structure

$$rD_r^2(rV) + \frac{1}{\sin\theta}D_\theta(\sin\theta D_\theta V) + \frac{1}{\sin^2\theta}D_\phi^2 V = 0 \qquad (1)$$

[v. (XIII) Art. 1].

To get a specific arrangement of (1) we will expect as expected that V is a result of capabilities every one of which includes yet a solitary variable.

Let $V = R.\Theta.\Phi$; where R includes r just, Θ includes θ just, and Φ ϕ as it were. Substitute in (1) and we get

$$\frac{r}{R}\frac{d^2(rR)}{dr^2} + \frac{1}{\Theta\sin\theta}\frac{d\left(\sin\theta\frac{d\Theta}{d\theta}\right)}{d\theta} + \frac{1}{\Phi\sin^2\theta}\frac{d^2\Phi}{d\phi^2} = 0 \qquad (2)$$

or

$$\frac{r\sin^2\theta}{R}\frac{d^2(rR)}{dr^2} + \frac{\sin\theta}{\Theta}\frac{d\left(\sin\theta\frac{d\Theta}{d\theta}\right)}{d\theta} = -\frac{1}{\Phi}\frac{d^2\Phi}{d\phi^2}.$$

As the primary part doesn't contain ϕ the subsequent part can't contain ϕ, and as it contains no other variable it should be consistent; call it n^2. Condition (2) is then comparable to the two conditions

$$\frac{d^2\Phi}{d\phi^2} + n^2\Phi = 0 \qquad (3)$$

and

$$\frac{r}{R}\frac{d^2(rR)}{dr^2} + \frac{1}{\Theta\sin\theta}\frac{d\left[\sin\theta\frac{d\Theta}{d\theta}\right]}{d\theta} - \frac{n^2}{\sin^2\theta} = 0 \qquad (4)$$

(3) has been addressed previously and gives us

$$\Phi = A\cos n\phi + B\sin n\phi \qquad (5)$$

[v. Art. 13(*a*)].

The initial term of (4) doesn't include θ and the second and third terms do not include r.

---

[1] See note, page 12.

SPHERICAL HARMONICS.   160

$\frac{r}{R}\frac{d^2(rR)}{dr^2}$ must, then, at that point, be a consistent; we will call it $m(m+1)$ as in Art. 13(c). Then (4) separates into

$$r\frac{d^2(rR)}{dr^2} = m(m+1)R \tag{6}$$

and
$$\frac{1}{\sin\theta}\frac{d\left[\sin\theta\frac{d\Theta}{d\theta}\right]}{d\theta} + \left[m(m+1) - \frac{n^2}{\sin^2\theta}\right]\Theta = 0. \tag{7}$$

(6) was addressed in Art. 13(c) and gives

$$R = A_1 r^m + B_1 r^{-m-1}. \tag{8}$$

If in (7) we supplant $\cos\theta$ by μ we get

$$\frac{d}{d\mu}\left[(1-\mu^2)\frac{d\Theta}{d\mu}\right] + \left[m(m+1) - \frac{n^2}{1-\mu^2}\right]\Theta = 0, \tag{9}$$

what might be compared to

$$(1-x^2)\frac{d^2z}{dx^2} - 2x\frac{dz}{dx} + \left[m(m+1) - \frac{n^2}{1-x^2}\right]z = 0, \tag{10}$$

[v. (17) Art. 85], which was addressed in Art. 85 for the situation where m and n are positive numbers and $n < m+1$. v. (18) and (19) Art. 85.

From (19) Art. 85 we get as a specific arrangement of (9)

$$\Theta = (1-\mu^2)^{\frac{n}{2}}\frac{d^n P_m(\mu)}{d\mu^n} = \sin^n\theta\frac{d^n P_m(\mu)}{d\mu^n}, \tag{11}$$

in the event that we confine ourselves to entire positive upsides of m and n, as we will do from now on except if the opposite is unequivocally expressed, and assume m not less than n.

A second however less helpful specific arrangement of (9) is

$$\Theta = (1-\mu^2)^{\frac{n}{2}}\frac{d^n Q_m(\mu)}{d\mu^n}.$$

Joining our outcomes we have as significant specific arrangements of (1)

$$V = r^m(A\cos n\phi + B\sin n\phi)\sin^n\theta\frac{d^n P_m(\mu)}{d\mu^n}, \tag{12}$$

and
$$V = \frac{1}{r^{m+1}}(A\cos n\phi + B\sin n\phi)\sin^n\theta\frac{d^n P_m(\mu)}{d\mu^n}, \tag{13}$$

where m and n are positive numbers and $n < m+1$.

102. $\sin^n\theta\frac{d^n P_m(\mu)}{d\mu^n}$ or $(1-\mu^2)^{\frac{n}{2}}\frac{d^n P_m(\mu)}{d\mu^n}$ is another capability of μ, that is of $\cos\theta$, and we will address it by $P_m^n(\mu)^2$ and will consider it a *associated function* of the nth request and mth degree. It is a worth of Θ fulfilling condition (9) Art. 101.

By separating the worth of $P_m(x)$ yielded (9) Art. 74 we get the recipe

$$P_m^n(\mu) = \frac{(2m)!\sin^n\theta}{2^m m!(m-n)!}\left[\mu^{m-n} - \frac{(m-n)(m-n-1)}{2.(2m-1)}\mu^{m-n-2}\right.$$
$$\left. + \frac{(m-n)(m-n-1)(m-n-2)(m-n-3)}{2.4.(2m-1)(2m-3)}\mu^{m-n-4} - \cdots\right] \tag{1}$$

---
[2]Most of the English journalists address this capability by $T_m^n(\mu)$.

SPHERICAL HARMONICS.

the articulation in the enclosure finishing with the term including μ^0 if $m - n$ is even and with the term including μ if $m - n$ is odd.

For comfort of reference we give on the following page a table from which $P_m^n(\mu)$ can be promptly acquired for upsides of m and n from 1 to 8.

$\cos n\phi P_m^n(\mu)$ and $\sin n\phi P_m^n(\mu)$, that is to say,

$$\cos n\phi \sin^n \theta \frac{d^n P_m(\mu)}{d\mu^n} \quad \text{and} \quad \sin n\phi \sin^n \theta \frac{d^n P_m(\mu)}{d\mu^n}$$

are called *Tesseral Harmonics* of the mth degree and nth request, and are upsides of V which fulfill the condition

$$m(m+1)V + \frac{1}{\sin\theta}D_\theta(\sin\theta D_\theta V) + \frac{1}{\sin^2\theta}D_\phi^2 V = 0 \tag{2}$$

or its equivalent

$$m(m+1)V + D_\mu[(1-\mu^2)D_\mu V] + \frac{1}{1-\mu^2}D_\phi^2 V = 0. \tag{3}$$

There are clearly $2m+1$ Tesseral Sounds of the mth degree, specifically

$$P_m(\mu), \qquad \cos\phi\sin\theta\frac{dP_m(\mu)}{d\mu}, \qquad \text{and} \sin\phi\sin\theta\frac{dP_m(\mu)}{d\mu}$$

$$\cos 2\phi\sin^2\theta\frac{d^2 P_m(\mu)}{d\mu^2}, \qquad \text{and} \sin 2\phi\sin^2\theta\frac{d^2 P_m(\mu)}{d\mu^2}$$

$$\cos 3\phi\sin^3\theta\frac{d^3 P_m(\mu)}{d\mu^3}, \qquad \text{and} \sin 3\phi\sin^3\theta\frac{d^3 P_m(\mu)}{d\mu^3}$$

$$\cdot \quad \cdot \quad \cdot \quad \cdot \quad \cdot \quad \cdot \quad \cdot \quad \cdot \quad \cdot \quad \cdot$$

$$\cos m\phi\sin^m\theta\frac{d^m P_m(\mu)}{d\mu^m}, \qquad \text{and} \sin m\phi\sin^m\theta\frac{d^m P_m(\mu)}{d\mu^m}$$

If each of these is increased by a steady and their total taken, this aggregate is known as a *Surface Circular Harmonic* of the mth degree, and is an answer of conditions (2) and (3). We will address it by $Y_m(\mu, \phi)$ or by $Y_m(\theta, \phi)$.

SPHERICAL HARMONICS.   162

$$\text{Table for } \csc^n \theta P_m^n(\mu) = \frac{d^n P_m(\mu)}{d\mu^n}.$$

m and $n = 1$.	$n = 2$.
1 and 1	
2 and 3μ	3
3 and $\frac{3}{2}(5\mu^2 - 1)$	15μ
4 and $\frac{5}{2}(7\mu^3 - 3\mu)$	$\frac{15}{2}(7\mu^2 - 1)$
5 and $\frac{15}{8}(21\mu^4 - 14\mu^2 + 1)$	$\frac{105}{2}(3\mu^3 - \mu)$
6 and $\frac{21}{8}(33\mu^5 - 30\mu^3 + 5\mu)$	$\frac{105}{8}(33\mu^4 - 18\mu^2 + 1)$
7 and $\frac{7}{16}(429\mu^6 - 495\mu^4 + 135\mu^2 - 5)$	$\frac{63}{8}(143\mu^5 - 110\mu^3 + 15\mu)$
8 and $\frac{9}{16}(715\mu^7 - 1001\mu^5 + 385\mu^3 - 35\mu)$	$\frac{315}{16}(143\mu^6 - 143\mu^4 + 33\mu^2 - 1)$

m and $n = 3$.	$n = 4$.
1 and	
2 and	
3 and 15	
4 and 105μ	105
5 and $\frac{105}{2}(9\mu^2 - 1)$	945μ
6 and $\frac{315}{2}(11\mu^3 - 3\mu)$	$\frac{945}{2}(11\mu^2 - 1)$
7 and $\frac{315}{8}(143\mu^4 - 66\mu^2 + 3)$	$\frac{3465}{2}(13\mu^3 - 3\mu)$
8 and $\frac{3465}{8}(39\mu^5 - 26\mu^3 + 3\mu)$	$\frac{10395}{8}(65\mu^4 - 26\mu^2 + 1)$

m	$n = 5$.	$n = 6$.	$n = 7$.	$n = 8$.
1				
2				
3				
4				
5	945			
6	10395μ	10395		
7	$\frac{10395}{2}(13\mu^2 - 1)$	135135μ	135135	
8	$\frac{135135}{2}(5\mu^3 - \mu)$	$\frac{135135}{2}(15\mu^2 - 1)$	2027025μ	2027025

$r^m Y_m(\mu, \phi)$ and $\frac{1}{r^{m+1}} Y_m(\mu, \phi)$ are called *Solid Round Harmonics* of the mth degree, and are arrangements of Laplace's Situation (1) Art. 101.

To form:- - -

$$Y_m(\mu, \phi) = \sum_{n=0}^{n=m} \left[A_n \cos n\phi \sin^n \theta \frac{d^n P_m(\mu)}{d\mu^n} + B_n \sin n\phi \sin^n \theta \frac{d^n P_m(\mu)}{d\mu^n} \right] \quad (4)$$

or
$$Y_m(\mu, \phi) = A_0 P_m(\mu) + \sum_{n=1}^{n=m} [A_n \cos n\phi P_m^n(\mu) + B_n \sin n\phi P_m^n(\mu)] \, and \quad (5)$$

is a Surface Round Symphonious of the mth degree.

A Tesseral Consonant is an exceptional instance of a Surface Circular Symphonious, and a Zonal Consonant an exceptional instance of a Tesseral Symphonious; $P_m(\mu)$ being the Tesseral Consonant of the zeroth request and the mth degree; it very well may be composed $P_m^0(\mu)$.

SPHERICAL HARMONICS.

EXAMPLES.

1. Show that
$$(1-x^2)\frac{d^2z}{dx^2} - 2x\frac{dz}{dx} + \left[m(m+1) - \frac{n^2}{1-x^2}\right]z = 0$$

diminishes to
$$(1-x^2)\frac{d^2y}{dx^2} - 2(n+1)x\frac{dy}{dx} + [m(m+1) - n(n+1)]y = 0$$

assuming we substitute $(1-x^2)^{\frac{n}{2}}y$ for z, in any event, when m and n are unrestricted.

2. Show that assuming in the second condition of Ex. 1 we let $y = \sum a_k x^k$ we get
$$a_{k+2} = -\frac{(m-n-k)(m+n+1+k)}{(k+1)(k+2)}a_k \qquad \text{(v. Art. 16)}$$

whence $z = p_m^n(x)$ and $z = q_m^n(x)$ are arrangements of the principal condition of Ex. 1, regardless of what the upsides of m and n, if

$$p_m^n(x) = (1-x^2)^{\frac{n}{2}}\left[1 - \frac{(m-n)(m+n+1)}{2!}x^2 \right.$$
$$\left. + \frac{(m-n)(m-n-2)(m+n+1)(m+n+2)}{4!}x^4 - \cdots\right]$$

furthermore,

$$q_m^n(x) = (1-x^2)^{\frac{n}{2}}\left[x - \frac{(m-n-1)(m+n+2)}{3!}x^3 \right.$$
$$\left. + \frac{(m-n-1)(m-n-3)(m+n+2)(m+n+4)}{5!}x^5 - \cdots\right].$$

In the event that $m - n$ is a positive number, $p_m^n(x)$ or $q_m^n(x)$ will end with the term including x^{m-n}, and all things considered

$$z = (1-x^2)^{\frac{n}{2}}\left[x^{m-n} - \frac{(m-n)(m-n-1)}{2.(2m-1)}x^{m-n-2} \right.$$
$$\left. + \frac{(m-n)(m-n-1)(m-n-2)(m-n-3)}{2.4.(2m-1)(2m-3)}x^{m-n-4} - \cdots\right].$$

the enclosure finishing with a term including x^0 if $m - n$ is even and x if $m - n$ is odd, is an answer of the primary condition of Ex. 1. If m and n are numbers this worth of z is $\frac{2^m m!(m-n)!}{(2m)!}P_m^n(x)$.

103. We have found in the last part that in numerous issues it is significant to have the option to communicate a given capability of $\cos\theta$, that is of μ, as far as Zonal Music of μ. So it is frequently attractive to communicate a given capability of μ and ϕ as far as Tesseral Music of μ and ϕ.

If, for instance, we are attempting to track down the *Potential Function* due to certain powers and have the worth of the capability given for some given worth of r, that is, on the outer layer of some given circle whose middle is at the beginning of coÃ¶rdinates, obviously the given worth will be a component of θ and ϕ and if we can communicate it as far as Round Music of θ and ϕ we have just to increase each term by the legitimate force of r to get the necessary arrangement of the issue. For we will then have a worth of V fulfilling Laplace's Condition and diminishing to the given capability of θ and ϕ on the outer layer of the given circle.

SPHERICAL HARMONICS.

104. Assume that we have a component of μ and ϕ given for all focuses on the unit circle, or at least, for all upsides of μ from -1 to 1 and for all upsides of ϕ from 0 to 2π, μ and ϕ being free factors, and that we wish to express it regarding Surface Round Sounds.

Expect that

$$f(\mu,\phi) = \sum_{m=0}^{m=\infty} \left[A_{0,m} P_m(\mu) + \sum_{n=1}^{n=m} \left(A_{n,m} \cos n\phi P_m^n(\mu) + B_{n,m} \sin n\phi P_m^n(\mu) \right) \right]. \qquad (1)$$

Allow us to look at first as a limited case, and endeavor to decide the coefficients so that

$$f(\mu,\phi) = \sum_{m=0}^{m=p} \left[A_{0,m} P_m(\mu) + \sum_{n=1}^{n=m} \left(A_{n,m} \cos n\phi P_m^n(\mu) + B_{n,m} \sin n\phi P_m^n(\mu) \right) \right] \qquad (2)$$

will hold great at whatever number places of the circle as could be expected under the circumstances. The articulation in sections in the second individual from (2) is a Surface Circular Symphonious of the mth degree and contains $2m + 1$ steady coefficients. The entire number of coefficients to be resolved is then the amount of an Arithmetical Movement of $p + 1$ terms the initial term of which is 1 and the latter is $2p + 1$, and is along these lines equivalent to $(p+1)^2$.

Let the span from $\mu = -1$ to $\mu = 1$ be separated into $p+2$ parts each of which is $\Delta\mu$ with the goal that $(p+2)\Delta\mu = 2$, and let the stretch from $\phi = 0$ to $\phi = 2\pi$ be separated into $p+2$ parts every one of which is $\Delta\phi$ with the goal that $(p+2)\Delta\phi = 2\pi$.

Then, at that point, assuming we substitute in condition (2) thusly the qualities $(-1+\Delta\mu, \Delta\phi)$, $(-1+2\Delta\mu, \Delta\phi), \cdots [-1+(p+1)\Delta\mu, \Delta\phi]; (-1+\Delta\mu, 2\Delta\phi), (-1+2\Delta\mu, 2\Delta\phi), \cdots [-1+(p+1)\Delta\mu, 2\Delta\phi]; \cdots [-1+\Delta\mu, (p+1)\Delta\phi], [-1+2\Delta\mu, (p+1)\Delta\phi], \cdots [-1+(p+1)\Delta\mu, (p+1)\Delta\phi]$; starting from the first part for each situation will be realized we will have $(p+1)^2$ conditions of the first degree containing no obscure with the exception of the $(p+1)^2$ coefficients, and from them the coefficients not set in stone. At the point when they are subbed in condition (2) it will hold great at the $(p+1)^2$ marks of the unit circle where $p+1$ circles of scope whose planes are equidistant cross $p+1$ meridians what partition the equator into equivalent curves. If presently p is endlessly expanded the restricting upsides of the coefficients will be the coefficients in condition (1), furthermore (1) will hold great all around the outer layer of the unit circle.

To decide a specific steady we duplicate each of our $(p+1)^2$ conditions by $\Delta\mu\Delta\phi$ times the coefficient of the consistent being referred to in that condition and add the conditions and afterward examine the restricting structure moved toward by the subsequent condition as p is endlessly expanded.

As p is endlessly expanded the summation being referred to will approach an incorporation; and since $d\mu d\phi = -\sin\theta.d\theta d\phi$ is the component of surface of the unit circle, and as the cutoff points -1 and 1 of μ compare to π and 0 of θ the incorporation is a *surface integration* over the outer layer of the unit circle.

In deciding any coefficient as $A_{n,m}$ in (1) the principal individual from the restricting type of our subsequent condition will be

$$\int_0^{2\pi} d\phi \int_{-1}^{1} f(\mu,\phi) \cos n\phi P_m^n(\mu) d\mu.$$

In the second part we will go over terms of the structures

$$\int_0^{2\pi} d\phi \int_{-1}^{1} \sin l\phi \cos n\phi P_m^l(\mu) P_m^n(\mu) d\mu, \qquad \int_0^{2\pi} d\phi \int_{-1}^{1} \cos l\phi \cos n\phi P_m^l(\mu) P_m^n(\mu) d\mu,$$

$$\int_0^{2\pi} d\phi \int_{-1}^{1} \sin n\phi \cos n\phi [P_m^n(\mu)]^2 d\mu, \qquad \int_0^{2\pi} d\phi \int_{-1}^{1} \cos^2 n\phi [P_m^n(\mu)]^2 d\mu,$$

SPHERICAL HARMONICS.

what's more, different terms all of which go under the structure

$$\int_0^{2\pi} d\phi \int_{-1}^1 Y_l(\mu,\phi) Y_m(\mu,\phi) d\mu,$$

where $Y_m(\mu,\phi)$ and $Y_l(\mu,\phi)$ are Surface Round Sounds of various degrees.

Assuming we are deciding a coefficient $B_{n,m}$ the main contrast is that $\sin n\phi$ also, $\cos n\phi$ will be exchanged in the structures recently determined.

105. *The vital over the outer layer of the unit circle of the result of two Surface Round Music of various degrees is zero.*

That is
$$\int_0^{2\pi} d\phi \int_{-1}^1 Y_l(\mu,\phi) Y_m(\mu,\phi) d\mu = 0. \tag{1}$$

For as we have seen $U = r^l Y_l(\mu,\phi)$ and $V = r^m Y_m(\mu,\phi)$ are arrangements of Laplace's Condition. Thus by *Green's Theorem*

$$\int (U D_n V - V D_n U) ds = 0 \qquad \text{v. Art. 92.}$$
$$D_n V = D_r V = m r^{m-1} Y_m(\mu,\phi),$$
$$D_n U = D_r U = l r^{l-1} Y_l(\mu,\phi);$$
$$U D_n V - V D_n U = (m-l) r^{l+m-1} Y_l(\mu,\phi) Y_m(\mu,\phi),$$
$$= (m-l) Y_l(\mu,\phi) Y_m(\mu,\phi)$$

on the outer layer of the unit circle; and

$$(m-l) \int Y_l(\mu,\phi) Y_m(\mu,\phi) ds = (m-l) \int_0^{2\pi} d\phi \int_1^1 Y_l(\mu,\phi) Y_m(\mu,\phi) d\mu = 0.$$

Consequently except if $l = m$

$$\int_0^{2\pi} d\phi \int_{-1}^1 Y_l(\mu,\phi) Y_m(\mu,\phi) d\mu = 0.$$

EXAMPLES.

1. Get (1) Art. 105 straightforwardly from the situation

$$m(m+1) Y_m(\mu,\phi) + D_\mu[(1-\mu^2) D_\mu Y_m(\mu,\phi)] + \frac{1}{1-\mu^2} D_\phi^2 Y_m(\mu,\phi) = 0$$

v. (3) Art. 102, and Art. 91.

2. Show that the basic over the outer layer of the unit circle of the item of two Tesseral Sounds of a similar degree yet of various orders is zero.

Suggestion:

$$\int_0^{2\pi} \sin k\phi \cos l\phi . d\phi = \int_0^{2\pi} \sin k\phi \sin l\phi . d\phi = \int_0^{2\pi} \cos k\phi \cos l\phi . d\phi = 0.$$

SPHERICAL HARMONICS.

106.
$$\int_{-1}^{1} P_l^n(\mu) P_m^n(\mu) d\mu = 0 \quad \text{unless} \quad l = m$$
$$= \frac{2}{2m+1} \frac{(m+n)!}{(m-n)!} \quad \text{if} \quad l = m.$$

For
$$\int_{-1}^{1} P_l^n(\mu) P_m^n(\mu) d\mu = \int_{-1}^{1} (1-\mu^2)^n \frac{d^n P_l(\mu)}{d\mu^n} \cdot \frac{d^n P_m(\mu)}{d\mu^n} d\mu$$
$$= (1-\mu^2)^n \frac{d^n P_m(\mu)}{d\mu^n} \cdot \frac{d^{n-1} P_l(\mu)}{d\mu^{n-1}} \bigg]_{-1}^{1}$$
$$- \int_{-1}^{1} \frac{d^{n-1} P_l(\mu)}{d\mu^{n-1}} \cdot \frac{d}{d\mu} \left[(1-\mu^2)^n \frac{d^n P_m(\mu)}{d\mu^n} \right] d\mu,$$
$$= - \int_{-1}^{1} \frac{d^{n-1} P_l(\mu)}{d\mu^{n-1}} \cdot \frac{d}{d\mu} \left[(1-\mu^2)^n \frac{d^n P_m(\mu)}{d\mu^n} \right] d\mu;$$

by *integration by parts*.

Supplanting n by $n-1$ in condition (2) Art. 84 and recalling that $\frac{d^{n-1} P_m(x)}{dx^{n-1}}$ is a potential worth of $z^{(n-1)}$ we get

$$(1-\mu^2) \frac{d^{n+1} P_m(\mu)}{d\mu^{n+1}} - 2n\mu \frac{d^n P_m(\mu)}{d\mu^n} + [m(m+1) - n(n-1)] \frac{d^{n-1} P_m(\mu)}{d\mu^{n-1}} = 0,$$

or then again assuming we duplicate by $(1-\mu^2)^{n-1}$

$$(1-\mu^2)^n \frac{d^{n+1} P_m(\mu)}{d\mu^{n+1}} - 2n\mu(1-\mu^2)^{n-1} \frac{d^n P_m(\mu)}{d\mu^n}$$
$$+ (m+n)(m-n+1)(1-\mu^2)^{n-1} \frac{d^{n-1} P_m(\mu)}{d\mu^{n-1}} = 0,$$

or on the other hand
$$\frac{d}{d\mu} \left[(1-\mu^2)^n \frac{d^n P_m(\mu)}{d\mu^n} \right] = -(m+n)(m-n+1)(1-\mu^2)^{n-1} \frac{d^{n-1} P_m(\mu)}{d\mu^{n-1}}.$$

Thus follows the *reduction formula*
$$\int_{-1}^{1} (1-\mu^2)^n \frac{d^n P_l(\mu)}{d\mu^n} \cdot \frac{d^n P_m(\mu)}{d\mu^n} d\mu$$
$$= (m+n)(m-n+1) \int_{-1}^{1} (1-\mu^2)^{n-1} \frac{d^{n-1} P_l(\mu)}{d\mu^{n-1}} \cdot \frac{d^{n-1} P_m(\mu)}{d\mu^{n-1}} d\mu.$$

Utilizing this recipe n times we get
$$\int_{-1}^{1} P_l^n(\mu) P_m^n(\mu) d\mu = \frac{(m+n)!}{(m-n)!} \int_{-1}^{1} P_l(\mu) P_m(\mu) d\mu$$
$$= 0 \quad \text{unless} \quad l = m$$
$$= \frac{2}{2m+1} \frac{(m+n)!}{(m-n)!} \quad \text{if} \quad l = m$$

v. Art. 89 (4) and (5).

SPHERICAL HARMONICS.

107. We are currently ready to finish the arrangement of the issue in Art. 104 furthermore, since $\int_0^{2\pi} \cos^2 n\phi\, d\phi = \int_0^{2\pi} \sin^2 n\phi\, d\phi = \pi$ and $\int_0^{2\pi} d\phi = 2\pi$ we get as the coefficients in (1) Art. 104

$$A_{0,m} = \frac{2m+1}{4\pi} \int_0^{2\pi} d\phi \int_{-1}^{1} f(\mu,\phi) P_m(\mu)\, d\mu, \tag{1}$$

$$A_{n,m} = \frac{2m+1}{2\pi} \cdot \frac{(m-n)!}{(m+n)!} \int_0^{2\pi} d\phi \int_{-1}^{1} f(\mu,\phi) \cos n\phi\, P_m^n(\mu)\, d\mu, \tag{2}$$

$$B_{n,m} = \frac{2m+1}{2\pi} \cdot \frac{(m-n)!}{(m+n)!} \int_0^{2\pi} d\phi \int_{-1}^{1} f(\mu,\phi) \sin n\phi\, P_m^n(\mu)\, d\mu, \tag{3}$$

whence

$$f(\mu,\phi) = \sum_{m=0}^{m=\infty} \left[A_{0,m} P_m(\mu) + \sum_{n=1}^{n=m} (A_{n,m} \cos n\phi + B_{n,m} \sin n\phi) P_m^n(\mu) \right] \tag{4}$$

also, the improvement holds great for all upsides of μ and ϕ comparing to focuses on the unit circle, gave just that the given capability fulfills the conditions that would need to be fulfilled if it somehow managed to be formed into a *Fourier's Series*.

On the off chance that we use μ_1 and ϕ_1 instead of μ and ϕ in (1), (2), and (3), we can compose (4) in the structure

$$f(\mu,\phi) = \frac{1}{2\pi} \sum_{m=0}^{m=\infty} (2m+1) \left[\frac{1}{2} \int_0^{2\pi} d\phi_1 \int_{-1}^{1} f(\mu_1,\phi_1) P_m(\mu) P_m(\mu_1)\, d\mu_1 \right.$$
$$\left. + \sum_{n=1}^{n=m} \frac{(m-n)!}{(m+n)!} \int_0^{2\pi} d\phi_1 \int_{-1}^{1} f(\mu_1,\phi_1) P_m^n(\mu) P_m^n(\mu_1) \cos n(\phi-\phi_1)\, d\mu_1 \right]. \tag{5}$$

Equations (1), (2), (3), and (4) are advantageous for genuine work; (5) is fairly all the more minimalistically composed.

108. As an illustration let us express $\sin^2\theta \cos^2\theta \sin\phi \cos\phi$ as far as Surface Round Sounds.

Here
$$f(\mu,\phi) = \frac{1}{2}\mu^2(1-\mu^2) \sin 2\phi.$$

$$A_{0,m} = \frac{2m+1}{8\pi} \int_{-1}^{1} \mu^2(1-\mu^2) P_m(\mu)\, d\mu \int_0^{2\pi} \sin 2\phi\, d\phi = 0,$$

$$A_{n,m} = \frac{2m+1}{4\pi} \cdot \frac{(m-n)!}{(m+n)!} \int_{-1}^{1} \mu^2(1-\mu^2) P_m^n(\mu)\, d\mu \int_0^{2\pi} \sin 2\phi \cos n\phi\, d\phi = 0,$$

$$B_{n,m} = \frac{2m+1}{4\pi} \cdot \frac{(m-n)!}{(m+n)!} \int_0^{1} \mu^2(1-\mu^2) P_m^n(\mu)\, d\mu \int_0^{2\pi} \sin 2\phi \sin n\phi\, d\phi,$$
$$= 0 \quad \text{unless} \quad n = 2.$$

If $n = 2$ and $\int_0^{2\pi} \sin 2\phi \sin n\phi\, d\phi = \int_0^{2\pi} \sin^2 2\phi\, d\phi = \pi$, and

SPHERICAL HARMONICS. 168

$$B_{2,m} = \frac{2m+1}{4} \cdot \frac{(m-2)!}{(m+2)!} \int_{-1}^{1} \mu^2 (1-\mu^2)^2 \frac{d^2 P_m(\mu)}{d\mu^2} d\mu$$

$$= \frac{1}{2^m m!} \frac{2m+1}{4} \frac{(m-2)!}{(m+2)!} \int_{-1}^{1} \mu^2 (1-\mu^2)^2 \frac{d^{m+2}(\mu^2-1)^m}{d\mu^{m+2}} d\mu.$$

$$\int_{-1}^{1} \mu^2 (1-\mu^2)^2 \frac{d^{m+2}(\mu^2-1)^m}{d\mu^{m+2}} d\mu = 720 \int_{-1}^{1} \frac{d^{m-4}(\mu^2-1)^m}{d\mu^{m-4}} d\mu$$

by rehashed incorporation by parts,

$$= 0 \quad \text{if} \quad m > 4,$$

$$= 720 \int_{-1}^{1} (\mu^2-1)^4 d\mu = \frac{4096}{7} \quad \text{if} \quad m = 4,$$

and
$$B_{2,4} = \frac{1}{2^4 4!} \cdot \frac{9}{4} \cdot \frac{2!}{6!} \cdot \frac{4096}{7} = \frac{1}{105}.$$

By a like interaction we find

$$B_{2,3} = 0 \quad \text{and} \quad B_{2,2} = \frac{1}{42}. \qquad \text{Hence}$$

$$\sin^2\theta \cos^2\theta \sin\phi\cos\phi = \frac{1}{42} P_2^2(\mu) \sin 2\phi + \frac{1}{105} P_4^2(\mu) \sin 2\phi, \qquad (1)$$

$$= \frac{1}{42} \sin 2\phi \sin^2\theta \frac{d^2 P_2(\mu)}{d\mu^2} + \frac{1}{105} \sin 2\phi \sin^2\theta \frac{d^2 P_4(\mu)}{d\mu^2}, \qquad (2)$$

$$= \frac{1}{14} \sin^2\theta \sin 2\phi + \frac{1}{14} \sin^2\theta (7\mu^2 - 1) \sin 2\phi. \qquad (3)$$

The expected articulation could have been gotten without utilizing the equations of Art. 107, by an extremely basic gadget, as follows:

$$\sin^2\theta \cos^2\theta \sin\phi\cos\phi = \frac{1}{2}\mu^2 \sin^2\theta \sin 2\phi. \qquad (4)$$

On the off chance that now we can communicate μ^2 in the structure $\sum \frac{d^2 P_m(\mu)}{d\mu^2}$ the work will be finished.

$$\mu^2 = \frac{1}{4.3} \frac{d^2(\mu^4)}{d\mu^2},$$

$$\mu^4 = \frac{8}{35} P_4(\mu) + \frac{4}{7} P_2(\mu) + \frac{1}{5} P_0(\mu), \qquad (5) \text{ Art. 95.}$$

$$\frac{d^2(\mu^4)}{d\mu^2} = \frac{8}{35} \frac{d^2 P_4(\mu)}{d\mu^2} + \frac{4}{7} \frac{d^2 P_2(\mu)}{d\mu^2};$$

whence
$$\mu^2 = \frac{2}{105} \frac{d^2 P_4(\mu)}{d\mu^2} + \frac{1}{21} \frac{d^2 P_2(\mu)}{d\mu^2},$$

what's more, subbing this worth in (4) we get (2).

SPHERICAL HARMONICS.

EXAMPLES.

1. Show that

$$\cos^3\theta \sin^3\theta \sin\phi \cos^2\phi = \left[\frac{1}{6930}P_6^3(\mu) + \frac{1}{1540}P_4^3(\mu)\right]\sin 3\phi$$
$$- \left[\frac{2}{693}P_6^1(\mu) - \frac{1}{770}P_4^1(\mu) - \frac{1}{63}P_2^1(\mu)\right]\sin\phi.$$

2. Show that

$$\cos 2\phi = 2\cos 2\phi \left[\frac{5}{4!}P_2^2(\mu) + \frac{9.2!}{6!}P_4^2(\mu) + \frac{13.4!}{8!}P_6^2(\mu) + \cdots\right].$$

3. If in an issue on the Potential Capability $V = f(\mu,\phi)$ when $r = a$, we will clearly have

$$V = \sum_{m=0}^{m=\infty} \frac{r^m}{a^m}\left[A_{0,m}P_m(\mu) + \sum_{n=1}^{n=m}(A_{n,m}\cos n\phi + B_{n,m}\sin n\phi)P_m^n(\mu)\right]$$

at an interior point and

$$V = \sum_{m=0}^{m=\infty} \frac{a^{m+1}}{r^{m+1}}\left[A_{0,m}P_m(\mu) + \sum_{n=1}^{n=m}(A_{n,m}\cos n\phi + B_{n,m}\sin n\phi)P_m^n(\mu)\right]$$

at an outside point, where $A_{0,m}$, $A_{n,m}$, and $B_{n,m}$ have the qualities surrendered (1), (2), and (3) Art. 107.

4. Take care of issues (3), (4), and (5) of Art. 94 for the situation where V isn't even as for a pivot.

109. Any Strong Circular Symphonious $r^m Y_m(\mu,\phi)$ being a worth of V that fulfills Laplace's Condition in Round CoÃ¶rdinates will change into a capability of x, y, and z fulfilling $\nabla^2 V = 0$ in the event that we change to a bunch of rectangular tomahawks having a similar beginning and a similar pivot of X as the polar framework. Additionally the new capability will be a homogeneous normal fundamental Logarithmic capability of x, y, z, of the mth degree.

For each term of $r^m \cos n\phi P_m^n(\mu)$ is of the structure

$$Cr^m \cos^{n-2k}\phi \sin^{2k}\phi \sin^n\theta \cos^{m-2l-n}\theta$$

where $\qquad 2k < n+1 \quad \text{and} \quad 2l < m-n+1.$

This might be written

$$Cr^{2l}.r^{m-2l-n}\cos^{m-2l-n}\theta.r^{n-2k}\sin^{n-2k}\theta\cos^{n-2k}\phi.r^{2k}\sin^{2k}\theta\sin^{2k}\phi$$

which becomes $\qquad C(x^2+y^2+z^2)^l x^{m-2l-n} y^{n-2k} z^{2k},$

furthermore, is a homogeneous sane necessary Mathematical capability of x, y, and z of the mth degree. Exactly the same thing might be displayed of each term of $r^m \sin n\phi P_m^n(\mu)$. Therefore $r^m Y_m(\mu,\phi)$ is a homogeneous sane essential Mathematical capability of the mth degree in x, y, and z.

110. Any homogeneous objective fundamental Arithmetical capability $S_m(x,y,z)$ of the mth degree in x, y, and z, which is a worth of V fulfilling $\nabla^2 V = 0$ contains $2m+1$ inconsistent steady coefficients.

SPHERICAL HARMONICS.

For $S_m(x,y,z)$ will in everyday comprise of $\frac{(m+1)(m+2)}{2}$ terms and will in this way contain $\frac{(m+1)(m+2)}{2}$ coefficients.

$\nabla^2 S_m(x,y,z)$ will be homogeneous of the $(m-2)$d degree and will contain $\frac{m(m-1)}{2}$ coefficients, which, obviously, will be elements of the coefficients in $S_m(x,y,z)$. Since $\nabla^2 S_m(x,y,z) = 0$ freely of the mathematical qualities of x, y, and z the $\frac{m(m-1)}{2}$ coefficients in $\nabla^2 S_m(x,y,z)$ should be independently zero, and that reality will give us $\frac{m(m-1)}{2}$ conditions of condition between the $\frac{(m+1)(m+2)}{2}$ unique coefficients and will leave $\frac{(m+1)(m+2)}{2} - \frac{m(m-1)}{2}$ or on the other hand $2m+1$ of them unsure. $S_m(x,y,z)$ contains, then, at that point, a similar number of erratic coefficients as $r^m Y_m(\mu, \phi)$.

We can then pick the coefficients in $r^m Y_m(\mu, \phi)$ so it will change into some random $S_m(x,y,z)$.

Thus a Strong Round Consonant of the mth degree may be characterized as *a homogeneous levelheaded necessary Arithmetical capability of x, y, and z, $S_m(x,y,z)$, of the mth degree fulfilling the equation $\nabla^2 S_m(x,y,z) = 0$*; and a Surface Circular Consonant of the mth degree as such a capability isolated by $(x^2+y^2+z^2)^{\frac{m}{2}}$, that is by r^m.

EXAMPLES.

1. Show that in the event that $S_m(x,y,z)$ is a Strong Round Symphonious of the mth degree

$$\nabla^2[r^n S_m(x,y,z)] = n(2m+n+1)r^{n-2}S_m(x,y,z).$$

Suggestion:

$$\nabla^2 S_m = 0. \quad \nabla^2 r = \frac{2}{r}. \quad D_r S_m = \frac{m S_m}{r}. \quad (D_x r)^2 + (D_y r)^2 + (D_z r)^2 = 1.$$

2. Show that assuming $f_n(x,y,z)$ is a normal basic homogeneous capability of x, y, and z of the nth degree it very well may be communicated in the structure

$$f_n(x,y,z) = S_n(x,y,z) + r^2 S_{n-2}(x,y,z) + r^4 S_{n-4}(x,y,z) + \cdots, \tag{1}$$

ending with $r^{n-1}S_1(x,y,z)$ assuming that n is odd, and with $r^n S_0(x,y,z)$ assuming n is even.

Suggestion: If a term rS_{n-1} were available in the second individual from (1), and we were to work with ∇^2 on the two individuals we ought to by Ex. 1 have a term $\frac{2n}{r}S_{n-1}$ which would be nonsensical when the wide range of various terms of the subsequent condition were sane. No such term, then, at that point, could happen. Similarly it could be shown by working two times on (1) with ∇^2 that there can be no term $r^3 S_{n-3}$ in (1); and in this manner bit by bit we can arrive at the outcome figured out in (1).

3. Express $x^2 yz$ in the structure $S_4 + r^2 S_2 + r^4 S_0$.

Suggestion: let
$$x^2 yz = S_4 + r^2 S_2 + r^4 S_0$$

and take ∇^2 of the two individuals we get

$$2yz = 14 S_2 + 20 r^2 S_0.$$

Operate again with ∇^2. $\qquad 0 = 120 S_0.$ $\hfill$ Whence

$$S_0 = 0, \quad S_2 = \frac{1}{7} yz, \quad \text{and} \quad S_4 = \frac{1}{7}(6x^2 - y^2 - z^2) yz.$$

4. Express $\sin^2 \theta \cos^2 \theta \sin \phi \cos \phi$ with regards to Surface Circular Sounds.

Suggestion: $\qquad \sin^2 \theta \cos^2 \theta \sin \phi \cos \phi = \dfrac{x^2 yz}{r^4}.$

For result v. Art. 108 (3).

SPHERICAL HARMONICS.

111. A change of coördinates to another arrangement of tomahawks having something very similar beginning as the old set will change a given Surface Circular Symphonious into one more of a similar degree. For such a change doesn't change the type of Laplace's Situation $\nabla^2 V = 0$ in the event that the two arrangements of tomahawks are rectangular, what's more, it is affected by supplanting x, y, and z in the Strong Consonant comparing to the given Surface Consonant by $x\cos\alpha_1 + y\cos\alpha_2 + z\cos\alpha_3$, $x\cos\beta_1 + y\cos\beta_2 + z\cos\beta_3$ and $x\cos\gamma_1 + y\cos\gamma_2 + z\cos\gamma_3$ individually where the cosines are the *direction cosines* of the new tomahawks, and it will leave the capability a homogeneous capability of the mth degree in the new factors, also, on isolating this by the mth force of the unaltered range vector we will have a Surface Round Symphonious of the mth degree.

112. We have seen in Art. 75 that if (x_1, y_1, z_1) are the coördinates of a given point

$$V = \frac{1}{\sqrt{(x-x_1)^2 + (y-y_1)^2 + (z-z_1)^2}} \quad (1)$$

is an answer of Laplace's Situation $\nabla^2 V = 0$, and changing to circular coördinates that

$$V = \frac{1}{\sqrt{r^2 - 2rr_1[\cos\theta\cos\theta_1 + \sin\theta\sin\theta_1\cos(\phi-\phi_1)] + r_1^2}} \quad (2)$$

is an answer of

$$rD_r^2(rV) + \frac{1}{\sin\theta}D_\theta(\sin\theta D_\theta V) + \frac{1}{\sin^2\theta}D_\phi^2 V = 0. \quad (3)$$

In the event that γ is the point between the radii vectores r and r_1 of the focuses (x, y, z) furthermore, (x_1, y_1, z_1) (1) can be composed

$$V = \frac{1}{\sqrt{r^2 - 2rr_1\cos\gamma + r_1^2}} \quad (4)$$

which should be comparable to (2), and consequently

$$\cos\gamma = \cos\theta\cos\theta_1 + \sin\theta\sin\theta_1\cos(\phi-\phi_1).$$

(4) which is an answer of (3) is of a similar structure as (5) Art. 75 and by creating it as we created (5) Art. 75 we view that as

$$V = P_m(\cos\gamma)$$

is an answer of the situation

$$m(m+1)V + \frac{1}{\sin\theta}D_\theta(\sin\theta D_\theta V) + \frac{1}{\sin^2\theta}D_\phi^2 V = 0 \quad (5)$$

and that $\qquad V = r^m P_m(\cos\gamma) \quad \text{and} \quad V = \frac{1}{r^{m+1}} P_m(\cos\gamma)$

are arrangements of (3).

On the off chance that we change our coördinates keeping the beginning unaltered and taking as our new polar pivot the range vector of (x_1, y_1, z_1) γ turns into our new θ and $P_m(\cos\gamma)$ decreases to $P_m(\cos\theta)$, a Surface Zonal Consonant, or a *Legendrian*,[3] of the mth degree. It is then a Legendrian having for its hub not the first polar pivot however the span vector of (x_1, y_1, z_1). Since a Legendrian is a Surface Circular Symphonious,

$$P_m(\cos\gamma) = P_m[\cos\theta\cos\theta_1 + \sin\theta\sin\theta_1\cos(\phi-\phi_1)]$$

is a Surface Round Consonant of the mth degree.

It is, in any case, of extremely exceptional structure, since being a determinate capability of μ, ϕ, μ_1, and ϕ_1 it contains however two erratic constants in the event that we see it as a capability of μ and ϕ, rather than containing $2m+1$.

---

[3] v. Art. 74.

SPHERICAL HARMONICS. 172

It is known as a *Laplace's Coefficient*, or momentarily as a Laplacian, of the mth degree.

We will before long communicate it in the guideline type of a Surface Round Consonant.

The range vector of (x_1, y_1, z_1) is known as the pivot of the Laplacian and the point where the hub cuts the outer layer of the unit circle is the *pole* of the Laplacian.

We will address the Laplacian $P_m(\cos\gamma)$ by $L_m(\mu, \phi, \mu_1, \phi_1)$. Obviously $L_m(\mu, \phi, 1, \phi_1) = P_m(\mu) = P_m(\cos\theta)$, and is truly autonomous of ϕ.

113. *If the result of a Surface Circular Symphonious of the mth degree by a Laplacian of a similar degree is coordinated over the outer layer of the unit circle, the result is equivalent to $\frac{4\pi}{2m+1}$ duplicated by the worth of the Round Consonant at the post of the Laplacian.*

That is,

$$\int_0^{2\pi} d\phi \int_{-1}^1 Y_m(\mu,\phi) L_m(\mu,\phi,\mu_1,\phi_1) d\mu = \frac{4\pi}{2m+1} Y_m(\mu_1,\phi_1). \tag{1}$$

Change to the pivot of the Laplacian as another polar hub, and let $Z_m(\mu,\phi)$ be the changed Circular Symphonious. $L_m(\mu, \phi, \mu_1, \phi_1)$ will become $P_m(\mu)$, also (1) will be demonstrated in the event that we can show that

$$\int_0^{2\pi} d\phi \int_{-1}^1 Z_m(\mu,\phi) P_m(\mu) d\mu = \frac{4\pi}{2m+1} Z_m(1,0). \tag{2}$$

$$Z_m(\mu,\phi) P_m(\mu) = A_0 [P_m(\mu)]^2 + \sum_{n=1}^{n=m} (A_n \cos n\phi + B_n \sin n\phi) P_m^n(\mu) P_m(\mu)$$

(v. (5) Art. 102).

$$\int_0^{2\pi} Z_m(\mu,\phi) P_m(\mu) d\phi = 2\pi A_0 [P_m(\mu)]^2 \qquad \text{and}$$

$$\int_{-1}^1 d\mu \int_0^{2\pi} Z_m(\mu,\phi) P_m(\mu) d\phi = \frac{4\pi}{2m+1} A_0 \qquad \text{(v. (5) Art. 89)}.$$

Yet, $Z_m(1,0) = A_0$, since $P_m(1) = 1$ and $P_m^n(1)$ contains $(1-1)^{\frac{n}{2}}$ as a component what's more, is equivalent to nothing.

Thus (2) is demonstrated.

114. We can now communicate a Laplacian in the guideline structure as a Round Symphonious, by the equations of Art. 107.

$$L_m(\mu,\phi,\mu_1,\phi_1) = P_m(\cos\gamma) = P_m[\cos\theta\cos\theta_1 + \sin\theta\sin\theta_1\cos(\phi-\phi_1)]$$

$$= \sum_{k=0}^{k=\infty} \left[A_{0,k} P_k(\mu) + \sum_{n=1}^{n=k} (A_{n,k}\cos n\phi + B_{n,k}\sin n\phi) P_k^n(\mu) \right]$$

where
$$A_{0,m} = \frac{2m+1}{4\pi} \int_0^{2\pi} d\phi \int_{-1}^1 L_m(\mu,\phi,\mu_1,\phi_1) P_m(\mu) d\mu.$$

$$= \frac{2m+1}{4\pi} \frac{4\pi}{2m+1} P_m(\mu_1) = P_m(\mu_1) \qquad \text{by (1) Art. 113,}$$

SPHERICAL HARMONICS.

$$A_{n,m} = \frac{2m+1}{2\pi}\frac{(m-n)!}{(m+n)!}\int_0^{2\pi} d\phi \int_{-1}^1 L_m(\mu,\phi,\mu_1,\phi_1)\cos n\phi P_m^n(\mu)d\mu$$

$$= \frac{2(m-n)!}{(m+n)!}\cos n\phi_1 P_m^n(\mu_1) \qquad \text{by (1) Art. 113, and}$$

$$B_{n,m} = \frac{2m+1}{2\pi}\frac{(m-n)!}{(m+n)!}\int_0^{2\pi} d\phi \int_{-1}^1 L_m(\mu,\phi,\mu_1,\phi_1)\sin n\phi P_m^n(\mu)d\mu$$

$$= \frac{2(m-n)!}{(m+n)!}\sin n\phi_1 P_m^n(\mu_1) \qquad \text{by (1) Art. 113,}$$

what's more, $A_{0,k} = A_{n,k} = B_{n,k} = 0$ by Art. 105 except if $k = m$. Thus

$$L_m(\mu,\phi,\mu_1,\phi_1) =$$

$$P_m(\mu)P_m(\mu_1) + 2\sum_{n=1}^{n=m}\left[\frac{(m-n)!}{(m+n)!}P_m^n(\mu)P_m^n(\mu_1)\cos n(\phi-\phi_1)\right]. \quad (1)$$

Each term of a Laplacian includes a mathematical coefficient, a component which is an element of μ, a second variable which is a similar capability of μ_1, and a third factor which is of the structure $\cos k(\phi - \phi_1)$. We give under a table of the initial not many Laplacians, taken from Minchin's Statics, precluding in each term for curtness the capability of μ_1.

By the guide of (1) we can compose (5) Art. 107 all the more minimalistically. It becomes

$$f(\mu,\phi) = \frac{1}{4\pi}\sum_{m=0}^{m=\infty}(2m+1)\int_0^{2\pi}d\phi_1\int_{-1}^1 f(\mu_1,\phi_1)L_m(\mu,\phi,\mu_1,\phi_1)d\mu_1 \qquad (2)$$

or

$$F(\theta,\phi) = \frac{1}{4\pi}\sum_{m=0}^{m=\infty}(2m+1)\int_0^{2\pi}d\phi_1\int_0^\pi F(\theta_1,\phi_1)P_m(\cos\gamma)\sin\theta_1 d\theta_1. \qquad (3)$$

LAPLACIANS.

	coef. of $\cos 0(\phi-\phi_1)$	coef. of $\cos(\phi-\phi_1)$	coef. of $\cos 2(\phi-\phi_1)$
L_0	1		
L_1	μ	$(1-\mu^2)^{\frac{1}{2}}$	
L_2	$\frac{1}{4}(3\mu^2-1)$	$3\mu(1-\mu^2)^{\frac{1}{2}}$	$\frac{3}{4}(1-\mu^2)$
L_3	$\frac{1}{4}(5\mu^3-3\mu)$	$\frac{3}{8}(1-\mu^2)^{\frac{1}{2}}(5\mu^2-1)$	$\frac{15}{4}\mu(1-\mu^2)$
L_4	$\frac{1}{64}(35\mu^4-30\mu^2+3)$	$\frac{5}{8}(1-\mu^2)^{\frac{1}{2}}(7\mu^3-3\mu)$	$\frac{5}{16}(1-\mu^2)(7\mu^2-1)$

SPHERICAL HARMONICS.   174

	coef. of $\cos 3(\phi - \phi_1)$	coef. of $\cos 4(\phi - \phi_1)$
L_0		
L_1		
L_2		
L_3	$\frac{5}{8}(1-\mu^2)^{\frac{3}{2}}$	
L_4	$\frac{35}{8}\mu(1-\mu^2)^{\frac{3}{2}}$	$\frac{35}{64}(1-\mu^2)^2$

EXAMPLE.

Figure out the issues of Art. 108 and Art. 108 Exs. 1 and 2 by the guide of (3) Art. 114.

115. Such issues as we have taken care of in Arts. 98 and 99, and furthermore issues contrasting from them in not having roundabout evenness about a pivot, can presently be settled by direct combination.

For example let it be expected to track down the worth at an outside place of the likely capability because of the fascination of a strong circle whose thickness at any point is relative to the result of any force of the sweep vector by a Surface Circular Symphonious.

Let $$\rho = Cr_1^k Y_m(\mu_1, \phi_1).$$

Then, at that point, utilizing our customary documentation we have

$$V = \int_0^a dr_1 \int_0^{2\pi} d\phi_1 \int_{-1}^1 \frac{Cr_1^k Y_m(\mu_1,\phi_1) r_1^2 d\mu_1}{\sqrt{r^2 - 2rr_1 \cos\gamma + r_1^2}}.$$

Yet, by (3) Art. 77

$$\frac{1}{\sqrt{r^2 - 2rr_1\cos\gamma + r_1^2}} = \frac{1}{r}\left[P_0(\cos\gamma) + \frac{r_1}{r}P_1(\cos\gamma) \right.$$
$$\left. + \frac{r_1^2}{r^2}P_2(\cos\gamma) + \cdots + \frac{r_1^m}{r^m}P_m(\cos\gamma) + \cdots\right]$$

if $r > r_1$.

Therefore since

$$\int_0^{2\pi} d\phi_1 \int_{-1}^1 Y_m(\mu_1,\phi_1) Y_n(\mu_1,\phi_1) d\mu_1 = 0,$$

V lessens to the single term

$$V = \frac{C}{r^{m+1}} \int_0^a r_1^{m+k+2} dr_1 \int_0^{2\pi} d\phi_1 \int_{-1}^1 Y_m(\mu_1,\phi_1) P_m(\cos\gamma) d\mu_1$$

$$= \frac{C}{r^{m+1}} \int_0^a r_1^{m+k+2} \left(\frac{4\pi}{2m+1} Y_m(\mu,\phi)\right) dr_1 \qquad \text{by Art. 113.}$$

$$V = \frac{4\pi C}{2m+1} \cdot \frac{a^{m+k+3}}{m+k+3} \cdot \frac{Y_m(\mu,\phi)}{r^{m+1}}.$$

EXAMPLES.

1. Tackle by direct combination the issues worked in Arts. 98 and 99 and Models 1, 2, 3, and 4 of Art. 99.

2. The thickness of a strong circle is relative to the result of the squares of the good ways from two commonly opposite diametral planes; track down the worth of the expected capability at an outside point.

Ans.
$$\rho = kr_1^4 \cos^2\theta_1 \sin^2\theta_1 \cos^2\phi_1$$
$$= kr_1^4 \left[\frac{1}{15} P_0(\mu_1) + \frac{1}{21} P_2(\mu_1) + \frac{1}{42} \cos 2\phi_1 P_2^2(\mu_1) \right.$$
$$\left. - \frac{4}{35} P_4(\mu_1) + \frac{1}{105} \cos 2\phi_1 P_4^2(\mu_1) \right].$$

$$V = \frac{M}{a}\left[\frac{a}{r} + \frac{a^3}{r^3}\left(\frac{1}{9}P_2(\mu) + \frac{1}{18}\cos 2\phi P_2^2(\mu)\right) \right.$$
$$\left. - \frac{a^5}{r^5}\left(\frac{4}{33}P_4(\mu) - \frac{1}{99}\cos 2\phi P_4^2(\mu)\right)\right].$$

3. Address Model 2 by an augmentation of the strategy for Arts. 98 and 99.

4. A leading circle of range a associated with the ground by a wire is set in the field of power because of an electric place where m units of power are concentrated. Track down the worth of the likely capability due to the instigated charge.

Suggestion: Let V_1 be the likely capability because of the point, and V_2 that because of the incited charge, and let b be the distance of the point from the focus of the circle. Then

$$V_1 = \frac{m}{\sqrt{b^2 - 2br\cos\theta + r^2}}$$
$$= \frac{m}{b}\left[P_0(\cos\theta) + \frac{r}{b}P_1(\cos\theta) + \frac{r^2}{b^2}P_2(\cos\theta) + \cdots\right] \quad \text{if} \quad r < b.$$
$$= \frac{m}{r}\left[P_0(\cos\theta) + \frac{b}{r}P_1(\cos\theta) + \frac{b^2}{r^2}P_2(\cos\theta) + \cdots\right] \quad \text{if} \quad r > b.$$
$$V_2 = A_0 P_0(\cos\theta) + A_1 \frac{r}{a} P_1(\cos\theta) + A_2 \frac{r^2}{a^2} P_2(\cos\theta) + \cdots \quad \text{if} \quad r < a.$$
$$= A_0 \frac{a}{r} P_0(\cos\theta) + A_1 \frac{a^2}{r^2} P_1(\cos\theta) + A_2 \frac{a^3}{r^3} P_2(\cos\theta) + \cdots \quad \text{if} \quad r > a.$$

When $r = a$ $V_1 + V_2 = 0$. Consequently

$$A_0 = -\frac{m}{b}, \quad A_1 = -\frac{ma}{b^2}, \quad A_2 = -\frac{ma^2}{b^3}, \cdots$$

also,

$$V_2 = -\frac{m}{b}\left[P_0(\cos\theta) + \frac{r}{b}P_1(\cos\theta) + \frac{r^2}{b^2}P_2(\cos\theta) + \cdots\right] \quad \text{if} \quad r < a$$
$$= -\frac{ma}{br}\left[P_0(\cos\theta) + \frac{a^2}{br}P_1(\cos\theta) + \frac{a^4}{b^2r^2}P_2(\cos\theta) + \cdots\right] \quad \text{if} \quad r > a.$$

Subsequently the impact of the initiated charge is unequivocally something very similar at an outer point as though the circle were supplanted by $\frac{ma}{b}$ units of negative power concentrated at the point $r = \frac{a^2}{b}$, $\theta = 0$. v. Peirce, Newt. Pot. Func., § 66.

SPHERICAL HARMONICS.

116. On the off chance that the two focuses P and P' are taken on the line OH whose bearing cosines are λ, μ, and ν, and if u and u' are the qualities at P and P' of any nonstop capability of the space coÃ¶rdinates, then $\lim_{PP' \doteq 0} \left[\dfrac{u' - u}{PP'} \right]$ is known as the *partial derivative* of u along the line OH and will be addressed by $D_h u$.

Let x, y, z be the coÃ¶rdinates of P and $x + \Delta x$, $y + \Delta y$, $z + \Delta z$ the coÃ¶rdinates of P'; then, at that point,

$$u' - u = D_x u . \Delta x + D_y u . \Delta y + D_z u . \Delta z + \epsilon$$

where ϵ is a minute of higher request than the first if Δx, Δy, and Δz are minute (v. Dif. Cal. Art. 198).

Hence $\quad \dfrac{u' - u}{PP'} = D_x u . \dfrac{\Delta x}{PP'} + D_y u . \dfrac{\Delta y}{PP'} + D_z u . \dfrac{\Delta z}{PP'} + \dfrac{\epsilon}{PP'}.$

But $\quad \dfrac{\Delta x}{PP'} = \lambda, \quad \dfrac{\Delta y}{PP'} = \mu, \quad \text{and} \quad \dfrac{\Delta z}{PP'} = \nu.$

Therefore $\quad D_h u = \lambda D_x u + \mu D_y u + \nu D_z u . and$ \hfill (1)

In the event that $\nabla^2 u = 0$, $D_x^p D_y^q D_z^r u$ is an answer of Laplace's Situation.

For $\quad \nabla^2(D_x^p D_y^q D_z^r u) = D_x^p D_y^q D_z^r(\nabla^2 u) = 0.$

Subsequently if $\nabla^2 u = 0$ $D_h u$ is an answer of Laplace's Situation, and if OH_1, OH_2, OH_3, $\cdots$ are a bunch of lines through the beginning $D_{h_1} D_{h_2} D_{h_3} \cdots u$ is an answer of Laplace's Situation.

117. On the off chance that H_k is a reasonable necessary homogeneous Mathematical capability of x, y, also, z of the kth degree

$$D_x \left(\dfrac{H_k}{r^l} \right) = D_r \left(\dfrac{H_k}{r^l} \right) D_x r + \dfrac{1}{r^l} D_x(H_k)$$
$$= -\dfrac{lx H_k}{r^{l+2}} + \dfrac{H_{k-1}}{r^l} = -\dfrac{lx H_k}{r^{l+2}} + \dfrac{r^2 H_{k-1}}{r^{l+2}},$$

furthermore, is of the structure $\dfrac{H_{k+1}}{r^{l+2}}$.

Exactly the same thing can be demonstrated of $D_y \left(\dfrac{H_k}{r^l} \right)$ and $D_z \left(\dfrac{H_k}{r^l} \right)$ and in this manner holds great of $D_h \left(\dfrac{H_k}{r^l} \right)$.

On the off chance that u is a homogeneous capability of x, y, and z of the degree $-m - 1$ and $\nabla^2 u = 0$ then $\nabla^2(r^{2m+1} u) = 0$.

$$\nabla^2(r^{2m+1} u) = (2m+1)(2m+2) r^{2m-1} u$$
$$+ 2(2m+1) r^{2m-1}(x D_x u + y D_y u + z D_z u) + r^{2m+1} \nabla^2 u$$
$$= 0,$$

since $\quad x D_x u + y D_y u + z D_z u = -(m+1) u$

by Euler's Hypothesis (v. Dif. Cal. Art. 220).

118. $\dfrac{M}{r} = \dfrac{M}{\sqrt{x^2 + y^2 + z^2}}$ is an answer of Laplace's Situation (v. Art. 75) what's more, is of the structure $\dfrac{H_0}{r}$.

$D_{h_1} D_{h_2} D_{h_3} \cdots D_{h_m} \left(\dfrac{M}{r} \right)$ is then an answer of Laplace's Situation by Workmanship. 116; it is of the structure $\dfrac{H_m}{r^{2m+1}}$ by Art. 117 and is a homogeneous capability of the degree $-m - 1$.

In this manner $r^{2m+1} D_{h_1} D_{h_2} D_{h_3} \cdots D_{h_m} \left(\dfrac{M}{r} \right)$ is an answer of Laplace's Situation, what's more, is a reasonable essential homogeneous Logarithmic capability of x, y, and z of the mth

SPHERICAL HARMONICS. 177

degree, and is subsequently a Strong Round Consonant of the mth degree (v. Art. 110); and $r^{m+1}D_{h_1}D_{h_2}D_{h_3}\cdots D_{h_m}\left(\frac{M}{r}\right)$ is a Surface Round Consonant of the mth degree.

Additionally since the course of every one of the lines $OH_1, OH_2, \cdots OH_m$ depends upon two points which might be taken at delight, these points and M are $2m+1$ inconsistent constants and might be decided to such an extent that $r^{m+1}D_{h_1}D_{h_2}D_{h_3}\cdots D_{h_m}\left(\frac{M}{r}\right)$ might be any given Surface Circular Consonant. Subsequently any given Surface Round Symphonious might be viewed as shaped by separating $\frac{M}{r}$ progressively along m determinate lines $OH_1, OH_2 \cdots OH_m$, and is given aside from the unsure variable M when these lines are given.

The lines $OH_1, OH_2, OH_3, \cdots OH_m$ are known as the *axes* of the Symphonious, and the places where they meet the outer layer of the unit circle the *poles* of the Consonant. The m tomahawks of a Zonal Consonant concur with the pivot of coÃ¶rdinates (v. Art. 86) and subsequently the m tomahawks of a Laplacian concur with what we have called the hub of the Laplacian (v. Art. 112).

119. Any *Surface Zonal Harmonic* $P_m(\mu)$ is equivalent to zero for m genuine and unmistakable upsides of μ which lie between -1 and 1; and any *Associated Function* $P_m^n(\mu)$ is equivalent to zero for $m-n$ genuine and unmistakable upsides of μ, which lie between -1 and 1.

$$P_m(\mu) = \frac{1}{2^m m!} \cdot \frac{d^m(\mu^2-1)^m}{d\mu^m}.\text{and}v.\text{ Art. 83 (1).}$$

$\frac{d^k(\mu^2-1)^m}{d\mu^k}$ contains $(\mu^2-1)^{m-k}$ as a variable. v. Art. 89.

From Rolle's Hypothesis, "On the off chance that $f(x)$ is consistent and single-esteemed and is equivalent to zero for the genuine qualities a and b of x, $\frac{df(x)}{dx}$ is equivalent to zero for something like one genuine worth of x among a and b," (v. Dif. Cal. Art. 126) it follows that since $(\mu^2-1)^m = 0$ when $\mu = -1$ and when $\mu = 1$ $\frac{d(\mu^2-1)^m}{d\mu} = 0$ for at any rate one worth of μ between -1 and 1. $\frac{d(\mu^2-1)^m}{d\mu}$ can't be equivalent to zero for more than one worth of μ between -1 and 1, for it contains $(\mu^2-1)^{m-1}$ as a factor and is an objective Mathematical polynomial of the $2m-1$st degree.

Similarly we can show that $\frac{d^2(\mu^2-1)^m}{d\mu^2} = 0$ has $m-2$ roots equivalent to -1, $m-2$ roots equivalent to 1 and two genuine roots between -1 and 1 which separate the three particular underlying foundations of $\frac{d(\mu^2-1)^m}{d\mu} = 0$; and overall if $k < m+1$ that $\frac{d^k(\mu^2-1)^m}{d\mu^k} = 0$ has $m-k$ attaches equivalent to -1, $m-k$ attaches equivalent to 1, what's more, k genuine roots isolating the $k+1$ particular foundations of $\frac{d^{k-1}(\mu^2-1)^m}{d\mu^{k-1}} = 0$.

Thus $P_m(\mu) = 0$ or $\frac{1}{2^m m!}\cdot\frac{d^m(\mu^2-1)^m}{d\mu^m} = 0$ has m genuine and particular roots between -1 and 1, and it has no more since it is of the mth degree.

The very thinking shows that $\frac{d^{m+n}(\mu^2-1)^m}{d\mu^{m+n}} = 0$ has $m-n$ unmistakable genuine roots between -1 and 1, and subsequently that $P_m^n(\mu)$ is equivalent to zero for $m-n$ particular genuine upsides of μ between -1 and 1. Since $P_m^n(\mu)$ contains $\sin^n\theta$ as a factor it is likewise equivalent to zero when $\mu = -1$ and when $\mu = 1$.

$\cos n\phi$ is equivalent to zero for $2n$ equidistant upsides of ϕ, and $\sin n\phi$ is equivalent to zero for $2n$ upsides of ϕ. Consequently any *Tesseral Harmonic* $\sin n\phi P_m^n(\mu)$ or $\cos n\phi P_m^n(\mu)$ is equivalent to zero for $2n$ equidistant upsides of ϕ, for $\mu = 1$, for $\mu = -1$, and for $m-n$ genuine and various upsides of μ between -1 and 1.

It follows that the worth of any Surface Zonal Consonant $P_m(\mu)$ at a point on the outer layer of the unit circle will have a similar sign insofar as the point stays on one of the *zones* into which the outer layer of the circle is partitioned by the m circles of scope comparing to the m underlying foundations of $P_m(\mu) = 0$, and will change sign at whatever point the point passes from one of these zones into a connecting one; and that the worth of any Tesseral Symphonious $\sin n\phi P_m^n(\mu)$ at a point on the outer layer of the unit circle will have a similar sign insofar as the point stays on any of the *tesserae* into which the outer layer of the circle is isolated by the $m-n$ circles of scope relating to the foundations of $P_m^n(\mu) = 0$ furthermore, the $2n$ meridians relating to the foundations of $\sin n\phi = 0$, and will change sign at whatever point the point passes from one of these tesserae into a connecting one.

Chapter 7

CYLINDRICAL HARMONICS

120. In Arts. 11 and 17 we got

$$z = AJ_0(x) + BK_0(x) \qquad (1)$$

as the overall arrangement of *Fourier's Equation*

$$\frac{d^2z}{dx^2} + \frac{1}{x}\frac{dz}{dx} + z = 0, \qquad (2)$$

where

$$J_0(x) = 1 - \frac{x^2}{2^2} + \frac{x^4}{2^2.4^2} - \frac{x^6}{2^2.4^2.6^2} + \cdots \qquad (3)$$

what's more, is known as a *Cylindrical Harmonic* or *Bessel's Function* of the zeroth request; furthermore, where

$$K_0(x) = J_0(x)\log x + \frac{x^2}{2^2} - \frac{x^4}{2^2.4^2}\left(\frac{1}{1} + \frac{1}{2}\right) + \frac{x^6}{2^2.4^2.6^2}\left(\frac{1}{1} + \frac{1}{2} + \frac{1}{3}\right) - \cdots \qquad (4)$$

furthermore, is known as a *Cylindrical Harmonic* or *Bessel's Function* of the Subsequent Kind, furthermore, of the zeroth request.

In Art. 17 we found that $z = J_n(x)$ is a specific arrangement of *Bessel's Equation*

$$\frac{d^2z}{dx^2} + \frac{1}{x}\frac{dz}{dx} + \left(1 - \frac{n^2}{x^2}\right)z = 0, \qquad (5)$$

where assuming n is unlimited in esteem

$$J_n(x) = \frac{x^n}{2^n \Gamma(n+1)}\left[1 - \frac{x^2}{2^2(n+1)} + \frac{x^4}{2^4.2!(n+1)(n+2)} - \frac{x^6}{2^6.3!(n+1)(n+2)(n+3)} + \cdots\right] \qquad (6)$$

what's more, is known as a *Cylindrical Harmonic* or *Bessel's Function* of the nth request; also, that except if n is a whole number

$$z = AJ_n(x) + BJ_{-n}(x)$$

is the overall arrangement of Bessel's Situation.

---

[0]The understudy should re-read cautiously Arts. 11, 17, also, 18(d) prior to starting this chapter.

CYLINDRICAL HARMONICS.

Assuming n is a whole number it tends to be shown that

$$J_n(x) = (-1)^n J_{-n}(x),$$

(v. Forsyth's Diff. Eq. Art. 102), and afterward

$$z = AJ_n(x) + B\{K_n(x)\}$$

is the overall arrangement of Bessel's Situation and

$$\{K_n(x)\} = J_n(x)\log x - \frac{1}{2}\left(\frac{x}{2}\right)^{-n}\sum_{k=0}^{k=n-1}\frac{(n-k-1)!}{k!}\left(\frac{x}{2}\right)^{2k}$$
$$-\frac{1}{2}\left(\frac{x}{2}\right)^{n}\sum_{k=0}^{k=\infty}\frac{(-1)^k}{(n+k)!k!}\left[1+\frac{1}{2}+\frac{1}{3}+\cdots+\frac{1}{k}\right.$$
$$\left.+1+\frac{1}{2}+\frac{1}{3}+\cdots+\frac{1}{n+k}\right]\left(\frac{x}{2}\right)^{2k} \quad (7)$$

v. M. BÃ´cher, Ann. Math. Vol. VI, No. 4.

121. A valuable articulation for $J_n(x)$ as an unequivocal indispensable can be gotten without trouble from Bessel's Situation [(5) Art. 120] by a slight change of the strategy given by Forsyth (Diff. Eq. Art. 136).

It was displayed in Art. 17 that $z = x^n v$ is an answer of Bessel's Situation if v fulfills the condition

$$\frac{d^2v}{dx^2} + \frac{2n+1}{x}\frac{dv}{dx} + v = 0. \quad (1)$$

Assume
$$v = \int_a^b T\cos(xt)dt \quad (2)$$

where x and t are free, T is an obscure capability of t, and a and b are at present dubious.

Then
$$\frac{dv}{dx} = -\int_a^b tT\sin(xt)dt$$

and
$$\frac{d^2v}{dx^2} = -\int_a^b t^2 T\cos(xt)dt.$$

Subbing in (1) subsequent to duplicating through by x, we have

$$\int_a^b (1-t^2)Tx\cos(xt)dt - \int_a^b (2n+1)tT\sin(xt)dt = 0. \quad (3)$$

By *integration by parts* we view that as

$$\int_a^b (1-t^2)Tx\cos(xt)dt = \left[(1-t^2)T\sin(xt)\right]_a^b$$
$$-\int_a^b \left[(1-t^2)\frac{dT}{dt} - 2tT\right]\sin(xt)dt,$$

CYLINDRICAL HARMONICS.

what's more (3) lessens to

$$\left[(1-t^2)T\sin(xt)\right]_a^b - \int_a^b \left[(1-t^2)\frac{dT}{dt} + (2n-1)tT\right]\sin(xt)dt = 0. \tag{4}$$

Assuming we decide T so that

$$(1-t^2)\frac{dT}{dt} + (2n-1)tT = 0, \tag{5}$$

and a and b so that

$$\left[(1-t^2)T\sin(xt)\right]_a^b = 0 \tag{6}$$

(4) will be fulfilled and our concern will be settled. (5) gives

$$T = C(1-t^2)^{n-\frac{1}{2}}, \tag{7}$$

furthermore (6) will clearly be fulfilled if $a = -1$ and $b = 1$.

Hence $\quad$ and $v = C\int_{-1}^{1} \frac{(1-t^2)^n \cos(xt)dt}{\sqrt{1-t^2}}$ is an answer of (1),

and and $z = Cx^n \int_{-1}^{1} \frac{(1-t^2)^n \cos(xt)dt}{\sqrt{1-t^2}}$ and (8)

is an answer of Bessel's Situation.

In the event that we let $t = \cos\phi$ in (8) we get

$$z = Cx^n \int_0^\pi \sin^{2n}\phi \cos(x\cos\phi)d\phi.$$

Extend $\cos(x\cos\phi)$ into a series including powers of $x\cos\phi$, coordinate term by term by the guide of the recipes

$$\int_0^{\frac{\pi}{2}} \sin^n x\,dx = \frac{\sqrt{\pi}}{2}\cdot\frac{\Gamma\left(\frac{n+1}{2}\right)}{\Gamma\left(\frac{n}{2}+1\right)} \qquad \text{[Int. Cal. (1) Art. 99]},$$

$$\int_0^{\frac{\pi}{2}} \sin^n x \cos^m x\,dx = \frac{\Gamma\left(\frac{m+1}{2}\right)\Gamma\left(\frac{n+1}{2}\right)}{2\Gamma\left(\frac{m+n}{2}+1\right)}$$

(Int. Cal. Art. 99 Ex. 2), and contrast and (6) Art. 120, and we get

$$J_n(x) = \frac{x^n}{2^n\sqrt{\pi}\,\Gamma\left(n+\frac{1}{2}\right)} \int_0^\pi \sin^{2n}\phi \cos(x\cos\phi)d\phi. \tag{9}$$

On the off chance that n is a positive number (9) decreases to

$$J_n(x) = \frac{1}{\pi}\cdot\frac{x^n}{1.3.5.\cdots(2n-1)} \int_0^\pi \sin^{2n}\phi \cos(x\cos\phi)d\phi. \tag{10}$$

Let $n = 0$ in (9) or (10) and we get

$$J_0(x) = \frac{1}{\pi}\int_0^\pi \cos(x\cos\phi)d\phi. \tag{11}$$

CYLINDRICAL HARMONICS.

EXAMPLES.

1. Get Recipe (11) straightforwardly from Fourier's Situation, (2) Art. 120.

2. Demonstrate by *integration by parts* that if $n > -\frac{1}{2}$

$$\int_0^\pi \sin^{2n}\phi \cos\phi \sin(x\cos\phi)d\phi = \frac{x}{2n+1}\int_0^\pi \sin^{2n+2}\phi \cos(x\cos\phi)d\phi.$$

3. Demonstrate by *integration by parts* that if $n > \frac{1}{2}$

$$\int_0^\pi \sin^{2n}\phi \cos\phi \sin(x\cos\phi)d\phi$$
$$= \frac{1}{x}\int_0^\pi [2n\sin^{2n}\phi - (2n-1)\sin^{2n-2}\phi]\cos(x\cos\phi)d\phi.$$

122. We can now promptly get various helpful equations. Separate (11) Art. 121 regarding x and we get

$$\frac{dJ_0(x)}{dx} = -\frac{1}{\pi}\int_0^\pi \cos\phi \sin(x\cos\phi)d\phi$$

$$= -\frac{x}{\pi}\int_0^\pi \sin^2\phi \cos(x\cos\phi)d\phi \qquad \text{by Ex. 2 Art. 121.}$$

Thus by (10) Art. 121 $\qquad \dfrac{dJ_0(x)}{dx} = -J_1(x). \qquad (1)$

Likewise by the guide of Exs. 3 and 2, Art. 121, we can acquire the relations

$$\frac{d[x^n J_n(x)]}{dx} = x^n J_{n-1}(x) \qquad (2)$$

in the event that $n > \frac{1}{2}$,

$$\frac{d[x^{-n} J_n(x)]}{dx} = -x^{-n} J_{n+1}(x) \qquad (3)$$

if $n > -\frac{1}{2}$.

(2) can be composed

$$\int_0^x x^n J_{n-1}(x)dx = x^n J_n(x) \qquad (4)$$

if $n > \frac{1}{2}$.

CYLINDRICAL HARMONICS.

(2) and (3) can be composed

$$x^n \frac{dJ_n(x)}{dx} + nx^{n-1}J_n(x) = x^n J_{n-1}(x)$$

and

$$x^{-n}\frac{dJ_n(x)}{dx} - nx^{-n-1}J_n(x) = -x^{-n}J_{n+1}(x),$$

or and $\dfrac{dJ_n(x)}{dx} = J_{n-1}(x) - \dfrac{n}{x}J_n(x)$ (5)

and and $\dfrac{dJ_n(x)}{dx} = -J_{n+1}(x) + \dfrac{n}{x}J_n(x);$ (6)

whence $2\dfrac{dJ_n(x)}{dx} = J_{n-1}(x) - J_{n+1}(x)$ (7)

and $\dfrac{2n}{x}J_n(x) = J_{n-1}(x) + J_{n+1}(x).$ (8)

The rehashed utilization of recipe (8) will empower us to get from $J_0(x)$ and $J_1(x)$ any of Bessel's Capabilities whose request is a positive number. For instance, we have

$$J_2(x) = \frac{2}{x}J_1(x) - J_0(x)$$
$$J_3(x) = \left(\frac{8}{x^2} - 1\right)J_1(x) - \frac{4}{x}J_0(x).$$

From a table giving the upsides of $J_0(x)$ and $J_1(x)$, then, at that point, tables for the elements of higher request are promptly built. Such a table taken from Rayleigh's Sound (Vol. I., page 265) will be tracked down in the Supplement (Table VI.).

By the guide of (5) and (6) any subsidiary of $J_n(x)$ can be communicated in wording of $J_n(x)$ and $J_{n+1}(x)$. For instance

$$\frac{d^2J_n(x)}{dx^2} = \left[\frac{n(n-1)}{x^2} - 1\right]J_n(x) + \frac{1}{x}J_{n+1}(x).$$

On the off chance that we compose $J_0(x)$ for z in Fourier's Situation [(2) Art. 120], duplicate through by xdx and coordinate from zero to x, working on the subsequent condition by *integration by parts*, we get

$$x\frac{dJ_0(x)}{dx} + \int_0^x xJ_0(x)dx = 0;$$

whence by (1) $\displaystyle\int_0^x xJ_0(x)dx = xJ_1(x).$ (9)

In the event that we compose $J_0(x)$ for z in Fourier's Situation, duplicate through by $x^2 \dfrac{dJ_0(x)}{dx} dx$ and incorporate from zero to x, streamlining by *integration by parts* we get

$$\frac{x^2}{2}\left[\left(\frac{dJ_0(x)}{dx}\right)^2 + (J_0(x))^2\right] - \int_0^x x(J_0(x))^2 dx = 0;$$

whence by (1) $\displaystyle\int_0^x x(J_0(x))^2 dx = \frac{x^2}{2}[(J_0(x))^2 + (J_1(x))^2].$ (10)

CYLINDRICAL HARMONICS.

Similarly we can get from Bessel's Situation [(5) Art. 120] the equation

$$\int_0^x x(J_n(x))^2 dx = \frac{1}{2}\left[x^2\left(\frac{dJ_n(x)}{dx}\right)^2 + (x^2 - n^2)(J_n(x))^2\right] \quad (11)$$

which (6) empowers us to lessen to the structure

$$\int_0^x x(J_n(x))^2 dx = \frac{x^2}{2}[(J_n(x))^2 + (J_{n+1}(x))^2] - nxJ_n(x)J_{n+1}(x). \quad (12)$$

Recipes (9), (10), (11), and (12) will demonstrate valuable when we endeavor to foster with regards to *Cylindrical Harmonics*.

Upsides of $J_n(x)$ for bigger upsides of x than those given in Table VI., Supplement, might be figured effectively from the recipe

$$J_n(x) = \sqrt{\frac{2}{\pi x}}\left[1 - \frac{(1^2 - 4n^2)(3^2 - 4n^2)}{2!(8x)^2}\right.$$
$$\left. + \frac{(1^2 - 4n^2)(3^2 - 4n^2)(5^2 - 4n^2)(7^2 - 4n^2)}{4!(8x)^4} - \cdots\right]\cos\left(x - \frac{\pi}{4} - n\frac{\pi}{2}\right)$$
$$+ \sqrt{\frac{2}{\pi x}}\left[\frac{1^2 - 4n^2}{1!8x}\right.$$
$$\left. - \frac{(1^2 - 4n^2)(3^2 - 4n^2)(5^2 - 4n^2)}{3!(8x)^3} + \cdots\right]\sin\left(x - \frac{\pi}{4} - n\frac{\pi}{2}\right). \quad (13)$$

v. Lommel, Studien Ã¼ber bite the dust Bessel'schen Functionen, page 59.

The series ends assuming that $2n$ is an odd whole number, however in any case it is dissimilar. It tends to be demonstrated, notwithstanding, that regardless the amount of m terms contrasts from $J_n(x)$ by not exactly the last term included, and subsequently the equation can securely be utilized for mathematical calculation.

EXAMPLES.

1. Affirm (1), (2), and (3), Art. 122, by acquiring them from (3) and (6), Art. 120.

2. Affirm (1), Art. 122, by showing that Fourier's Condition will separate into the exceptional structure accepted by Bessel's Situation when $n = 1$.

3. Show that (9), Art. 122, is a unique instance of (4), Art. 122.

4. Show that the breaking point drew closer by $J_n(x)$ as n increments endlessly is zero, and by the guide of this reality and of (8), Art. 122, demonstrate that

$$J_{n-1}(x) = \frac{2}{x}[nJ_n(x) - (n+2)J_{n+2}(x) + (n+4)J_{n+4}(x) + \cdots].$$

5. Demonstrate that

$$\frac{dJ_n(x)}{dx} = \frac{2}{x}[\tfrac{1}{2}nJ_n(x) - (n+2)J_{n+2}(x) + (n+4)J_{n+4}(x) - \cdots].$$

6. Show that the replacement of $\left(1 - \frac{y^2}{n^2}\right)^{\frac{1}{2}}$ for x in Legendre's Situation will lessen it to the structure

$$\left(1 - \frac{y^2}{n^2}\right)\frac{d^2z}{dy^2} + \left(\frac{1}{y} - \frac{2y}{n^2}\right)\frac{dz}{dy} + \left(1 + \frac{1}{n}\right)z = 0,$$

CYLINDRICAL HARMONICS. 185

furthermore, that the restricting structure drew closer by this situation as n is endlessly expanded is Fourier's Condition, and consequently that $J_0(x)$ can be viewed as some consistent variable increased by the restricting worth drew nearer by $P_n\left(1 - \frac{x^2}{n^2}\right)^{\frac{1}{2}}$ as n is endlessly expanded.

123. To finish the arrangement of the drumhead issue taken up in Workmanship. 11, we found that it would be important to foster a given capability of r in the structure

$$f(r) = A_1 J_0(\mu_1 r) + A_2 J_0(\mu_2 r) + A_3 J_0(\mu_3 r) + \cdots$$

where μ_1, μ_2, μ_3, &c., are the foundations of the supernatural condition $J_0(\mu a) = 0$; furthermore, in Art. 11, Ex. the improvement of solidarity in a progression of exactly the same structure was needed.

(*a*) Let us think about another issue.

The raised surface and one base of a chamber of range a and length b are kept at the steady temperature zero, the temperature at each mark of the other base is a given capability of the distance of the point from the focal point of the base; required the temperature of any mark of the chamber after the long-lasting temperatures have been laid out.

Here we need to tackle Laplace's Condition in Barrel shaped Coördinates ([XIV] Art. 1).

$$D_r^2 u + \frac{1}{r} D_r u + \frac{1}{r^2} D_\phi^2 u + D_z^2 u = 0 \qquad (1)$$

dependent upon the circumstances

$$u = 0 \quad \text{when} \quad z = 0$$
$$u = 0 \quad " \quad r = a$$
$$u = f(r) \quad " \quad z = b,$$

also, from the balance of the issue we know that $D_\phi^2 u = 0$.

Expecting as expected $u = R.Z$ we break (1) up into the situations

$$\frac{d^2 Z}{dz^2} - \mu^2 Z = 0$$

$$\frac{d^2 R}{dr^2} + \frac{1}{r}\frac{dR}{dr} + \mu^2 R = 0,$$

whence $\qquad u = \sinh(\mu z) J_0(\mu r) \qquad (2)$

and $\qquad u = \cosh(\mu z) J_0(\mu r) \qquad (3)$

are specific arrangements of (1).

On the off chance that μ_k is a root of $J_0(\mu a) = 0$

$$\qquad \qquad (4)$$

$$u = \sinh(\mu_k z) J_0(\mu_k r)$$

fulfills (1) and two of the three conditions of condition.

If then $\qquad f(r) = A_1 J_0(\mu_1 r) + A_2 J_0(\mu_2 r) + A_3 J_0(\mu_3 r) + \cdots \qquad (5)$

CYLINDRICAL HARMONICS.

μ_1, μ_2, μ_3, &c., being underlying foundations of (4),

$$u = A_1 \frac{\sinh(\mu_1 z)}{\sinh(\mu_1 b)} J_0(\mu_1 r) + A_2 \frac{\sinh(\mu_2 z)}{\sinh(\mu_2 b)} J_0(\mu_2 r) + A_3 \frac{\sinh(\mu_3 z)}{\sinh(\mu_3 b)} J_0(\mu_3 r) + \cdots \quad (6)$$

fulfills (1) and the conditions of condition, and is all the required solution.

(b) If rather than keeping the raised surface of the chamber at the temperature zero we encompass it by a coat impenetrable to warm, the condition of condition $u = 0$ when $r = a$ will be supplanted by $D_r u = 0$ when $r = a$, or on the other hand if

and $u = \sinh(\mu z) J_0(\mu r)$,

by $\qquad \dfrac{dJ_0(\mu r)}{dr} = 0 \qquad$ when $r = a$,

that is by $\qquad \mu J_0'(\mu a) = 0^1 \qquad$ or (v. (1) Art. 122)

by and $J_1(\mu a) = 0$. $\hfill (7)$

In the event that now in (5) and (6) μ_1, μ_2, μ_3, &c., are underlying foundations of (7), (6) will be the arrangement of our new problem.

(c) If rather than keeping the curved surface of the chamber at the temperature zero we permit it to cool in air at the temperature zero, the condition $u = 0$ when $r = a$ will be supplanted by $D_r u + hu = 0$ when $r = a$, or on the other hand if

$$u = \sinh(\mu z) J_0(\mu r)$$

by $\qquad \mu J_0'(\mu r) + h J_0(\mu r) = 0 \qquad$ when $r = a$

that is by $\qquad \mu a J_0'(\mu a) + ah J_0(\mu a) = 0 \qquad$ or (v. (1) Art. 122)

by $\qquad \mu a J_1(\mu a) - ah J_0(\mu a) = 0$. $\hfill (8)$

If presently in (5) and (6) μ_1, μ_2, μ_3, &c., are foundations of (8), (6) will be the arrangement of our current issue.

124. It very well may be shown that $\qquad J_0(x) = 0 \hfill (1)$

$$J_1(x) = 0 \hfill (2)$$

and and $x J_0'(x) + \lambda \qquad J_0(x) = 0 \hfill (3)$

have each a limitless number of genuine positive roots (v. Riemann, Par. Dif. Gl., § 97). The prior foundations of these situations can be registered without serious trouble from the table for the upsides of $J_0(x)$ (Table VI., Index).

The initial twelve foundations of $J_0(x) = 0$ and $J_1(x) = 0$ are given in Table IV., Reference section, a table due to Stirs up. Enormous foundations of $J_0(x) = 0$ and of $J_1(x) = 0$ might be handily figured from the equations

$$\frac{x_0^{(s)}}{\pi} = s - .25 + \frac{.050661}{4s-1} - \frac{.053041}{(4s-1)^3} + \frac{.262051}{(4s-1)^5} - \cdots \quad (4)$$

$$\frac{x_1^{(s)}}{\pi} = s + .25 - \frac{.151982}{4s+1} + \frac{.015399}{(4s+1)^3} - \frac{.245270}{(4s+1)^5} + \cdots \quad (5)$$

given by Stirs up in Camb. Phil. Trans., Vol. IX., $x_0^{(s)}$ addressing the sth root of $J_0(x) = 0$, and $x_1^{(s)}$ the sth foundation of $J_1(x) = 0$.

---

[1] We will find it advantageous to utilize the recognizable documentation of $f'(x) = \frac{df(x)}{dx}$ (v. Dif. Cal., p. 119).

CYLINDRICAL HARMONICS.

125. We have seen in Art. 123 that $U = \sinh(\mu_k z) J_0(\mu_k r)$ and $V = \sinh(\mu_l z) J_0(\mu_l r)$ are arrangements of $\nabla^2 U = 0$ also, $\nabla^2 V = 0$ assuming we express Laplace's Condition as far as Barrel shaped CoÃ¶rdinates (v. (1) Art. 123).

Thus, if $\int dS$ addresses the surface basic over any shut surface, we have

$$\int (U D_n V - V D_n U) dS = 0$$

by Green's Hypothesis (v. Art. 92).

In the event that we accept the chamber of Art. 123 as our surface, and play out the mixes and improve on the subsequent condition, we find

$$\int_0^a r J_0(\mu_k r) J_0(\mu_l r) dr = \frac{-1}{\mu_k^2 - \mu_l^2}[\mu_k a J_0(\mu_l a) J_0'(\mu_k a) - \mu_l a J_0(\mu_k a) J_0'(\mu_l a)]$$

$$= \frac{-1}{\mu_l^2 - \mu_k^2}[\mu_k a J_0(\mu_l a) J_1(\mu_k a) - \mu_l a J_0(\mu_k a) J_1(\mu_l a)]. \quad (1)$$

Thus if μ_k and μ_l are various foundations of

$$J_0(\mu a) = 0,$$

or of $and J_1(\mu a)\qquad = 0,$

or of $\qquad\qquad \mu a J_1(\mu a) - \lambda J_0(\mu a) = 0,$

then $and \int_0^a r J_0(\mu_k r) J_0(\qquad\qquad \mu_l r) dr = 0. \quad (2)$

EXAMPLE.

Get (1) Art. 125 straightforwardly from Fourier's Situation

$$\frac{d^2 J_0(\mu r)}{dr^2} + \frac{1}{r}\frac{d J_0(\mu r)}{dr} + \mu^2 J_0(\mu r) = 0.$$

126. We are presently ready to acquire the advancements called for in Art. 123.

Let $\qquad f(r) = A_1 J_0(\mu_1 r) + A_2 J_0(\mu_2 r) + A_3 J_0(\mu_3 r) + \cdots \qquad (1)$

μ_1, μ_2, μ_3, &c., being underlying foundations of $J_0(\mu a) = 0$, or of $J_1(\mu a) = 0$, or of

$$\mu a J_1(\mu a) - \lambda J_0(\mu a) = 0.$$

To decide any coefficient A_k increase (1) by $r J_0(\mu_k r) dr$ and incorporate from zero to a. The primary part will turn into

$$\int_0^a r f(r) J_0(\mu_k r) dr.$$

Each term of the subsequent part will evaporate by (2) Art. 125 with the exception of the term

$$A_k \int_0^a r (J_0(\mu_k r))^2 dr.$$

$$\int_0^a r (J_0(\mu_k r))^2 dr = \frac{1}{\mu_k^2} \int_0^{\mu_k a} x (J_0(x))^2 dx = \frac{a^2}{2}[(J_0(\mu_k a))^2 + (J_1(\mu_k a))^2]$$

by (10) Art. 122.

Hence
$$A_k = \frac{2}{a^2[(J_0(\mu_k a))^2 + (J_1(\mu_k a))^2]} \int_0^a rf(r) J_0(\mu_k r) dr. \qquad (2)$$

The turn of events (1) holds great from $r = 0$ to $r = a$ (v. Arts. 24, 25, and 88).

On the off chance that μ_1, μ_2, μ_3, &c., are foundations of $J_0(\mu a) = 0$, (2) diminishes to

$$A_k = \frac{2}{a^2 (J_1(\mu_k a))^2} \int_0^a rf(r) J_0(\mu_k r) dr. \qquad (3)$$

On the off chance that μ_1, μ_2, μ_3, &c., are foundations of $J_1(\mu a) = 0$, (2) diminishes to

$$A_k = \frac{2}{a^2 (J_0(\mu_k a))^2} \int_0^a rf(r) J_0(\mu_k r) dr. \qquad (4)$$

On the off chance that μ_1, μ_2, μ_3, &c., are foundations of $\mu a J_1(\mu a) - \lambda J_0(\mu a) = 0$, (2) diminishes to

$$A_k = \frac{2\mu_k^2}{(\lambda^2 + \mu_k^2 a^2)(J_0(\mu_k a))^2} \int_0^a rf(r) J_0(\mu_k r) dr. \qquad (5)$$

For the significant situation where $f(r) = 1$

$$\int_0^a rf(r) J_0(\mu_k r) dr = \int_0^a r J_0(\mu_k r) dr = \frac{1}{\mu_k^2} \int_0^{\mu_k a} x J_0(x) dx = \frac{a}{\mu_k} J_1(\mu_k a) \qquad (6)$$

by (9) Art. 122, and (3) lessens to

$$A_k = \frac{2}{\mu_k a J_1(\mu_k a)}, \qquad (7)$$

(4) lessens to $A_k = 0$ with the exception of $k = 1$ when $\mu_k = 0$ and we have $A_1 = 1$,

(5) lessens to
$$A_k = \frac{2\lambda}{(\lambda^2 + \mu_k^2 a^2) J_0(\mu_k a)}. \qquad (8)$$

EXAMPLES.

1. Show that in (12) Art. 11 any coefficient A_k has the worth yielded (3) Art. 126; and in the solution to Art. 11, Ex. the esteem yielded (7) Art. 126.

2. Show that assuming a drumhead be at first mutilated so it has round balance, it won't in everyday give a melodic note; that it could be at first twisted to give a melodic note; that for this situation the vibration will be a *steady* vibration; that the frequencies of the different melodic notes that can be considering when the contortion has roundabout evenness are corresponding to the roots of $J_0(x) = 0$; that the potential hubs for such vibrations are concentric circles whose radii are corresponding to the underlying foundations of $J_0(x) = 0$.

3. A chamber of sweep one meter and height one meter has its upper surface kept at the temperature 100Â°, and its base and curved surface at the temperature 15Â°, until the *stationary temperature* is set up. Track down the temperature at focuses on the hub 25 cm., 50 cm., and 75

CYLINDRICAL HARMONICS.

cm. from the base, and furthermore at a point 25 cm. from the base and 50 cm. from the axis.
$$\text{Ans., } 29°.6; 47°.6; 71°.2; 25°.8.$$

4. An iron chamber one meter long and twenty centimeters in distance across has its raised surface covered with a supposed non-directing concrete one centimeter thick. One end and the arched surface of the chamber in this way covered are kept at the temperature zero, the opposite end at the temperature of 100°. Find to the closest 10th of a degree the temperature of the center place of the pivot, and of the places of the hub twenty centimeters from each end after the temperatures have quit evolving. Considering that the conductivity of iron is 0.185 and of concrete 0.000162 in C. G. S. units. Track down likewise the temperature of a point on the surface halfway between the closures, and of focuses on a superficial level twenty centimeters from each end. Track down the temperatures of the three places of the pivot, assuming the covering an ideal non-guide, and once more, assuming the covering missing. Disregard the shape of the coating.
$$\text{Ans., } 15°.4; 40°.85; 72°.8; 15°.3; 40°.7; 72°.5; 0°.0; 0°.0; 1°.3.$$

127. If rather than considering the cooling of a chamber as in Art. 123 we need to manage a round and hollow shell whose bended surfaces are co-hub chambers, we are obliged to utilize the Bessel's Elements of the subsequent kind. Leave our conditions of condition alone

$$u = 0 \quad \text{when} \quad z = 0, \qquad u = 0 \quad \text{when} \quad r = a,$$
$$u = f(r) \quad " \quad z = b, \qquad u = 0 \quad " \quad r = c.$$

Then (v. Art. 123)

$$u = \sinh(\mu_k z) \left[J_0(\mu_k r) - \frac{J_0(\mu_k c)}{K_0(\mu_k c)} K_0(\mu_k r) \right]$$

where μ_k is a base of the situation

$$J_0(\mu a) - \frac{J_0(\mu c)}{K_0(\mu c)} K_0(\mu a) = 0 \tag{1}$$

will fulfill Laplace's Condition [(1) Art. 123] and every one of the situations of condition aside from the second.

$$\text{Hence} \quad u = \sum_{k=1}^{k=\infty} A_k \frac{\sinh(\mu_k z)}{\sinh(\mu_k b)} \left[J_0(\mu_k r) - \frac{J_0(\mu_k c)}{K_0(\mu_k c)} K_0(\mu_k r) \right] \tag{2}$$

is the expected arrangement if

$$f(r) = \sum_{k=1}^{k=\infty} A_k \left[J_0(\mu_k r) - \frac{J_0(\mu_k c)}{K_0(\mu_k c)} K_0(\mu_k r) \right]. \tag{3}$$

The turn of events (3) is effectively gotten.

Call the bracket for curtness $B_0(\mu_k r)$. Then, at that point, by the technique of Art. 125 we get in the event that we coordinate over our round and hollow shell

$$\int_a^c r B_0(\mu_k r) B_0(\mu_l r) dr = 0 \tag{4}$$

on the off chance that μ_k and μ_l are foundations of (1); and by a simple augmentation of (10) Art. 122

$$\int_a^c r [B_0(\mu_k r)]^2 dr = \tfrac{1}{2} \{ c^2 [B_0'(\mu_k c)]^2 - a^2 [B_0'(\mu_k a)]^2 \}. \tag{5}$$

CYLINDRICAL HARMONICS.

Deciding the coefficients in (3) as in Art. 124 and disentangling by the help of (4) we have

$$A_k = \frac{2\int_a^c rf(r)\left[J_0(\mu_k r) - \frac{J_0(\mu_k c)}{K_0(\mu_k c)}K_0(\mu_k r)\right]dr}{c^2\left[J_0'(\mu_k c) - \frac{J_0(\mu_k c)}{K_0(\mu_k c)}K_0'(\mu_k c)\right]^2 - a^2\left[J_0'(\mu_k a) - \frac{J_0(\mu_k c)}{K_0(\mu_k c)}K_0'(\mu_k a)\right]^2}. \quad (6)$$

EXAMPLE.

In the event that a film limited by concentric circles of range a and span b, and secured at the edges, is at first mutilated into a structure balanced with deference to the middle, and afterward permitted to vibrate

$$y = \sum_{k=1}^{k=\infty} A_k \cos(\mu_k ct)\left[J_0(\mu_k r) - \frac{J_0(\mu_k b)}{K_0(\mu_k b)}K_0(\mu_k r)\right]$$

where A_k is gotten from (6) Art. 127 by supplanting c by b.

128. If in the cooling of a chamber $u = 0$ when $z = 0$, $u = 0$ when $z = b$, furthermore, $u = f(z)$ when $r = a$, the issue is effortlessly tackled.

If in (2) and (3) Art. 123 μ is supplanted by μi we can promptly get

$$z = \sin(\mu z)J_0(\mu r i)$$

and

$$z = \cos(\mu z)J_0(\mu r i)$$

as specific arrangements of Laplace's Situation [(1) Art. 123]; and

$$J_0(xi) = 1 + \frac{x^2}{2^2} + \frac{x^4}{2^2.4^2} + \frac{x^6}{2^2.4^2.6^2} + \cdots \quad (1)$$

also, is genuine.

$$f(z) = \sum_{k=1}^{k=\infty} A_k \sin\frac{k\pi z}{b}$$

where

$$A_k = \frac{2}{b}\int_a^b f(z)\sin\frac{k\pi z}{b}dz \quad (2)$$

by Art. 31 (7) and (8).

Hence

$$u = \sum_{k=1}^{k=\infty} A_k \sin\frac{k\pi z}{b}\frac{J_0\left(\frac{k\pi r i}{b}\right)}{J_0\left(\frac{k\pi a i}{b}\right)} \quad (3)$$

is our required solution.

EXAMPLES.

CYLINDRICAL HARMONICS. 191

1. Assuming the chamber is empty and we have $u = 0$ when $z = 0$, $u = 0$ when $z = b$, $u = 0$ when $r = c$, and $u = f(z)$ when $r = a$; then

$$u = \sum_{k=1}^{k=\infty} A_k \sin \frac{k\pi z}{b} \left[\frac{J_0\left(\frac{k\pi ri}{b}\right)}{J_0\left(\frac{k\pi ci}{b}\right)} - \frac{\overline{K_0}\left(\frac{k\pi ri}{b}\right)}{\overline{K_0}\left(\frac{k\pi ci}{b}\right)} \right] \div \left[\frac{J_0\left(\frac{k\pi ai}{b}\right)}{J_0\left(\frac{k\pi ci}{b}\right)} - \frac{\overline{K_0}\left(\frac{k\pi ai}{b}\right)}{\overline{K_0}\left(\frac{k\pi ci}{b}\right)} \right]$$

where A_k has the worth yielded (2) Art. 128, and

$$\overline{K_0}(xi) = K_0(xi) - J_0(xi) \log i$$
$$= J_0(xi) \log x - \frac{x^2}{2^2} - \frac{x^4}{2^2 \cdot 4^2}\left(\frac{1}{1}+\frac{1}{2}\right) - \frac{x^6}{2^2 \cdot 4^2 \cdot 6^2}\left(\frac{1}{1}+\frac{1}{2}+\frac{1}{3}\right) - \cdots$$

[v. (4) Art. 120], and is real.

2. An empty chamber 6 feet in length whose inward surface has the span 3 inches, also, whose external surface has the span 1 foot, has its bases and external surface kept at the temperature 0Â°, and its internal surface at the temperature 100Â°, until the extremely durable condition of temperatures is laid out; track down the temperatures of two focuses in a plane lined up with the bases and somewhere between them, one of which is 6 inches and the other 9 crawls from the pivot. *Ans.*, 49Â°.6; 20Â°.2.

129. If in the issue of Art. 123 the temperatures of the places of the upper base of the chamber are unsymmetrical so that $u = f(r,\theta)$ when $z = b$, we need to get specific arrangements of Laplace's Situation [(1) Art. 123] for the situation where $D_\phi^2 u$ isn't equivalent to nothing. We promptly see that as

$$u = \sinh(\mu z)[A \cos n\phi + B \sin n\phi] J_n(\mu r)$$

and

$$u = \cosh(\mu z)[A \cos n\phi + B \sin n\phi] J_n(\mu r)$$

are such arrangements, and that

$$u = \sum_{n=0}^{n=\infty} \sum_{k=1}^{k=\infty} \frac{\sinh \mu_k z}{\sinh \mu_k b}[A_{n,k} \cos n\phi + B_{n,k} \sin n\phi] J_n(\mu_k r) \qquad (1)$$

is the arrangement of the given issue if

$$f(r,\phi) = \sum_{n=0}^{n=\infty} \sum_{k=1}^{k=\infty} (A_{n,k} \cos n\phi + B_{n,k} \sin n\phi) J_n(\mu_k r) \qquad (2)$$

where μ_k is a foundation of the situation

$$\frac{J_n(\mu a)}{\mu^n a^n} = 0. \qquad (3)$$

CYLINDRICAL HARMONICS.

EXAMPLES.

1. Show that

$$\int_0^a r J_n(\mu_k r) J_n(\mu_l r) dr$$
$$= \frac{a}{\mu_k^2 - \mu_l^2}[\mu_l J_n(\mu_k a) J_n'(\mu_l a) - \mu_k J_n(\mu_l a) J_n'(\mu_k a)]$$
$$= \frac{a}{\mu_k^2 - \mu_l^2}[\mu_k J_n(\mu_l a) J_{n+1}(\mu_k a) - \mu_l J_n(\mu_k a) J_{n+1}(\mu_l a)].$$

2. Show that

$$\int_0^a r[J_n(\mu_k r)]^2 dr$$
$$= \frac{1}{2}\left[a^2 (J_n'(\mu_k a))^2 + \left(a^2 - \frac{n^2}{\mu_k^2}\right)(J_n(\mu_k a))^2\right]$$
$$= \frac{a^2}{2}[(J_n(\mu_k a))^2 + (J_{n+1}(\mu_k a))^2] - \frac{na}{\mu_k} J_n(\mu_k a) J_{n+1}(\mu_k a).$$

3. Show that in Art. 129

$$A_{0,k} = \frac{1}{\pi} \frac{\int_0^{2\pi} d\phi \int_0^a r f(r,\phi) J_0(\mu_k r) dr}{a^2 [J_1(\mu_k a)]^2},$$

$$B_{0,k} = 0,$$

$$A_{n,k} = \frac{2}{\pi} \frac{\int_0^{2\pi} d\phi \int_0^a r f(r,\phi) \cos n\phi J_n(\mu_k r) dr}{a^2 [J_{n+1}(\mu_k a)]^2},$$

$$B_{n,k} = \frac{2}{\pi} \frac{\int_0^{2\pi} d\phi \int_0^a r f(r,\phi) \sin n\phi J_n(\mu_k r) dr}{a^2 [J_{n+1}(\mu_k a)]^2}.$$

4. Get the coefficients for the situation where the curved surface of the chamber is impenetrable to heat.

5. Acquire the coefficients for the situation where the curved surface of the chamber is presented to air at the temperature zero.

6. Show that assuming that in a drumhead issue of Art. 11 the underlying contortion is unsymmetrical, so we need to address the condition [XI] Art. 1 subject to the circumstances $z = f(r, \phi)$ when $t = 0$, $D_t z = 0$ when $t = 0$, $z = 0$ when $r = a$, the arrangement is

$$z = \sum_{n=0}^{n=\infty} \sum_{k=1}^{k=\infty} \cos(\mu_k ct)(A_{n,k} \cos n\phi + B_{n,k} \sin n\phi) J_n(\mu_k r)$$

where $A_{0,k}$, $B_{0,k}$, $A_{n,k}$, and $B_{n,k}$ have the qualities given in Ex. 3.

7. What alterations do the assertions made in Ex. 2, Art. 126, need to cause them to apply to the unsymmetrical case treated in Ex. 6?

Show that any conceivable nodal framework in Ex. 6 is made out of concentric circles and of radii whose external furthest points are equidistant. v. Rayleigh's Sound, Vol. I., Arts. (202-207).

CYLINDRICAL HARMONICS.

8. Take care of the issue of Art. 127 and of Art. 127. Ex. for the unsymmetrical case. *Suggestion:* $AJ_n(x) + BK_n(x)$ is an answer of Bessel's Equation.

9. Take care of the issue of Art. 128 and of Art. 128. Ex. 1. for the situation where $u = f(z, \phi)$ when $r = a$. *Suggestion:* $u = \sin \mu z (A \cos n\phi + B \sin n\phi) J_n(\mu r i)$ is an answer of Laplace's Situation, and $f(z, \phi)$ can be formed into a twofold Fourier's Series [v. (15) Art. 71].

10. Show that in managing a wedge cut from a chamber via planes gone through the pivot, or with a film as a round area, it could be important to utilize Bessel's Elements of fragmentary or incommensurable orders.

11. *Bernouilli's Problem* (v. Part IX). In taking into account little cross over vibrations of a uniform, weighty, adaptable, inelastic string secured toward one side furthermore, at first misshaped into some given bend, we need to tackle the condition $D_t^2 y = c^2(x D_x^2 y + D_x y)$, dependent upon the circumstances $D_t y = 0$ when $t = 0$, $y = f(x)$ when $t = 0$, $y = 0$ when $x = a$; the beginning being taken at the distance a underneath the place of suspension and the hub of X taken vertical.

Show that
$$y = \sum_{k=1}^{k=\infty} A_k \cos \mu_k ct \, B_0(\mu_k^2 x),$$

where and $B_0(x) = 1 - \frac{x}{1^2} + \frac{x^2}{1^2 \cdot 2^2} - \frac{x^3}{1^2 \cdot 2^2 \cdot 3^2} + \cdots$

$$= J_0(2\sqrt{x})$$

furthermore, μ_k is a base of the situation

$$B_0(\mu^2 a) = J_0(2\mu \sqrt{a}) = 0,$$

and
$$A_k = \frac{\int_0^a f(x) B_0(\mu_k^2 x) dx}{\mu^2 a^2 [B_0'(\mu_k^2 a)]^2} = \frac{\int_0^a f(x) J_0(2\mu_k \sqrt{x}) dx}{a [J_1(2\mu_k \sqrt{a})]^2}.$$

12. As a basic case under Model 10 think about the vibrations of a roundabout film secured at the border and furthermore along a sweep and afterward at first misshaped (v. Rayleigh's Sound, Art. 207). For this situation we should change the recipe given in Ex. 6 by exiting the terms including $\cos n\phi$ and by taking $n = \frac{m}{2}$. The necessary arrangement is

$$z = \sum_{m=1}^{m=\infty} \sum_{k=1}^{k=\infty} B_{m,k} \cos \mu_k ct \sin \frac{m\phi}{2} J_{\frac{m}{2}}(\mu_k r)$$

where μ_k is a root of
$$\frac{J_{\frac{m}{2}}(\mu a)}{\mu^{\frac{m}{2}} a^{\frac{m}{2}}} = 0$$

and
$$B_{m,k} = \frac{2}{\pi} \frac{\int_0^{2\pi} d\phi \int_0^a r f(r, \phi) \sin \frac{m\phi}{2} J_{\frac{m}{2}}(\mu_k r) dr}{a^2 [J_{\frac{m}{2}}'(\mu_k a)]^2}.$$

CYLINDRICAL HARMONICS.

For the terms in which m is odd, $J_{\frac{m}{2}}(x)$ can be promptly acquired from (13) Art. 122, which will turn into a limited total.

For instance, (13) Art. 122 gives the qualities

$$J_{\frac{1}{2}}(x) = \sqrt{\frac{2}{\pi x}} \sin x; \quad J_{\frac{3}{2}}(x) = \sqrt{\frac{2}{\pi x}} \left(\frac{1}{x} \sin x - \cos x \right);$$

$$J_{\frac{5}{2}}(x) = -\sqrt{\frac{2}{\pi x}} \left[\left(1 + \frac{3}{x^2} \right) \sin x + \frac{3}{x} \cos x \right]; \quad \&c.$$

13. The topic of the progression of intensity in three aspects includes an issue similar to the last.

Assume the underlying temperatures of all places in a circle of range c given, furthermore, let the surface be kept at the temperature zero. Then, at that point, we need to tackle the condition

$$D_t u = \frac{a^2}{r^2} \left[D_r(r^2 D_r u) + \frac{1}{\sin \theta} D_\theta (\sin \theta D_\theta u) + \frac{1}{\sin^2 \theta} D_\phi^2 u \right] \qquad (1)$$

([IV] Art. 1) dependent upon the circumstances

$$u = 0 \quad \text{when} \quad r = c,$$
$$u = f(r, \theta, \phi) \quad \text{when} \quad t = 0.$$

In the event that we expect $u = T.R.V$ where T is an element of t just, R of r just, and V of θ and ϕ no one but, (1) can be separated into

$$\frac{dT}{dt} + a^2 \alpha^2 T = 0 \qquad (2)$$

$$m(m+1)V + \frac{1}{\sin \theta} D_\theta (\sin \theta D_\theta V) + \frac{1}{\sin^2 \theta} D_\phi^2 V = 0 \qquad (3)$$

and

$$\frac{d^2 R}{dr^2} + \frac{2}{r} \frac{dR}{dr} + \left[\alpha^2 - \frac{m(m+1)}{r^2} \right] R = 0. \qquad (4)$$

Subsequently $T = e^{-a^2 \alpha^2 t}$, $V = Y_m(\mu, \phi)$ [v. Art. 102 (2)], and R is still to be found. In the event that in (4) we let $x = \alpha r$ and $z = R\sqrt{\alpha r}$ it becomes

$$\frac{d^2 z}{dx^2} + \frac{1}{x} \frac{dz}{dx} + \left[1 - \frac{(m+\frac{1}{2})^2}{x^2} \right] z = 0$$

which is fulfilled by $z = J_{m+\frac{1}{2}}(x)$. (v. Art. 17.)

Therefore $$R = \frac{1}{\sqrt{\alpha r}} J_{m+\frac{1}{2}}(\alpha r).$$

$$f(r, \theta, \phi) = \frac{1}{4\pi} \sum_{m=0}^{m=\infty} (2m+1) \int_0^{2\pi} d\phi_1 \int_0^\pi f(r, \theta_1, \phi_1) P_m(\cos \gamma) \sin \theta_1 d\theta_1 \qquad \text{by (3) Art. 114,}$$

$$= \sum_{m=0}^{m=\infty} \sum_{n=0}^{n=m} [A_{m,n} f_{m,n}(r) \cos n\phi + B_{m,n} F_{m,n}(r) \sin n\phi] P_m^n(\mu).$$

$$\sqrt{r} f_{m,n}(r) = \sum_{k=0}^{k=\infty} C_{m,n,k} J_{m+\frac{1}{2}}(\alpha_k r)$$

CYLINDRICAL HARMONICS.

where α_k is a base of the situation

$$\frac{J_{m+\frac{1}{2}}(\alpha c)}{(\alpha c)^{m+\frac{1}{2}}} = 0$$

and

$$C_{m,n,k} = \frac{2\int_0^c r^{\frac{3}{2}} f_{m,n}(r) J_{m+\frac{1}{2}}(\alpha_k r) dr}{c^2 [J'_{m+\frac{1}{2}}(\alpha_k c)]^2}.$$

$$\sqrt{r} F_{m,n}(r) = \sum_{m=0}^{m=\infty} D_{m,n,k} J_{m+\frac{1}{2}}(\alpha_k r)$$

where

$$D_{m,n,k} = \frac{2\int_0^c r^{\frac{3}{2}} F_{m,n}(r) J_{m+\frac{1}{2}}(\alpha_k r) dr}{c^2 [J'_{m+\frac{1}{2}}(\alpha_k c)]^2}.$$

The last arrangement is

$$u = \frac{1}{\sqrt{r}} \sum_{m=0}^{m=\infty} \sum_{n=0}^{n=m} \left[P_m^n(\mu) \sum_{k=1}^{k=\infty} (A_{m,n} C_{m,n,k} \cos n\phi + B_{m,n} D_{m,n,k} \sin n\phi) e^{-a^2 \alpha_k^2 t} J_{m+\frac{1}{2}}(\alpha_k r) \right]$$

cf. Riemann, Par. Dif. Gl., §§ 72 and 73.

Chapter 8

LAPLACE'S EQUATION

130. *Orthogonal Curvilinear Coördinates.*

If
$$\left.\begin{array}{l}F_1(x,y,z)=\rho_1\\F_2(x,y,z)=\rho_2\\F_3(x,y,z)=\rho_3\end{array}\right\} \quad (1)$$

are the conditions in rectangular coördinates of three surfaces that are commonly opposite regardless of what the upsides of ρ_1, ρ_2, and ρ_3, the boundaries ρ_1, ρ_2, and ρ_3, might be viewed as a bunch of coördinates for a mark of crossing point of the three surfaces, as in when ρ_1, ρ_2, ρ_3 are given the point in not set in stone, and when the fact is given the comparing values of ρ_1, ρ_2, ρ_3, can be found.

From conditions (1) x, y, and z can be communicated as far as ρ_1, ρ_2, and ρ_3. Assume this done. On the off chance that now x, y, z are the rectangular coördinates of the point $\rho_1 = a$, $\rho_2 = b$, $\rho_3 = c$, the rectangular coördinates of the places $\rho_1 = a + d\rho_1$, $\rho_2 = b$, $\rho_3 = c$, are clearly $x + D_{\rho_1}x.d\rho_1 + \epsilon_1$, $y + D_{\rho_1}y.d\rho_1 + \epsilon_2$, $z + D_{\rho_1}z.d\rho_1 + \epsilon_3$, where ϵ_1, ϵ_2, and ϵ_3 are infinitesimals of higher request than $d\rho_1$. Consequently the square of the distance between the focuses will contrast by an minuscule of higher request than that of $d\rho_1^2$ from dn_1^2 where

$$dn_1^2 = [(D_{\rho_1}x)^2 + (D_{\rho_1}y)^2 + (D_{\rho_1}z)^2]d\rho_1^2.$$

Let
$$\left.\begin{array}{l}\dfrac{1}{h_1^2} = (D_{\rho_1}x)^2 + (D_{\rho_1}y)^2 + (D_{\rho_1}z)^2\\[4pt]\dfrac{1}{h_2^2} = (D_{\rho_2}x)^2 + (D_{\rho_2}y)^2 + (D_{\rho_2}z)^2\\[4pt]\dfrac{1}{h_3^2} = (D_{\rho_3}x)^2 + (D_{\rho_3}y)^2 + (D_{\rho_3}z)^2.\end{array}\right\} \quad (2)$$

Then, at that point, assuming dn_1 is the component of length ordinary to the surface $\rho_1 = a$, dn_2 typical to $\rho_2 = b$, and dn_3 ordinary to $\rho_3 = c$

$$dn_1 = \frac{d\rho_1}{h_1}, \quad dn_2 = \frac{d\rho_2}{h_2}, \quad dn_3 = \frac{d\rho_3}{h_3}. \quad (3)$$

The component of surface dS_1 on a superficial level $\rho_1 = a$ is effectively seen to be

$$dS_1 = \frac{d\rho_2 d\rho_3}{h_2 h_3}; \quad (4)$$

ELLIPSOIDAL HARMONICS.

furthermore, the component of volume dv is

$$dv = \frac{d\rho_1 d\rho_2 d\rho_3}{h_1 h_2 h_3}. \tag{5}$$

EXAMPLE.

Show that
$$h_1^2 = (D_x\rho_1)^2 + (D_y\rho_1)^2 + (D_z\rho_1)^2$$
$$h_2^2 = (D_x\rho_2)^2 + (D_y\rho_2)^2 + (D_z\rho_2)^2$$
$$h_3^2 = (D_x\rho_3)^2 + (D_y\rho_3)^2 + (D_z\rho_3)^2.$$

Suggestion: On the off chance that h_1 has the worth just given $\frac{D_x\rho_1}{h_1}$, $\frac{D_y\rho_1}{h_1}$, $\frac{D_z\rho_1}{h_1}$ are the bearing cosines of the typical at some random mark of $\rho_1 = a$. (v. Int. Cal. page 161.) Then, at that point,

$$dn_1 = \frac{D_x\rho_1}{h_1}dx + \frac{D_y\rho_1}{h_1}dy + \frac{D_z\rho_1}{h_1}dz = \frac{1}{h_1}d\rho_1.$$

131. *Laplace's Condition in symmetrical curvilinear coÃ¶rdinates.*
Assuming we apply the exceptional type of Green's Hypothesis

$$\iiint \nabla^2 V \, dx\,dy\,dz = \int D_n V \, dS \qquad \text{(v. Art. 98)}$$

to the space limited by the surfaces $\rho_1 = a$, $\rho_2 = b$, $\rho_3 = c$, $\rho_1 = a + d\rho_1$, $\rho_2 = b + d\rho_2$, $\rho_3 = c + d\rho_3$, we have

$$\frac{\nabla^2 V d\rho_1 d\rho_2 d\rho_3}{h_1 h_2 h_3} =$$

$$-h_1 D_{\rho_1} V \frac{d\rho_2 d\rho_3}{h_2 h_3} + h_1 D_{\rho_1} V \frac{d\rho_2 d\rho_3}{h_2 h_3} + D_{\rho_1}\left(\frac{h_1}{h_2 h_3}D_{\rho_1}V\right) d\rho_1 d\rho_2 d\rho_3$$
$$-h_2 D_{\rho_2} V \frac{d\rho_3 d\rho_1}{h_3 h_1} + h_2 D_{\rho_2} V \frac{d\rho_3 d\rho_1}{h_3 h_1} + D_{\rho_2}\left(\frac{h_2}{h_3 h_1}D_{\rho_2}V\right) d\rho_1 d\rho_2 d\rho_3$$
$$-h_3 D_{\rho_3} V \frac{d\rho_1 d\rho_2}{h_1 h_2} + h_3 D_{\rho_3} V \frac{d\rho_1 d\rho_2}{h_1 h_2} + D_{\rho_3}\left(\frac{h_3}{h_1 h_2}D_{\rho_3}V\right) d\rho_1 d\rho_2 d\rho_3;$$

whence

$$\nabla^2 V = h_1 h_2 h_3 \left[D_{\rho_1}\left(\frac{h_1}{h_2 h_3}D_{\rho_1}V\right) + D_{\rho_2}\left(\frac{h_2}{h_3 h_1}D_{\rho_2}V\right) + D_{\rho_3}\left(\frac{h_3}{h_1 h_2}D_{\rho_3}V\right)\right], \tag{6}$$

furthermore, Laplace's Condition in our curvilinear framework is

$$h_1 h_2 h_3 \left[D_{\rho_1}\left(\frac{h_1}{h_2 h_3}D_{\rho_1}V\right) + D_{\rho_2}\left(\frac{h_2}{h_3 h_1}D_{\rho_2}V\right) + D_{\rho_3}\left(\frac{h_3}{h_1 h_2}D_{\rho_3}V\right)\right] = 0. \tag{7}$$

Assuming it happens that $\nabla^2 \rho_1 = 0$, $V = \rho_1$ will fulfill (7) and we will have $h_1 h_2 h_3 D_{\rho_1}\left(\frac{h_1}{h_2 h_3}\right) = 0$. Likewise if $\nabla^2 \rho_2 = 0$ we have $D_{\rho_2}\left(\frac{h_2}{h_3 h_1}\right) = 0$, also, if $\nabla^2 \rho_3 = 0$ we have $D_{\rho_3}\left(\frac{h_3}{h_1 h_2}\right) = 0$; and consequently (7) lessens to

$$h_1^2 D_{\rho_1}^2 V + h_2^2 D_{\rho_2}^2 V + h_3^2 D_{\rho_3}^2 V = 0 \tag{8}$$

when $\nabla^2 \rho_1 = 0$, $\nabla^2 \rho_2 = 0$, and $\nabla^2 \rho_3 = 0$.

ELLIPSOIDAL HARMONICS.

132. If rather than having the worth of the Potential Capability V given on the outer layer of a circle as in our Circular Consonant issue, we have it given at every one of the focuses on the outer layer of a *oblate spheroid*, and are expected to find its worth at any inside or outer point, we can without much of a stretch get an answer by techniques in no fundamental regard unique in relation to those generally utilized, if by some stroke of good luck we properly pick our arrangement of coÃ¶rdinates.

On the off chance that we take a circle and a hyperbola having similar foci, and rotate them about the minor pivot of the oval, we will get a couple of surfaces which are opposite together; a plane through the pivot of insurgency will cut both the *spheroid* and the *hyperboloid* symmetrically.

The conditions of the three surfaces can be composed:- - -

$$F_1(x,y,z,\lambda) = \frac{x^2}{\lambda^2} + \frac{y^2}{\lambda^2 - b^2} + \frac{z^2}{\lambda^2} - 1 = 0 \tag{1}$$

$$F_2(x,y,z,\mu) = \frac{x^2}{\mu^2} + \frac{y^2}{\mu^2 - b^2} + \frac{z^2}{\mu^2} - 1 = 0 \tag{2}$$

$$F_3(x,y,z,\nu) = z - \nu x = 0, \tag{3}$$

where $\lambda^2 > b^2 > \mu^2$, $2b$ being the distance between the foci.

For all upsides of λ, μ, and ν reliable with the disparity above composed the surfaces (1), (2), (3) converge in genuine focuses and cut symmetrically.

λ, μ, and ν can be decided to such an extent that the surfaces will converge in any given point, and in this manner can be taken as a bunch of curvilinear coÃ¶rdinates, and Laplace's Condition can be communicated with regards to them by the guide of Equation [xv] Art. 1.

From (1), (2), and (3) we promptly get

$$\left. \begin{array}{l} x^2 = \dfrac{\lambda^2 \mu^2}{b^2(1+\nu^2)} \\[4pt] y^2 = \dfrac{(\lambda^2 - b^2)(b^2 - \mu^2)}{b^2} \\[4pt] z^2 = \dfrac{\lambda^2 \mu^2 \nu^2}{b^2(1+\nu^2)}; \end{array} \right\} \tag{4}$$

whence $\quad D_\lambda x = \frac{\mu}{b\sqrt{1+\nu^2}}, \quad D_\lambda y = \frac{\lambda}{b}\sqrt{\frac{b^2-\mu^2}{\lambda^2-b^2}}, \quad D_\lambda z = \frac{\mu\nu}{b\sqrt{1+\nu^2}};$

andand $\dfrac{1}{h_1^2} = \dfrac{\lambda^2 - \mu^2}{\lambda^2 - b^2} \tag{5}$

[v. 130 (2)]. In this way we get

$$\frac{1}{h_2^2} = \frac{\lambda^2 - \mu^2}{b^2 - \mu^2} \tag{6}$$

andand $\dfrac{1}{h_3^2} = \dfrac{\lambda^2 \mu^2}{b^2(1+\nu^2)^2}, \tag{7}$

what's more, [xv] Art. 1 becomes

$$\frac{\mu}{b(1+\nu^2)\sqrt{b^2-\mu^2}} D_\lambda[\lambda\sqrt{\lambda^2 - b^2}.D_\lambda V]$$
$$+ \frac{\lambda}{b(1+\nu^2)\sqrt{\lambda^2 - b^2}} D_\mu[\mu\sqrt{b^2 - \mu^2}.D_\mu V]$$
$$+ \frac{b(\lambda^2 - \mu^2)}{\lambda\mu\sqrt{(\lambda^2 - b^2)(b^2 - \mu^2)}} D_\nu[(1+\nu^2)D_\nu V] = 0, \tag{8}$$

ELLIPSOIDAL HARMONICS.

which is Laplace's Condition as far as our *Spheroidal Coördinates* λ, μ, and ν.

In the event that now instead of λ, μ, and ν we can present some capability of λ, some capability of μ, and some capability of ν which, subsequently, will address the same arrangement of symmetrical surfaces, and in the event that we can pick these capabilities α, β, also, γ, which obviously are elements of x, y, and z, so that $\nabla^2\alpha = 0$, $\nabla^2\beta = 0$, and $\nabla^2\gamma = 0$, condition (8) should decrease to the straightforward and balanced structure given in [XVI] Art. 1.

These capabilities α, β, and γ are effortlessly found. Condition (8) is $\nabla^2 V = 0$ communicated as far as λ, μ, and ν. Expect that V is an element of λ as it were; then $D_\mu V = 0$, and $D_\nu V = 0$, and (8) decreases to

$$D_\lambda[\lambda\sqrt{\lambda^2-b^2}\cdot D_\lambda V] = 0$$

whence
$$\lambda\sqrt{\lambda^2-b^2}\,\frac{dV}{d\lambda} = c_1,$$

$$dV = \frac{c_1 d\lambda}{\lambda\sqrt{\lambda^2-b^2}},$$

and
$$V = \frac{c_1}{b}\sec^{-1}\frac{\lambda}{b},$$

furthermore, is a component of λ which fulfills Laplace's Condition.

Accept this as α leaving c_1 at present dubious, so that

$$d\alpha = \frac{c_1 d\lambda}{\lambda\sqrt{\lambda^2-b^2}} \quad \text{and} \quad \alpha = \frac{c_1}{b}\sec^{-1}\frac{\lambda}{b}.$$

Similarly we get

$$d\beta = \frac{c_2 d\mu}{\mu\sqrt{b^2-\mu^2}} \quad \text{and} \quad \beta = \frac{c_2}{b}\text{sech}^{-1}\frac{\mu}{b},$$

(v. Int. Cal. Art. 46, Ex.)

$$d\gamma = \frac{c_3 d\nu}{1+\nu^2}, \quad \text{and} \quad \gamma = c_3\tan^{-1}\nu.$$

Subbing these qualities in (8) and taking $c_1 = -c_2 = b$, and $c_3 = 1$, (8) decreases immediately to

$$\frac{D_\alpha^2 V}{\lambda^2} + \frac{D_\beta^2 V}{\mu^2} + \frac{\lambda^2-\mu^2}{\lambda^2\mu^2}D_\gamma^2 V = 0, \tag{9}$$

or since
$$\lambda = b\sec\alpha, \quad \mu = b\,\text{sech}\,\beta, \quad \text{and} \quad \nu = \tan\gamma, \tag{10}$$

to
$$\cos^2\alpha\, D_\alpha^2 V + \cosh^2\beta\, D_\beta^2 V + (\cosh^2\beta - \cos^2\alpha)D_\gamma^2 V = 0 \tag{11}$$

which is Laplace's Condition as far as what we might call *Normal Oblate Spheroidal Coördinates*.

In utilizing (11) it is to be noticed that the point whose coördinates are (α,β,γ) is the place of convergence of an oblate spheroid whose semi-tomahawks are $b\sec\alpha$ furthermore, $b\tan\alpha$, an unparted hyperboloid of upheaval whose semi-tomahawks are $b\,\text{sech}\,\beta$ and $b\tanh\beta$, and a plane containing the pivot of the framework and making the point γ with a proper plane; and that assuming the hub of upset is the pivot of Y and the decent plane is the plane of XY, the rectangular coördinates of (α,β,γ) are

$$x = b\sec\alpha\,\text{sech}\,\beta\cos\gamma, \quad y = b\tan\alpha\tanh\beta, \quad z = b\sec\alpha\,\text{sech}\,\beta\sin\gamma \tag{12}$$

[v. (4)].

Assuming that now we let α range from 0 to $\frac{\pi}{2}$, β from $-\infty$ to ∞, and γ from 0 to 2π, we will have the option to address all focuses in space; and assuming we concur that negative upsides of β will have a place with focuses under a plane through the beginning and opposite to the pivot of insurgency and positive upsides of β to focuses over that plane, not exclusively will we have no vagueness, yet in addition the rectangular coördinates of any point as yielded (12) will have their legitimate signs.

ELLIPSOIDAL HARMONICS.

EXAMPLES.

1. On the off chance that the spheroid is a *prolate* spheroid, the circle and confocal hyperbola should be spun about the significant hub of the circle, and the plane should contain that hub. Instead of conditions (1), (2), and (3) of Art. 132 we have, then, at that point,

$$\frac{x^2}{\lambda^2} + \frac{y^2}{\lambda^2 - b^2} + \frac{z^2}{\lambda^2 - b^2} - 1 = 0$$

$$\frac{x^2}{\mu^2} + \frac{y^2}{\mu^2 - b^2} + \frac{z^2}{\mu^2 - b^2} - 1 = 0$$

$$z - \nu y = 0$$

where and $\lambda^2 > b^2 > \mu^2$.

$$h_1^2 = \frac{\lambda^2 - b^2}{\lambda^2 - \mu^2}, \quad h_2^2 = \frac{b^2 - \mu^2}{\lambda^2 - \mu^2}, \quad h_3^2 = \frac{b^2(1 + \nu^2)^2}{(\lambda^2 - b^2)(b^2 - \mu^2)}.$$

Laplace's Condition becomes

$$\frac{1}{b^2(1+\nu^2)} D_\lambda[(\lambda^2 - b^2)D_\lambda V] + \frac{1}{b^2(1+\nu^2)} D_\mu[(b^2 - \mu^2)D_\mu V]$$
$$+ \frac{\lambda^2 - \mu^2}{(\lambda^2 - b^2)(b^2 - \mu^2)} D_\nu[(1+\nu^2)D_\nu V] = 0. \qquad (1)$$

(1) decreases to $\qquad \dfrac{D_\alpha^2 V}{\lambda^2 - b^2} + \dfrac{D_\beta^2 V}{b^2 - \mu^2} + \dfrac{\lambda^2 - \mu^2}{(\lambda^2 - b^2)(b^2 - \mu^2)} D_\gamma^2 V = 0, \qquad (2)$

where $\qquad d\alpha = -\dfrac{b\, d\lambda}{\lambda^2 - b^2}, \quad d\beta = \dfrac{b\, d\mu}{b^2 - \mu^2}, \quad d\gamma = \dfrac{d\nu}{1+\nu^2},$

$$\alpha = \operatorname{ctnh}^{-1} \frac{\lambda}{b}, \quad \beta = \tanh^{-1} \frac{\mu}{b}, \quad \text{and} \quad \gamma = \tan^{-1} \nu.$$

Since and $\lambda = b \operatorname{ctnh} \alpha, \quad \mu = b \tanh \beta, \quad \text{and} \quad \nu = \tan \gamma$

(2) can be decreased to

$$\sinh^2 \alpha D_\alpha^2 V + \cosh^2 \beta D_\beta^2 V + (\sinh^2 \alpha + \cosh^2 \beta) D_\gamma^2 V = 0. \qquad (3)$$

In utilizing (3) it is to be noticed that the point (α, β, γ) is the mark of convergence of a prolate spheroid whose semi-tomahawks are $b \operatorname{ctnh} \alpha$ and $b \operatorname{csch} \alpha$, a biparted hyperboloid of upset whose semi-tomahawks are $b \tanh \beta$ and $b \operatorname{sech} \beta$, furthermore, a plane containing the hub of insurgency and making the point γ with a fixed plane.

Assuming the decent plane is that of (XY) the rectangular coÃ¶rdinates of any point (α, β, γ) are

$$x = b \operatorname{ctnh} \alpha \tanh \beta, \quad y = b \operatorname{csch} \alpha \operatorname{sech} \beta \cos \gamma, \quad z = b \operatorname{csch} \alpha \operatorname{sech} \beta \sin \gamma,$$

also, α may go from ∞ to 0, β from $-\infty$ to ∞, and γ from 0 to 2π. Negative upsides of β are to be taken for directs lying toward the left of a plane through the beginning opposite to the hub of revolution.

2. Change Laplace's Condition in Round CoÃ¶rdinates [XIII] Art. 1 to the balanced structure

$$\alpha^2 D_\alpha^2 V + \cosh^2 \beta D_\beta^2 V + \cosh^2 \beta D_\gamma^2 V = 0$$

ELLIPSOIDAL HARMONICS.

where
$$\alpha = \frac{1}{r}, \quad \beta = \log\tan\frac{\theta}{2}, \quad \text{and} \quad \gamma = \phi.$$

3. Change Laplace's Condition in Tube shaped Coördinates [XIV] Art. 1 to the balanced structure
$$D_\alpha^2 V + D_\beta^2 V + e^{2\alpha} D_\gamma^2 V = 0$$

where
$$\alpha = \log r, \quad \beta = \phi, \quad \text{and} \quad \gamma = z.$$

133. In every one of the cases we have thought of, it has been not difficult to pass from Laplace's Situation as far as the picked coördinates addressing an symmetrical arrangement of surfaces to the even structure [XVI] Art. 1; and it is clear that our new coördinate α is a worth of V relating to such a dispersion that the surfaces acquired by giving specific qualities to ρ_1 are *equipotential* surfaces; that β is a worth of V comparing to such a appropriation that the surfaces got by giving specific qualities to ρ_2 are equipotential surfaces; and that γ is a worth of V comparing to such a circulation that the surfaces got by giving specific qualities to ρ_3 are equipotential surfaces. α, β, and γ are called by Lamé "*thermometric parameters.*"

The condition that these qualities ought to exist, for a given arrangement of surfaces, that will be, that the dispersion portrayed above ought to be conceivable, is promptly acquired. We will sort out it for α. It is just the condition that V in Laplace's Condition might be a component of ρ_1 alone.

In the event that V is an element of ρ_1 alone

$$D_x V = \frac{dV}{d\rho_1} D_x \rho_1, \quad D_y V = \frac{dV}{d\rho_1} D_y \rho_1, \quad D_z V = \frac{dV}{d\rho_1} D_z \rho_1,$$

$$D_x^2 V = \frac{d^2 V}{d\rho_1^2}(D_x \rho_1)^2 + \frac{dV}{d\rho_1} D_x^2 \rho_1$$

$$D_y^2 V = \frac{d^2 V}{d\rho_1^2}(D_y \rho_1)^2 + \frac{dV}{d\rho_1} D_y^2 \rho_1$$

$$D_z^2 V = \frac{d^2 V}{d\rho_1^2}(D_z \rho_1)^2 + \frac{dV}{d\rho_1} D_z^2 \rho_1.$$

In this way $[(D_x\rho_1)^2 + (D_y\rho_1)^2 + (D_z\rho_1)^2]\frac{d^2V}{d\rho_1^2} + [D_x^2\rho_1 + D_y^2\rho_1 + D_z^2\rho_1]\frac{dV}{d\rho_1} = 0$

whence
$$\frac{D_x^2\rho_1 + D_y^2\rho_1 + D_z^2\rho_1}{(D_x\rho_1)^2 + (D_y\rho_1)^2 + (D_z\rho_1)^2} = -\frac{d^2V}{d\rho_1^2} \div \frac{dV}{d\rho_1}.$$

or *and*
$$\frac{\nabla^2 \rho_1}{h_1^2} = F_1(\rho_1)$$

where $F_1(\rho_1)$ might be any capability of ρ_1 alone. Our necessary circumstances are then, at that point,

$$\left.\begin{array}{l}\dfrac{\nabla^2 \rho_1}{h_1^2} = F_1(\rho_1) \\[4pt] \dfrac{\nabla^2 \rho_2}{h_2^2} = F_2(\rho_2) \\[4pt] \dfrac{\nabla^2 \rho_3}{h_3^2} = F_3(\rho_3)\end{array}\right\} \tag{1}$$

also, when they are satisfied the first curvilinear coördinates ρ_1, ρ_2, ρ_3, compare to conceivable *equipotential* or *isothermal* surfaces, *thermometric parameters* α, β, and γ exist, and the decrease of Laplace's Situation to the balanced structure [XVI] Art. 1 is conceivable.

134. Getting back to our Oblate Spheroid issue of Art. 132 we can continue as expected to separate our condition (**11**) Art. 132.

Expect that $V = L.M.N$, where L is an element of α just, M of β as it were, what's more, N of γ as it were. (11) Art. 132 becomes

$$\frac{\cos^2\alpha}{L}\frac{d^2L}{d\alpha^2} + \frac{\cosh^2\beta}{M}\frac{d^2M}{d\beta^2} + \frac{[\cosh^2\beta - \cos^2\alpha]}{N}\frac{d^2N}{d\gamma^2} = 0$$

or

$$\frac{1}{L}\frac{\cos^2\alpha}{\cosh^2\beta - \cos^2\alpha}\frac{d^2L}{d\alpha^2} + \frac{1}{M}\frac{\cosh^2\beta}{\cosh^2\beta - \cos^2\alpha}\frac{d^2M}{d\beta^2} = -\frac{1}{N}\frac{d^2N}{d\gamma^2}.$$

The main part is free of γ, and the subsequent part is autonomous of α and β, and the two individuals are indistinguishably equivalent. The subsequent part is then autonomous of α, β, and γ and should be consistent; call it n^2. We have, then,

$$\frac{d^2N}{d\gamma^2} + n^2N = 0 \tag{1}$$

and

$$\frac{\cos^2\alpha}{L}\frac{d^2L}{d\alpha^2} + \frac{\cosh^2\beta}{M}\frac{d^2M}{d\beta^2} - n^2(\cosh^2\beta - \cos^2\alpha) = 0. \tag{2}$$

(1) gives us

$$N = A\cos n\gamma + B\sin n\gamma. \tag{3}$$

(2) can be composed

$$\frac{\cos^2\alpha}{L}\frac{d^2L}{d\alpha^2} + n^2\cos^2\alpha = n^2\cosh^2\beta - \frac{\cosh^2\beta}{M}\frac{d^2M}{d\beta^2} = m(m+1),$$

whence

$$\cos^2\alpha\frac{d^2L}{d\alpha^2} + [n^2\cos^2\alpha - m(m+1)]L = 0 \tag{4}$$

and

$$\cosh^2\beta\frac{d^2M}{d\beta^2} + [m(m+1) - n^2\cosh^2\beta]M = 0. \tag{5}$$

Assuming that we present $x = \tanh\beta$ in (5) it becomes

$$(1-x^2)\frac{d^2M}{dx^2} - 2x\frac{dM}{dx} + \left[m(m+1) - \frac{n^2}{1-x^2}\right]M = 0 \tag{6}$$

where since $x = \tanh\beta$ and β may have any worth from $-\infty$ to ∞, x may have any worth between -1 and 1. (6) is a natural condition having for a specific arrangement

$$M = (1-x^2)^{\frac{n}{2}}\frac{d^n P_m(x)}{dx^n} = P_m^n(x) = P_m^n(\tanh\beta). \tag{7}$$

(v. Arts. 101 and 102).

In the event that we present in (4) $x = \tan\alpha$ it lessens to

$$(1+x^2)\frac{d^2L}{dx^2} + 2x\frac{dL}{dx} + \left[\frac{n^2}{1+x^2} - m(m+1)\right]L = 0. \tag{8}$$

(8) is a new condition, yet it tends to be treated as (6) was dealt with in the event that we make careful arrangements to return to the start and follow the means of the treatment of Legendre's Situation.

This work can be saved, notwithstanding, by taking note of that assuming we let $x = \frac{y}{i}$ (8) becomes

$$(1-y^2)\frac{d^2L}{dy^2} - 2y\frac{dL}{dy} + \left[m(m+1) - \frac{n^2}{1-y^2}\right]L = 0$$

furthermore, is indistinguishable in structure with (6). Subsequently

$$L = P_m^n(y) \quad \text{and} \quad L = (1-y^2)^{\frac{n}{2}}\frac{d^n Q_m(y)}{dy^n} \quad \text{(v. Art. 101)},$$

ELLIPSOIDAL HARMONICS.

where $y = I\tan\alpha$, are specific arrangements of (4).

We can keep away from imaginaries in the event that we utilize the qualities

$$L = (-i)^{m-n} P_m^n(y) \quad \text{and} \quad L = i^{m+n+1}(1-y^2)^{\frac{n}{2}} \frac{d^n Q_m(y)}{dy^n}. \tag{9}$$

Since we accepted $V = L.M.N$ we have

$$V = (A\cos n\gamma + B\sin n\gamma)P_m^n(\tanh\beta)(-i)^{m-n}P_m^n(I\tan\alpha)$$

and

$$V = (A\cos n\gamma + B\sin n\gamma)P_m^n(\tanh\beta)i^{m+n+1}\sec^n\alpha\frac{d^n Q_m(I\tan\alpha)}{(d(i\tan\alpha))^n} \tag{10}$$

as specific arrangements of (**11**) Art. 132.

Assuming the issue is even concerning the hub of the spheroid $D_\gamma^2 V = 0$, $n^2 = 0$ and our specific arrangements (10) lessen to

$$V = (-i)^m P_m(I\tan\alpha)P_m(\tanh\beta)$$

and

$$V = i^{m+1} Q_m(I\tan\alpha)P_m(\tanh\beta). \tag{11}$$

On the off chance that, V is given on the outer layer of a spheroid as a component of β and γ, we should communicate it as an element of $\tanh\beta$ and γ, and will be obliged to foster it as far as *Spherical Harmonics* of $\tanh\beta$ and γ by the recipes of Part VII, involving the main condition in (10) for the worth of V at an inward point, and the second for the worth of V at an outer point. In the event that the issue is even, we should foster in Zonal Sounds of $\tanh\beta$ by the equations of Part VI.

A helpful structure for $Q_m(I\tan\alpha)$ is gotten from (2) Art. 100; it is

$$Q_m(I\tan\alpha) = -IP_m(i\tan\alpha)\int_{\tan\alpha}^{\infty}\frac{dx}{(1+x^2)[P_m(xi)]^2}. \tag{12}$$

Hence

$$Q_0(I\tan\alpha) = -i\int_{\tan\alpha}^{\infty}\frac{dx}{1+x^2} = -I\left(\frac{\pi}{2}-\alpha\right). \tag{13}$$

EXAMPLES.

1. A guide as an oblate spheroid whose semi-tomahawks are $b\sec\alpha_0$ and $b\tan\alpha_0$ is accused of power and is viewed as at potential V_0; track down the worth of the likely capability at any inner or outside point.

Here $V_0 = V_0 P_0(\tanh\beta)$. Thus at an inside point

$$V = V_0 \frac{P_0(I\tan\alpha)}{P_0(I\tan\alpha_0)} P_0(\tanh\beta) = V_0, \tag{1}$$

what's more, at an outer point

$$V = V_0 \frac{Q_0(I\tan\alpha)}{Q_0(I\tan\alpha_0)} P_0(\tanh\beta) = V_0 \frac{\left(\frac{\pi}{2}-\alpha\right)}{\left(\frac{\pi}{2}-\alpha_0\right)}. \tag{2}$$

Since V in (2) includes α just, the equipotential surfaces are spheroids confocal with the conductor.

2. The upper portion of an oblate spheroid whose semi-tomahawks are $b\sec\alpha_0$ and $b\tan\alpha_0$ is kept at the temperature solidarity, and the lower half at the temperature zero. Track down the long-lasting temperature at any interior point.

Ans. $u = \frac{1}{2} + \frac{3}{4}\frac{P_1(i\tan\alpha)}{P_1(i\tan\alpha_0)}P_1(\tanh\beta) - \frac{7}{8}\cdot\frac{1}{2}\frac{P_3(i\tan\alpha)}{P_3(i\tan\alpha_0)}P_3(\tanh\beta)+\cdots$

ELLIPSOIDAL HARMONICS.

(v. Art. 93). u might be communicated as far as x, y, and z without serious trouble [v. (12) Art. 132].

$$u = \frac{1}{2} + \frac{3}{4}\frac{y}{c} - \frac{7}{8}\cdot\frac{1}{2}\cdot\frac{1}{2}\frac{[25y^3 - 15y(x^2 + y^2 + z^2 - b^2) - 9b^2 y]}{5c^3 + 3b^2 c} + \cdots$$

if $2c = 2b\tan\alpha_0 =$ minor pivot of spheroid.

135. Let us currently track down the expected capability at an outer point due to the fascination of a strong homogeneous oblate spheroid, utilizing the technique utilized in Arts. 98 and 99.

Consider first the expected capability because of a shell limited by the spheroids for which $\alpha = \phi$ and $\alpha = \phi + d\phi$.

By (1) Art. 98 we have

$$4\pi\rho\kappa = [D_n V_1 - D_n V_2]_{\alpha=\phi}, \tag{1}$$

where ρ is the thickness and κ the thickness of the shell, V_1 the worth of the expected capability at an inward point, and V_2 the worth of the potential capability at an outside point.

Let
$$V_1 = \sum A_m (-i)^m P_m(i\tan\alpha) P_m(\tanh\beta)$$

and
$$V_2 = \sum B_m i^{m+1} Q_m(i\tan\alpha) P_m(\tanh\beta) \qquad \text{[v. (11) Art. 134]}.$$

Since V_1 and V_2 should have a similar worth when $\alpha = \phi$

$$A_m = B_m i^{2m+1}\frac{Q_m(i\tan\phi)}{P_m(i\tan\phi)} = (-1)^m B_m \int_{\tan\phi}^{\infty} \frac{dx}{(1+x^2)[P_m(xi)]^2} \tag{2}$$

[v. (12) Art. 134].

Hence
$$\left.\begin{aligned} V_1 &= \sum i^m B_m P_m(\tanh\beta) P_m(i\tan\alpha) \int_{\tan\phi}^{\infty} \frac{dx}{(1+x^2)[P_m(xi)]^2} \\ V_2 &= \sum i^m B_m P_m(\tanh\beta) P_m(i\tan\alpha) \int_{\tan\alpha}^{\infty} \frac{dx}{(1+x^2)[P_m(xi)]^2} \end{aligned}\right\} \tag{3}$$

$$D_n V_1 = D_\alpha V_1 . D_n\alpha. \qquad D_n V_2 = D_\alpha V_2 . D_n\alpha$$

$$[D_n V_1 - D_n V_2]_{\alpha=\phi} = [D_\alpha V_1 - D_\alpha V_2]_{\alpha=\phi}(D_n\alpha)_{\alpha=\phi}$$
$$= [D_\alpha(V_1 - V_2)]_{\alpha=\phi}[D_n\alpha]_{\alpha=\phi}.$$

$$V_1 - V_2 = \sum i^m B_m P_m(\tanh\beta) P_m(i\tan\alpha) \int_{\tan\phi}^{\tan\alpha} \frac{dx}{(1+x^2)[P_m(xi)]^2}.$$

$$D_\alpha(V_1 - V_2) = \sum i^m B_m P_m(\tanh\beta)\left[P_m(i\tan\alpha)\frac{\sec^2\alpha}{(1+\tan^2\alpha)[P_m(i\tan\alpha)]^2} \right.$$
$$\left. + \frac{dP_m(i\tan\alpha)}{d\alpha}\int_{\tan\phi}^{\tan\alpha} \frac{dx}{(1+x^2)[P_m(xi)]^2}\right].$$

$$D_\alpha[V_1 - V_2]_{\alpha=\phi} = \sum i^m B_m \frac{P_m(\tanh\beta)}{P_m(i\tan\phi)}.$$

$$D_n\alpha = \frac{d\alpha}{dn}$$

ELLIPSOIDAL HARMONICS.

$$dn = \frac{d\rho_1}{h_1} = \frac{d\lambda}{h_1} = \frac{\sqrt{\lambda^2 - \mu^2}}{\sqrt{\lambda^2 - b^2}} d\lambda = b \sec \alpha \sqrt{\tan^2 \alpha + \tanh^2 \beta}.d\alpha \qquad (4)$$

v. Art. 130 (3), and Art. 132 (5) and (10).

$$[D_n \alpha]_{\alpha=\phi} = \frac{1}{b \sec \phi \sqrt{\tan^2 \phi + \tanh^2 \beta}}.$$

Hence

$$[D_n V_1 - D_n V_2]_{\alpha=\phi} = \frac{1}{b \sec \phi \sqrt{\tan^2 \phi + \tanh^2 \beta}} \sum i^m B_m \frac{P_m(\tanh \beta)}{P_m(i \tan \phi)}.$$

$$\kappa = [dn]_{\alpha=\phi} = b \sec \phi \sqrt{\tan^2 \phi + \tanh^2 \beta}.d\phi$$

by (4), and (1) might be composed

$$4\pi \rho b^2 \sec^2 \phi (\tan^2 \phi + \tanh^2 \beta) d\phi = \sum i^m B_m \frac{P_m(\tanh \beta)}{P_m(i \tan \phi)}. \qquad (5)$$

Since $\qquad \tanh^2 \beta = \tfrac{1}{3} P_0(\tanh \beta) + \tfrac{2}{3} P_2(\tanh \beta)$

by (5) Art. 95, to fulfill (5) we should give m the qualities 0 and 2 and

and
$$B_0 = \tfrac{4}{3}\pi \rho b^2 \sec^2 \phi (3\tan^2 \phi + 1) d\phi$$
$$B_2 = \tfrac{4}{3}\pi \rho b^2 \sec^2 \phi (3\tan^2 \phi + 1) d\phi.$$

So that by (3)

$$V_1 = \tfrac{4}{3}\pi \rho b^2 \sec^2 \phi (3\tan^2 \phi + 1) d\phi \Bigg[\int_{\tan \phi}^{\infty} \frac{dx}{1+x^2}$$
$$- P_2(\tanh \beta) P_2(i \tan \alpha) \int_{\tan \phi}^{\infty} \frac{dx}{(1+x^2)[P_2(xi)]^2} \Bigg] \qquad (6)$$

and $\qquad V_2 = \tfrac{4}{3}\pi \rho b^2 \sec^2 \phi (3\tan^2 \phi + 1) d\phi [i Q_0(I \tan \alpha)$
$$+ i^3 P_2(\tanh \beta) Q_2(I \tan \alpha)]. \qquad (7)$$

The expected capability at an outer point because of the strong spheroid for which $\alpha = \alpha_0$ is

$$V = \int_{\phi=0}^{\phi=\alpha_0} V_2 = \tfrac{4}{3}\pi \rho b^2 \sec^2 \alpha_0 \tan \alpha_0 [i Q_0(I \tan \alpha) + i^3 P_2(\tanh \beta) Q_2(I \tan \alpha)]. \qquad (8)$$

Assuming $2a$ is the significant pivot and $2c$ the minor hub of the spheroid

$$\tfrac{4}{3}\pi \rho b^2 \sec^2 \alpha_0 \tan \alpha_0 = \tfrac{4}{3} \frac{\pi \rho a^2 c}{b} = \frac{M}{b}$$

where M is the mass of the spheroid. Accordingly

$$V = \frac{M}{b}[i Q_0(I \tan \alpha) + i^3 P_2(\tanh \beta) Q_2(I \tan \alpha)] \qquad (9)$$

is the necessary worth. (9) can be decreased to

$$V = \frac{M}{b}\left\{ \frac{\pi}{2} - \alpha + \frac{1}{4}\left[\left(\frac{\pi}{2} - \alpha\right)(3\tan^2 \alpha + 1) - 3\tan \alpha\right][3\tanh^2 \beta - 1] \right\}. \qquad (10)$$

EXAMPLES.

1. Separation the condition (3) Ex. 1, Art. 132, for the prolate spheroid, and acquire specific arrangements of the term

$$V = (A\cos n\gamma + B\sin n\gamma)P_m^n(\tanh\beta)P_m^n(\operatorname{ctnh}\alpha),$$

$$V = (A\cos n\gamma + B\sin n\gamma)P_m^n(\tanh\beta)(-1)^{\frac{n}{2}}\operatorname{csch}^n\alpha\frac{d^n Q_m(\operatorname{ctnh}\alpha)}{(d\operatorname{ctnh}\alpha)^n}.$$

2. Separation and settle the conditions of Exs. 2 and 3, Art. 132, and show that they lead to recognizable forms.

3. If in Ex. 1, Art. 132, the guide is a prolate spheroid whose semi-tomahawks are $b\operatorname{ctnh}\alpha_0$ and $b\operatorname{csch}\alpha_0$ show that

$$V = V_0 \text{ at an inner point.} \qquad V = V_0\frac{\alpha}{\alpha_0} \text{ at an outer point.}$$

4. Show that the possible capability at an outer point because of the fascination of a homogeneous strong prolate spheroid is

$$V = \frac{M}{b}[Q_0(\operatorname{ctnh}\alpha) - P_2(\tanh\beta)Q_2(\operatorname{ctnh}\alpha)].$$

Ellipsoidal Harmonics.

136. On the off chance that we are managing a *ellipsoid* rather than a spheroid, we can take as our symmetrical arrangement of surfaces a bunch of *confocal quadrics*;

$$\left.\begin{array}{l}\dfrac{x^2}{\lambda^2} + \dfrac{y^2}{\lambda^2 - b^2} + \dfrac{z^2}{\lambda^2 - c^2} - 1 = 0 \\[6pt] \dfrac{x^2}{\mu^2} + \dfrac{y^2}{\mu^2 - b^2} + \dfrac{z^2}{\mu^2 - c^2} - 1 = 0 \\[6pt] \dfrac{x^2}{\nu^2} + \dfrac{y^2}{\nu^2 - b^2} + \dfrac{z^2}{\nu^2 - c^2} - 1 = 0 \end{array}\right\} \quad (1)$$

where $\lambda^2 > c^2 > \mu^2 > b^2 > \nu^2$. Here the main surface is an ellipsoid, the second an unparted hyperboloid, and the third a biparted hyperboloid. Each of the three chief areas of the framework comprises of confocal conics, and it is notable and is effectively shown that the surfaces cut symmetrically. λ, μ, what's more, ν will be our curvilinear coÃ¶rdinates, and are known as Ellipsoidal CoÃ¶rdinates.

We find without trouble that

$$x^2 = \frac{\lambda^2\mu^2\nu^2}{b^2c^2}, \quad y^2 = \frac{(\lambda^2 - b^2)(\mu^2 - b^2)(b^2 - \nu^2)}{b^2(c^2 - b^2)},$$

$$z^2 = \frac{(\lambda^2 - c^2)(c^2 - \mu^2)(c^2 - \nu^2)}{c^2(c^2 - b^2)}, \quad (2)$$

$$h_1^2 = \frac{(\lambda^2 - b^2)(\lambda^2 - c^2)}{(\lambda^2 - \mu^2)(\lambda^2 - \nu^2)}, \quad h_2^2 = \frac{(\mu^2 - b^2)(c^2 - \mu^2)}{(\mu^2 - \nu^2)(\lambda^2 - \mu^2)},$$

$$h_3^2 = \frac{(b^2 - \nu^2)(c^2 - \nu^2)}{(\lambda^2 - \nu^2)(\mu^2 - \nu^2)}. \quad (3)$$

ELLIPSOIDAL HARMONICS.

To keep away from vagueness, we will guess that of the nine semi-tomahawks in (1) $\sqrt{c^2 - \mu^2}$ is to be taken with the positive sign for a point on the portion of the unparted hyperboloid on which z is positive, and with the negative sign for a point on the half on which z is negative; $\sqrt{b^2 - \nu^2}$ is to be taken with the positive sign for a point on the portion of the biparted hyperboloid on which y is positive, and with the negative sign for a point on the half on which y is negative; ν is to be taken positive for a point on the portion of the biparted hyperboloid on which x is positive, and negative for a point on the half on which x is negative, and that the leftover six are to be generally certain. It follows that our Ellipsoidal CoÃ¶rdinates have the inconvenience that to completely fix a point we want to know not only the upsides of its coÃ¶rdinates λ, μ, and ν, yet the indications of $\sqrt{c^2 - \mu^2}$, and $\sqrt{b^2 - \nu^2}$ too.

We will see later, Art. 139, when we come to present what we might call the *Normal Ellipsoidal CoÃ¶rdinates* α, β, and γ that they are liberated from this hindrance.

It is to be seen that λ may go from c to ∞, μ from b to c, and ν from $-b$ to b.

The component of length opposite to the Ellipsoid is

$$dn = \frac{d\lambda}{h_1} = \sqrt{\frac{(\lambda^2 - \mu^2)(\lambda^2 - \nu^2)}{(\lambda^2 - b^2)(\lambda^2 - c^2)}}.d\lambda. \qquad (4)$$

The component of Ellipsoidal surface is

$$dS = \frac{d\mu d\nu}{h_2 h_3} = (\mu^2 - \nu^2)\sqrt{\frac{(\lambda^2 - \mu^2)(\lambda^2 - \nu^2)}{(\mu^2 - b^2)(c^2 - \mu^2)(b^2 - \nu^2)(c^2 - \nu^2)}}.d\mu d\nu, \qquad (5)$$

furthermore, the component of volume is

$$dv = \frac{d\lambda d\mu d\nu}{h_1 h_2 h_3}$$

$$= \frac{(\lambda^2 - \mu^2)(\lambda^2 - \nu^2)(\mu^2 - \nu^2)}{\sqrt{(\lambda^2 - b^2)(\lambda^2 - c^2)(\mu^2 - b^2)(c^2 - \mu^2)(b^2 - \nu^2)(c^2 - \nu^2)}}d\lambda d\mu d\nu. \qquad (6)$$

The surface vital of some random capability of μ and ν assumed control over the ellipsoid is

$$\int f(\mu,\nu)dS = \int_{-b}^{b} d\nu \int_{b}^{c} [f_1(\mu,\nu) + f_2(\mu,\nu) + f_3(\mu,\nu)$$

$$+ f_4(\mu,\nu)](\mu^2 - \nu^2)\sqrt{\frac{(\lambda^2 - \mu^2)(\lambda^2 - \nu^2)}{(\mu^2 - b^2)(c^2 - \mu^2)(b^2 - \nu^2)(c^2 - \nu^2)}}.d\mu, \qquad (7)$$

where $f_1(\mu,\nu)$, $f_2(\mu,\nu)$, $f_3(\mu,\nu)$ and $f_4(\mu,\nu)$ are the upsides of the given capability on the four fourth of the ellipsoid into which it is separated by the planes of (XY) and (XZ).

Laplace's Condition demonstrates reducible to

$$(\mu^2 - \nu^2)D_\alpha^2 V + (\lambda^2 - \nu^2)D_\beta^2 V + (\lambda^2 - \mu^2)D_\gamma^2 V = 0 \qquad (8)$$

where

$$\alpha = c\int_{c}^{\lambda} \frac{d\lambda}{\sqrt{(\lambda^2 - b^2)(\lambda^2 - c^2)}}, \quad \beta = c\int_{b}^{\mu} \frac{d\mu}{\sqrt{(c^2 - \mu^2)(\mu^2 - b^2)}},$$

$$\gamma = c\int_{0}^{\nu} \frac{d\nu}{\sqrt{(b^2 - \nu^2)(c^2 - \nu^2)}}. \qquad (9)$$

α, β, and γ can be communicated as Elliptic Integrals of the top notch and are

$$\alpha = F\left(\frac{b}{c}, \frac{\pi}{2}\right) - F\left(\frac{b}{c}, \sin^{-1}\frac{c}{\lambda}\right), \quad \beta = F\left(\sqrt{1 - \frac{b^2}{c^2}}, \sin^{-1}\sqrt{\frac{1 - \frac{b^2}{\mu^2}}{1 - \frac{b^2}{c^2}}}\right),$$

$$\gamma = F\left(\frac{b}{c}, \sin^{-1}\frac{\nu}{b}\right); \qquad (10)$$

ELLIPSOIDAL HARMONICS.

whence
$$\lambda = \frac{c}{\operatorname{sn}(K-\alpha)}\left(\bmod \frac{b}{c}\right) = c\frac{\operatorname{dn}\alpha}{\operatorname{cn}\alpha}\left(\bmod \frac{b}{c}\right),$$

$$\mu = \frac{b}{\operatorname{dn}\beta}\left(\bmod \left(1-\frac{b^2}{c^2}\right)^{\frac{1}{2}}\right), \qquad \nu = b\operatorname{sn}\gamma\left(\bmod \frac{b}{c}\right) \tag{11}$$

(v. Int. Cal. Arts. 179, 192, and 196).

137. If in (8) Art. 136 we accept $V = L.M.N$ where L includes α as it were, M includes β just, and N includes γ no one but, (8) can be composed

$$\frac{\mu^2 - \nu^2}{L}\frac{d^2L}{d\alpha^2} + \frac{\lambda^2 - \nu^2}{M}\frac{d^2M}{d\beta^2} + \frac{\lambda^2 - \mu^2}{N}\frac{d^2N}{d\gamma^2} = 0. \tag{1}$$

(1) is too convoluted to be in any way separated by our typical technique.

In the event that, in any case, we let

$$\frac{1}{L}\frac{d^2L}{d\alpha^2} = \sum a_k\lambda^k, \qquad \frac{1}{M}\frac{d^2M}{d\beta^2} = \sum b_k\mu^k, \qquad \frac{1}{N}\frac{d^2N}{d\gamma^2} = \sum c_k\nu^k,$$

substitute in (1) and utilize the way that the outcome should be indistinguishably zero, we observe that the coefficients are zero for all upsides of k with the exception of $k = 0$ and $k = 2$, and that $a_0 = -b_0 = c_0$, and $a_2 = -b_2 = c_2$.

Thusly (1) can be separated into the three conditions

$$\frac{d^2L}{d\alpha^2} = (a_0 + a_2\lambda^2)L$$

$$\frac{d^2M}{d\beta^2} = -(a_0 + a_2\mu^2)M$$

$$\frac{d^2N}{d\gamma^2} = (a_0 + a_2\nu^2)N.$$

We will find it helpful to take a_2 as $m(m+1)$ and a_0 as $-(b^2 + c^2)p$; whence

$$\left.\begin{array}{l}\dfrac{d^2L}{d\alpha^2} and [m(m+1)\lambda^2 - (b^2 + c^2)p]L = 0 \\[2mm] \qquad\qquad \dfrac{d^2M}{d\beta^2} + [m(m+1)\mu^2 - (b^2 + c^2)p]M = 0 \\[2mm] \dfrac{d^2N}{d\gamma^2} and [m(m+1)\nu^2 - (b^2 + c^2)p]N = 0.\end{array}\right\} \tag{2}$$

On the off chance that now in (2) we supplant α, β, and γ by their qualities as far as λ, μ, and ν, we get

$$\left.\begin{array}{l}(\lambda^2 - b^2)(\lambda^2 - c^2)\dfrac{d^2L}{d\lambda^2} + \lambda(\lambda^2 - b^2 + \lambda^2 - c^2)\dfrac{dL}{d\lambda} \\[2mm] \qquad\qquad\qquad -[m(m+1)\lambda^2 - (b^2 + c^2)p]L = 0 \\[2mm] (\mu^2 - b^2)(\mu^2 - c^2)\dfrac{d^2M}{d\mu^2} + \mu(\mu^2 - b^2 + \mu^2 - c^2)\dfrac{dM}{d\mu} \\[2mm] \qquad\qquad\qquad -[m(m+1)\mu^2 - (b^2 + c^2)p]M = 0 \\[2mm] (\nu^2 - b^2)(\nu^2 - c^2)\dfrac{d^2N}{d\nu^2} + \nu(\nu^2 - b^2 + \nu^2 - c^2)\dfrac{dN}{d\nu} \\[2mm] \qquad\qquad\qquad -[m(m+1)\nu^2 - (b^2 + c^2)p]N = 0.\end{array}\right\} \tag{3}$$

ELLIPSOIDAL HARMONICS.

Whence if $L = E_m^p(\lambda)$, it follows that $M = E_m^p(\mu)$ and $N = E_m^p(\nu)$, and that

$$V = E_m^p(\lambda) E_m^p(\mu) E_m^p(\nu) \tag{4}$$

is an answer of Laplace's Situation, (8) Art. 136.

The condition

$$(x^2 - b^2)(x^2 - c^2)\frac{d^2z}{dx^2} + x(x^2 - b^2 + x^2 - c^2)\frac{dz}{dx}$$
$$- [m(m+1)x^2 - (b^2 + c^2)p]z = 0 \tag{5}$$

is known as LamÃ©'s Situation, and $E_m^p(x)$ as a *LamÃ©'s Function* or a *Ellipsoidal Harmonic*. We will assume m a positive number.

To get a specific arrangement of (5) let $z = \sum a_k x^k$. Substitute in (5) and decrease and we get

$$[k(k+1) - m(m+1)]a_k - (b^2 + c^2)[(k+2)^2 - p]a_{k+2}$$
$$+ b^2 c^2 (k+3)(k+4) a_{k+4} = 0. \tag{6}$$

We have now just to pick a grouping of coefficients fulfilling (6), and we may take any two continuous coefficients with no obvious end goal in mind.

(6) which is usually a connection interfacing three successive coefficients decreases to a connection between two when $k = m$, when $k = -3$, and when $k = -4$. Assuming we take $a_{m+2} = 0$, a_{m+4}, a_{m+6}, &c., will disappear. Let $a_m = 1$. On the off chance that m is even the coefficient of a_0 in (6) will be zero; if p has such a worth that a_{-2} is zero, a_{-4}, a_{-6}, &c., will be zero, and there will be no terms in the arrangement including negative powers of x.

Assuming we compose the upsides of a_{m-2}, a_{m-4}, &c., by the guide of (6) that's what we see a_{m-2} is of the main degree in p, a_{m-4} of the second degree in p, &c., and a_{-2} of the degree $\frac{m}{2}+1$ in p. There are then $\frac{m}{2}+1$ upsides of p which we will call p_1, p_2, p_3, &c., for which a_{-2} will disappear, and for which our answers will be of the structure

$$E_m^p(x) = x^m + a_{m-2}x^{m-2} + a_{m-4}x^{m-4} + \cdots + a_0$$

on the off chance that m is even.

In the event that m is odd, the coefficient of a_1 in (6) will evaporate and we can pick p so that a_{-1} will be zero, and afterward all coefficients of lower request will disappear. a_{-1} is of the degree $\frac{m+1}{2}$ in p, and there will be $\frac{m+1}{2}$ values p_1, p_2, p_3, &c., of p for which

$$E_m^p(x) = x^m + a_{m-2}x^{m-2} + a_{m-4}x^{m-4} + \cdots + a_1 x.$$

Following Heine we will call the arrangement just got $K_m^p(x)$ so that

$$K_m^p(x) = x^m + a_{m-2}x^{m-2} + a_{m-4}x^{m-4} + \cdots \tag{7}$$

ending with a_0 assuming that m is even, and with $a_1 x$ assuming m is odd. Assuming that m is even, there are $\frac{m}{2}+1$ of these capabilities $K_m^{p_1}(x)$, $K_m^{p_2}(x)$, &c., and there are $\frac{m+1}{2}$ of them assuming that m is odd. The coefficients can be processed by the guide of (6).

On the off chance that in LamÃ©'s Situation (5) we let $z = v\sqrt{x^2 - b^2}$ we get the condition

$$(x^2 - b^2)(x^2 - c^2)\frac{d^2v}{dx^2} + x[x^2 - b^2 + 3(x^2 - c^2)]\frac{dv}{dx}$$
$$- [(m+2)(m-1)x^2 + c^2 - (b^2 + c^2)p]v = 0. \tag{8}$$

Letting $v = \sum a_k x^k$ we acquire the connection

$$[k(k+3) - (m+2)(m-1)]a_k - \{(b^2 + c^2)[(k+2)^2 - p] + c^2(2k+5)\}a_{k+2}$$
$$+ b^2 c^2 (k+3)(k+4) a_{k+4} = 0. \tag{9}$$

ELLIPSOIDAL HARMONICS. 210

Continuing precisely as in the past, we observe that there are $\frac{m}{2}$ values q_1, q_2, q_3, &c., of p for which $v = x^{m-1} + a_{m-3}x^{m-3} + \cdots + a_1 x$ assuming that m is even, and $\frac{m+1}{2}$ values for which $v = x^{m-1} + a_{m-3}x^{m-3} + \cdots + a_0$ assuming m is odd.

Calling $v\sqrt{x^2 - b^2}\ L_m^p(x)$ so that

$$L_m^p(x) = \sqrt{x^2 - b^2}[x^{m-1} + a_{m-3}x^{m-3} + a_{m-5}x^{m-5} + \cdots], \qquad (10)$$

ending with $a_1 x$ assuming that m is even and with a_0 in the event that m is odd, we have $\frac{m}{2}$ upsides of $E_m^p(x)$, to be specific $L_m^{q_1}(x)$, $L_m^{q_2}(x)$, &c., of the structure (10) assuming m is even furthermore, $\frac{m+1}{2}$ values assuming m is odd.

By exchanging b and c in (8), (9), and (10) we might show that if

$$M_m^p(x) = \sqrt{x^2 - c^2}[x^{m-1} + a_{m-3}x^{m-3} + a_{m-5}x^{m-5} + \cdots] \qquad (11)$$

there are $\frac{m}{2}$ upsides of $E_m^p(x)$, specifically $M_m^{r_1}(x)$, $M_m^{r_2}(x)$, $M_m^{r_3}(x)$, &c., of the structure (11) assuming that m is even and $\frac{m+1}{2}$ qualities assuming m is odd.

At last on the off chance that in LamÃ©'s Situation (5) we let $z = v\sqrt{(x^2 - b^2)(x^2 - c^2)}$ we get

$$(x^2 - b^2)(x^2 - c^2)\frac{d^2 v}{dx^2} + 3x(x^2 - b^2 + x^2 - c^2)\frac{dv}{dx}$$
$$- [(m+3)(m-2)x^2 - (b^2 + c^2)(p-1)]v = 0. \qquad (12)$$

On the off chance that now we let $v = \sum a_k x^k$ we acquire the connection

$$[k(k+5) - (m-2)(m+3)]a_k$$
$$- (b^2 + c^2)[(k+2)(k+4) + 1 - p]a_{k+2} + b^2 c^2 (k+3)(k+4)a_{k+4} = 0. \qquad (13)$$

Continuing as before we observe that there are $\frac{m}{2}$ values s_1, s_2, s_3, &c., of p for which $v = x^{m-2} + a_{m-4}x^{m-4} + a_{m-6}x^{m-6} + \cdots + a_0$ assuming that m is even, and $\frac{m+1}{2}$ values for which $v = x^{m-2} + a_{m-4}x^{m-4} + \cdots + a_1 x$ in the event that m is odd.

Calling $v\sqrt{(x^2 - b^2)(x^2 - c^2)}\ N_m^p(x)$ so that

$$N_m^p(x) = \sqrt{(x^2 - b^2)(x^2 - c^2)}[x^{m-2} + a_{m-4}x^{m-4} + a_{m-6}x^{m-6} + \cdots] \qquad (14)$$

ending with a_0 assuming m is even and with $a_1 x$ on the off chance that m is odd, we have $\frac{m}{2}$ values of $E_m^p(x)$, to be specific $N_m^{s_1}(x)$, $N_m^{s_2}(x)$, $N_m^{s_3}(x)$, &c., of the structure (14) on the off chance that m is even and $\frac{m-1}{2}$ qualities assuming m is odd.

Summarizing our outcomes we see that there are $2m+1$ Ellipsoidal Music $E_m^p(x)$ every one of which is a limited amount of the mth degree in x, or in x and $\sqrt{x^2 - b^2}$, or then again in x and $\sqrt{x^2 - c^2}$, or in x and $\sqrt{x^2 - b^2}$ and $\sqrt{x^2 - c^2}$.

It was demonstrated by LamÃ© that the $2m+1$ upsides of p, specifically p_1, p_2, p_3, &c., q_1, q_2, q_3, &c., r_1, r_2, r_3, &c., s_1, s_2, s_3, &c., were all genuine, and by Liouville that they were all unique.

We give tables of the Ellipsoidal Sounds for $m = 0$, $m = 1$, $m = 2$, and $m = 3$. The coefficients were gotten by the guide of recipes (6), (9), also (13).

ELLIPSOIDAL HARMONICS.

$$E_0(x)$$

$$\boxed{\begin{aligned} K_0(x) &= 1 \\ L_0(x) &= 0 \\ M_0(x) &= 0 \\ N_0(x) &= 0 \end{aligned}}$$

$$E_1(x)$$

$$\boxed{\begin{aligned} K_1(x) &= x \\ L_1(x) &= \sqrt{x^2 - b^2} \\ M_1(x) &= \sqrt{x^2 - c^2} \\ N_1(x) &= 0 \end{aligned}}$$

$$E_2(x)$$

$$\boxed{\begin{aligned} K_2^{p_1}(x) &= x^2 - \tfrac{1}{3}[b^2 + c^2 - \sqrt{(b^2+c^2)^2 - 3b^2c^2}] \\ K_2^{p_2}(x) &= x^2 - \tfrac{1}{3}[b^2 + c^2 + \sqrt{(b^2+c^2)^2 - 3b^2c^2}] \\ L_2(x) &= x\sqrt{x^2 - b^2} \\ M_2(x) &= x\sqrt{x^2 - c^2} \\ N_2(x) &= \sqrt{(x^2 - b^2)(x^2 - c^2)} \end{aligned}}$$

$$E_3(x)$$

$$\boxed{\begin{aligned} K_3^{p_1}(x) &= x^3 - \tfrac{x}{5}[2(b^2 + c^2) - \sqrt{4(b^2+c^2)^2 - 15b^2c^2}] \\ K_3^{p_2}(x) &= x^3 - \tfrac{x}{5}[2(b^2 + c^2) + \sqrt{4(b^2+c^2)^2 - 15b^2c^2}] \\ L_3^{q_1}(x) &= \sqrt{x^2 - b^2}[x^2 - \tfrac{1}{5}(b^2 + 2c^2 - \sqrt{(b^2+2c^2)^2 - 5b^2c^2})] \\ L_3^{q_2}(x) &= \sqrt{x^2 - b^2}[x^2 - \tfrac{1}{5}(b^2 + 2c^2 + \sqrt{(b^2+2c^2)^2 - 5b^2c^2})] \\ M_3^{r_1}(x) &= \sqrt{x^2 - c^2}[x^2 - \tfrac{1}{5}(2b^2 + c^2 - \sqrt{(2b^2+c^2)^2 - 5b^2c^2})] \\ M_3^{r_2}(x) &= \sqrt{x^2 - c^2}[x^2 - \tfrac{1}{5}(2b^2 + c^2 + \sqrt{(2b^2+c^2)^2 - 5b^2c^2})] \\ N_3(x) &= x\sqrt{(x^2 - b^2)(x^2 - c^2)} \end{aligned}}$$

It is to be noticed that since in the arrangement (4) of Laplace's Situation,

$$V = E_m^p(\lambda) E_m^p(\mu) E_m^p(\nu),$$

we have the equivalent m and p in every one of the three elements, we will need to bargain only with items comprised of variables of a similar structure, for instance,

$$K_m^{p_k}(\lambda) K_m^{p_k}(\mu) K_m^{p_k}(\nu), \quad L_m^{q_k}(\lambda) L_m^{q_k}(\mu) L_m^{q_k}(\nu), \quad \&c.;$$

what's more, that in an answer of the structure

$$V = \sum A_{m,p} E_m^p(\lambda) E_m^p(\mu) E_m^p(\nu)$$

we will have for a given m just $2m + 1$ terms.

138. From the specific arrangement of Lamã©'s Situation [(5) Art. 137] $z = E_m^p(x)$, we can get by equation (5), Art. 18, the overall arrangement.

It is $and z = A E_m^p(x) + B E_m^p(x) \int_x^\infty \dfrac{dx}{\sqrt{(x^2 - b^2)(x^2 - c^2)} [E_m^p(x)]^2} .and$ \hfill (1)

ELLIPSOIDAL HARMONICS.

Causing $A = 0$ and $B = 2m+1$ we to get a second type of specific arrangement of LamÃ©'s Condition, $z = F_m^p(x)$ where

$$F_m^p(x) = (2m+1)E_m^p(x) \int_x^\infty \frac{dx}{\sqrt{(x^2-b^2)(x^2-c^2)}[E_m^p(x)]^2}. \qquad (2)$$

We will call $F_m^p(x)$ a LamÃ©'s Capability of the second kind.

It is effortlessly seen to move toward the worth zero as x is endlessly expanded.

EXAMPLES.

1. Assuming an ellipsoidal transmitter is accused of power, and is found to be at likely V_0, show that since $V_0 = V_0 K_0(\lambda)$,

$$V = V_0 K_0(\lambda) K_0(\mu) K_0(\nu) = V_0$$

at an inner point, and

$$V = V_0 K_0(\mu) K_0(\nu) \left[K_0(\lambda) \int_\lambda^\infty \frac{dx}{\sqrt{(x^2-b^2)(x^2-c^2)}[K_0(x)]^2} \right.$$

$$\left. \tilde{A} K_0(\lambda_0) \int_{\lambda_0}^\infty \frac{dx}{\sqrt{(x^2-b^2)(x^2-c^2)}[K_0(x)]^2} \right]$$

$$= V_0 \left[\int_\lambda^\infty \frac{dx}{\sqrt{(x^2-b^2)(x^2-c^2)}} \tilde{A} \int_{\lambda_0}^\infty \frac{dx}{\sqrt{(x^2-b^2)(x^2-c^2)}} \right] = V_0 \frac{F\left(\frac{b}{c}, \sin^{-1}\frac{c}{\lambda}\right)}{F\left(\frac{b}{c}, \sin^{-1}\frac{c}{\lambda_0}\right)},$$

whence and $V = V_0 \dfrac{\left(F\left(\frac{b}{c}, \frac{\pi}{2}\right) - \alpha\right)}{F\left(\frac{b}{c}, \frac{\pi}{2}\right) - \alpha_0}$ and \hfill v. (10) Art. 136.

2. Track down the worth of the possible capability at an outer point due to the fascination of a strong homogeneous ellipsoid (v. Art. 135).

See that

$$(l^2 - \mu^2)(l^2 - \nu^2) = \tfrac{1}{3}[3l^4 - 2(b^2+c^2)l^2 + b^2 c^2] K_0(\mu) K_0(\nu)$$

$$+ \tfrac{1}{2}\left[1 + \frac{b^2+c^2-3l^2}{\sqrt{(b^2+c^2)^2 - 3b^2c^2}}\right] K_2^{p_1}(\mu) K_2^{p_1}(\nu)$$

$$+ \tfrac{1}{2}\left[1 - \frac{b^2+c^2-3l^2}{\sqrt{(b^2+c^2)^2 - 3b^2c^2}}\right] K_2^{p_2}(\mu) K_2^{p_2}(\nu);$$

also, that

$$\int_0^{\lambda_0} \tfrac{4}{3}\pi\rho \frac{3l^4 - 2(b^2+c^2)l^2 + b^2 c^2}{\sqrt{(l^2-b^2)(l^2-c^2)}} dl = \tfrac{4}{3}\pi\rho\lambda_0 \sqrt{(\lambda_0^2 - b^2)(\lambda_0^2 - c^2)} = M$$

where M is the mass of the ellipsoid.

Ans. $$V = M \left\{ \int_\lambda^\infty \frac{dx}{\sqrt{(x^2-b^2)(x^2-c^2)}} - \frac{3}{2\sqrt{(b^2+c^2)^2 - 3b^2c^2}} \right.$$

$$\left[K_2^{p_1}(\mu) K_2^{p_1}(\nu) K_2^{p_1}(\lambda) \int_\lambda^\infty \frac{dx}{\sqrt{(x^2-b^2)(x^2-c^2)}.(K_2^{p_1}(x))^2} \right.$$

$$\left.\left. - K_2^{p_2}(\mu) K_2^{p_2}(\nu) K_2^{p_2}(\lambda) \int_\lambda^\infty \frac{dx}{\sqrt{(x^2-b^2)(x^2-c^2)}.(K_2^{p_2}(x))^2} \right] \right\}.$$

ELLIPSOIDAL HARMONICS.

139. If for quickness we address $\frac{b}{c}$ by k, and $\left(1 - \frac{b^2}{c^2}\right)^{\frac{1}{2}}$ by k' in the equations (11) Art. 136 we have

$$\lambda = c\frac{\operatorname{dn}\alpha}{\operatorname{cn}\alpha}(\operatorname{mod} k), \quad \mu = \frac{b}{\operatorname{dn}\beta(\operatorname{mod} k')}, \quad \nu = b\operatorname{sn}\gamma(\operatorname{mod} k) \tag{1}$$

what's more, from these we get easily (v. Int. Cal. Art. 192)

$$\left.\begin{array}{ll} \sqrt{\lambda^2 - b^2} = \dfrac{ck'}{\operatorname{cn}\alpha(\operatorname{mod} k)}, & \sqrt{\mu^2 - b^2} = \dfrac{bk'\operatorname{sn}\beta}{\operatorname{dn}\beta}(\operatorname{mod} k'), \\[2mm] \sqrt{b^2 - \nu^2} = b\operatorname{cn}\gamma(\operatorname{mod} k), & \sqrt{\lambda^2 - c^2} = \dfrac{ck'\operatorname{sn}\alpha}{\operatorname{cn}\alpha}(\operatorname{mod} k), \\[2mm] \sqrt{c^2 - \mu^2} = \dfrac{ck'\operatorname{cn}\beta}{\operatorname{dn}\beta}(\operatorname{mod} k'), & \sqrt{c^2 - \nu^2} = c\operatorname{dn}\gamma(\operatorname{mod} k). \end{array}\right\} \tag{2}$$

In the event that we let α range from 0 to K, and β from 0 to $2K'$, and γ from 0 to $4K$, where K and K' are the finished Elliptic Integrals $F\left(k, \frac{\pi}{2}\right)$ and $F\left(k', \frac{\pi}{2}\right)$ separately, (α, β, γ) may address any point in space, and there will be no vagueness in sign (v. Art. 136).

We might take note of that if $0 < \beta < K'$, z is positive; if $K' < \beta < 2K'$, z is negative; if $0 < \gamma < K$, x and y are both positive; if $K < \gamma < 2K$, x is positive and y negative; if $2K < \gamma < 3K$, x and y are both negative; and if $3K < \gamma < 4K$, x is negative and y positive (v. Art. 136).

We can compose the qualities in (4), (5), (6), and (7), Art. 136, all the more flawlessly by acquiring α, β, and γ. We get

$$dn = \frac{1}{c}\sqrt{(\lambda^2 - \mu^2)(\lambda^2 - \nu^2)}d\alpha, \tag{3}$$

$$dS = \frac{1}{c^2}(\mu^2 - \nu^2)\sqrt{(\lambda^2 - \mu^2)(\lambda^2 - \nu^2)}d\beta d\gamma, \tag{4}$$

$$dv = \frac{1}{c^3}(\lambda^2 - \mu^2)(\lambda^2 - \nu^2)(\mu^2 - \nu^2)d\alpha d\beta d\gamma. \tag{5}$$

For the vital of any capability of α, β, and γ more than the ellipsoid $\alpha = \alpha_0$, we will have

$$\int F(\alpha,\beta,\gamma)dS = \frac{1}{c^2}\int_0^{2K'} d\beta \int_0^{4K} F(\alpha_0,\beta,\gamma)(\mu^2 - \nu^2)\sqrt{(\lambda^2 - \mu^2)(\lambda^2 - \nu^2)}d\gamma. \tag{6}$$

140. Assuming that we utilize the recipe (2) Art. 92

$$\int (UD_nV - VD_nU)dS = 0 \tag{1}$$

furthermore, take as our shut surface any given ellipsoid, we can get a vital result.

If $\qquad U = E_m^p(\lambda)E_m^p(\mu)E_m^p(\nu) \quad \text{and} \quad V = E_n^q(\lambda)E_n^q(\mu)E_n^q(\nu)$

then $\qquad \nabla^2 U = \nabla^2 V = 0.$

$$D_nU = D_\alpha U D_n\alpha = E_m^p(\mu)E_m^p(\nu)\frac{dE_m^p(\lambda)}{d\alpha}\frac{c}{\sqrt{(\lambda^2 - \mu^2)(\lambda^2 - \nu^2)}},$$

and $\qquad D_nV = D_\alpha V D_n\alpha = E_n^q(\mu)E_n^q(\nu)\dfrac{dE_n^q(\lambda)}{d\alpha}\dfrac{c}{\sqrt{(\lambda^2 - \mu^2)(\lambda^2 - \nu^2)}},$

$$UD_nV - VD_nU = E_m^p(\mu)E_m^p(\nu)E_n^q(\mu)E_n^q(\nu)$$
$$\left(E_m^p(\lambda)\frac{dE_n^q(\lambda)}{d\alpha} - E_n^q(\lambda)\frac{dE_m^p(\lambda)}{d\alpha}\right)\frac{c}{\sqrt{(\lambda^2 - \mu^2)(\lambda^2 - \nu^2)}}.$$

ELLIPSOIDAL HARMONICS. 214

Incorporating $UD_nV - VD_nU$ over the entire ellipsoid, and composing the outcome equivalent to nothing, we have

$$\frac{1}{c} \int_0^{2K'} d\beta \int_0^{4K} E_m^p(\mu)E_m^p(\nu)E_n^q(\mu)E_n^q(\nu)$$

$$\left[E_m^p(\lambda)\frac{dE_n^q(\lambda)}{d\alpha} - E_n^q(\lambda)\frac{dE_m^p(\lambda)}{d\alpha} \right](\mu^2 - \nu^2)d\gamma = 0.$$

Hence and $\int_0^{2K'} d\beta \int_0^{4K} E_m^p(\mu)E_m^p(\nu) \qquad E_n^q(\mu)E_n^q(\nu)(\mu^2 - \nu^2)d\gamma = 0 \qquad (2)$

unless and $E_m^p(\lambda)\frac{dE_n^q(\lambda)}{d\alpha} \qquad -E_n^q(\lambda)\frac{dE_m^p(\lambda)}{d\alpha} = 0. \qquad (3)$

Yet, as our ellipsoid might be taken at delight, λ and α are unhindered, and in the event that (3) is valid it should be valid indistinguishably.

Assuming that we partition (3) by $[E_m^p(\lambda)]^2$ it becomes

$$\frac{d}{d\alpha}\left[\frac{E_n^q(\lambda)}{E_m^p(\lambda)}\right] = 0 \quad \text{and} \quad \frac{E_n^q(\lambda)}{E_m^p(\lambda)} = \text{a constant};$$

what's more, this clearly can't be valid except if $n = m$ and $q = p$.

EXAMPLES.

1. Show that it follows from (2) Art. 140 that

$$\int_{-K'}^{K'} d\beta \int_{-K}^{K} E_m^p(\mu)E_m^p(\nu)E_n^q(\mu)E_n^q(\nu)(\mu^2 - \nu^2)d\gamma = 0.$$

Suggestion:

$$\int_0^{2K'} E_m^p(\mu)E_n^q(\mu)(\mu^2 - \nu^2)d\beta = \int_0^{K'} E_m^p(\mu)E_n^q(\mu)(\mu^2 - \nu^2)d\beta$$

$$+ \int_{K'}^{2K'} E_m^p(\mu)E_n^q(\mu)(\mu^2 - \nu^2)d\beta.$$

In the event that in the last fundamental we supplant β by $\beta + 2K'$ it becomes

$$\hat{A} \pm \int_{-K'}^{0} E_m^p(\mu)E_n^q(\mu)(\mu^2 - \nu^2)d\beta$$

v. Arts. 136 and 139 and Int. Cal. Art. 196.

2. Show that

$$\int_0^{2K'} d\beta \int_0^{4K} [E_m^p(\mu)E_m^p(\nu)]^2(\mu^2 - \nu^2)d\gamma = 8 \int_0^{K'} d\beta \int_0^{K} [E_m^p(\mu)E_m^p(\nu)]^2(\mu^2 - \nu^2)d\gamma.$$

ELLIPSOIDAL HARMONICS.

141. We can now take care of the issue of tracking down the worth of V anytime in space when it is given at every one of the focuses on the outer layer of the ellipsoid $\alpha = \alpha_0$.

We have first to create in Quite a while a component of μ and ν or rather of α and β given at all focuses on the outer layer of the ellipsoid being referred to; what's more, this is presently effortlessly achieved by our standard strategy, which drives us to the outcome

$$f(\alpha_0, \beta, \gamma) = \sum_{m=0}^{m=\infty} \sum_{k=1}^{k=2m+1} A_{m,p_k} E_m^{p_k}(\mu) E_m^{p_k}(\nu), \tag{1}$$

whereand$A_{m,p_k} = \dfrac{\int_0^{2K'} d\beta \int_0^{4K} f(\alpha_0, \beta, \gamma) E_m^{p_k}(\mu) E_m^{p_k}(\nu)(\mu^2 - \nu^2) d\gamma}{8 \int_0^{K'} d\beta \int_0^{K} [E_m^{p_k}(\mu) E_m^{p_k}(\nu)]^2 (\mu^2 - \nu^2) d\gamma}$.and (2)

Our last arrangement is

$$V = \sum_{m=0}^{m=\infty} \sum_{k=1}^{k=2m+1} A_{m,p_k} \frac{E_m^{p_k}(\lambda)}{E_m^{p_k}(\lambda_0)} E_m^{p_k}(\mu) E_m^{p_k}(\nu) \tag{3}$$

at an inward point;

$$V = \sum_{m=0}^{m=\infty} \sum_{k=1}^{k=2m+1} A_{m,p_k} \frac{F_m^{p_k}(\lambda)}{F_m^{p_k}(\lambda_0)} E_m^{p_k}(\mu) E_m^{p_k}(\nu) \tag{4}$$

at an outer point.

LamÃ© has demonstrated rather cunningly that

$$\int_0^{K'} d\beta \int_0^{K} [E_m^{p_k}(\mu) E_m^{p_k}(\nu)]^2 (\mu^2 - \nu^2) d\gamma$$

can constantly be found and that it is equivalent to $\frac{\pi}{2}$ increased by a sane vital capability of the coefficients of $E_m^{p_k}(x)$ and of c^2 and $\left(\frac{b}{c}\right)^2$.

Obviously the work of getting even a couple of terms of the improvement of a capability that is at all confounded is gigantic.

142. Assuming in Laplace's Situation (8) Art. 136 we let $V = E_m^p(\lambda) U$ assuming U to be an element of β and γ just, we get subsequent to supplanting $\frac{1}{E_m^p(\lambda)} \frac{d^2 E_m^p(\lambda)}{d\alpha^2}$ by its worth $m(m+1)\lambda^2 - (b^2 + c^2)p$ [v. (2) Art. 137]

$$(\lambda^2 - \nu^2) D_\beta^2 U + (\lambda^2 - \mu^2) D_\gamma^2 U + (\mu^2 - \nu^2)[m(m+1)\lambda^2 - (b^2 + c^2)p] U = 0; \tag{1}$$

what's more, since by speculation U is free of λ, the coefficient of λ^2 in (1) should disappear. Subsequently

$$D_\beta^2 U + D_\gamma^2 U + (\mu^2 - \nu^2) m(m+1) U = 0. \tag{2}$$

Obviously $U = E_m^p(\mu) E_m^p(\nu)$ will fulfill (2).

EXAMPLES.

1. Substitute $U = E_m^p(\mu) E_m^p(\nu)$ in (2) Art. 142 and by the guide of (2) Art. 137 show that the condition (2) Art. 142 is satisfied.

ELLIPSOIDAL HARMONICS.

2. Acquire (2) Art. 140 straightforwardly from (2) Art. 142.

3. *Conical Coördinates.* Consider the arrangement of coördinates characterized by the conditions

$$\left.\begin{array}{c} x^2 + y^2 + z^2 = r^2 \\ \dfrac{x^2}{\mu^2} + \dfrac{y^2}{\mu^2 - b^2} + \dfrac{z^2}{\mu^2 - c^2} = 0 \\ \dfrac{x^2}{\nu^2} + \dfrac{y^2}{\nu^2 - b^2} + \dfrac{z^2}{\nu^2 - c^2} = 0 \end{array}\right\} \quad (1)$$

where $c^2 > \mu^2 > b^2 > \nu^2$.

Show that

$$x^2 = \frac{r^2 \mu^2 \nu^2}{b^2 c^2}, \quad y^2 = \frac{r^2(\mu^2 - b^2)(\nu^2 - b^2)}{b^2(b^2 - c^2)}, \quad z^2 = \frac{r^2(\mu^2 - c^2)(\nu^2 - c^2)}{c^2(c^2 - b^2)};$$

$$h_1^2 = \frac{(\mu^2 - b^2)(c^2 - \mu^2)}{r^2(\mu^2 - \nu^2)}, \quad h_2^2 = \frac{(b^2 - \nu^2)(c^2 - \nu^2)}{r^2(\mu^2 - \nu^2)}, \quad h_3^2 = 1.$$

Laplace's Condition is

$$D_\alpha^2 V + D_\beta^2 V + (\mu^2 - \nu^2) D_r(r^2 D_r V) = 0 \quad (2)$$

where $\alpha = \int_b^\mu \dfrac{d\mu}{\sqrt{(\mu^2 - b^2)(c^2 - \mu^2)}}$ and $\beta = \int_0^\nu \dfrac{d\nu}{\sqrt{(b^2 - \nu^2)(c^2 - \nu^2)}}$,

In the event that $V = U.R$ (2) separates into

$$\frac{d}{dr}\left(r^2 \frac{dR}{dr}\right) = m(m+1)R, \quad (3)$$

$$D_\alpha^2 U + D_\beta^2 U + m(m+1)(\mu^2 - \nu^2)U = 0. \quad (4)$$

(3) gives *and* $R \qquad = Ar^m + Br^{-m-1}$.

(4) gives $\qquad U = E_m^p(\mu) E_m^p(\nu)$ *and* $\qquad$ (v. Art. 142).

With the goal that an answer of (2) is

$$V = Ar^m E_m^p(\mu) E_m^p(\nu).$$

In any case, since (2) is Laplace's Condition, $V = Ar^m Y_m(\mu, \phi)$, whenever communicated in Funnel shaped Coördinates, should fulfill it, therefore $E_m^p(\mu) E_m^p(\nu)$ should be just a Round Symphonious of the mth degree.

Toroidal Coördinates.

143. Any sets of circles having a place with the symmetrical framework got and figured in Art. 46 can be addressed by the situations

$$\left.\begin{array}{c} \dfrac{2ax}{\sinh\alpha} = \dfrac{x^2 + y^2 + a^2}{\cosh\alpha} \\ \dfrac{2ay}{\sin\beta} = \dfrac{x^2 + y^2 - a^2}{\cos\beta} \end{array}\right\} \quad (1)$$

in the event that we take $2a$ rather than 2 as the distance between the focuses normal to the second arrangement of circles.

ELLIPSOIDAL HARMONICS.

In the event that we turn the framework about the hub of y we get a bunch of circles and a set of anchor rings what cut symmetrically. These and a bunch of planes through the hub of unrest will shape a symmetrical arrangement of surfaces, and the boundaries comparing to them might be taken as a bunch of curvilinear coördinates and might be called *Toroidal Coördinates*. In the event that we take the hub of the framework as the pivot of Z, the conditions of a bunch of the surfaces might be composed

$$\left. \begin{array}{c} \dfrac{4a^2(x^2+y^2)}{\sinh^2 \alpha} = \dfrac{[x^2+y^2+z^2+a^2]^2}{\cosh^2 \alpha} \\ \dfrac{2az}{\sin \beta} = \dfrac{x^2+y^2+z^2-a^2}{\cos \beta} \\ y = x \tan \gamma \end{array} \right\} \quad (2)$$

α, β, and γ being viewed as the coördinates of a mark of convergence of the three surfaces. Finding Laplace's Condition in the typical way we get

$$x = \frac{a \sinh \alpha \cos \gamma}{\cosh \alpha \mp \cos \beta}, \quad y = \frac{a \sinh \alpha \sin \gamma}{\cosh \alpha \mp \cos \beta}, \quad z = \frac{a \sin \beta}{\cosh \alpha \mp \cos \beta},$$

$$r = \sqrt{x^2+y^2} = \frac{a \sinh \alpha}{\cosh \alpha \mp \cos \beta}, \quad a + z \operatorname{ctn} \beta = \frac{a \cosh \alpha}{\cosh \alpha \mp \cos \beta};$$

$$h_1 = \frac{\cosh \alpha \mp \cos \beta}{a}, \quad h_2 = \frac{\cosh \alpha \mp \cos \beta}{a}, \quad h_3 = \frac{\cosh \alpha \mp \cos \beta}{a \sinh \alpha};$$

what's more, Laplace's Condition becomes

$$D_\alpha \left[\frac{a \sinh \alpha}{\cosh \alpha \mp \cos \beta} D_\alpha V \right] + D_\beta \left[\frac{a \sinh \alpha}{\cosh \alpha \mp \cos \beta} D_\beta V \right]$$
$$+ D_\gamma \left[\frac{a}{\sinh \alpha (\cosh \alpha \mp \cos \beta)} D_\gamma V \right] = 0, \quad (1)$$

or
$$D_\alpha(r D_\alpha V) + D_\beta(r D_\beta V) + \frac{1}{\sinh^2 \alpha} r D_\gamma^2 V = 0. \quad (2)$$

We can't continue further by our standard strategy, for the suspicion that V is a component of α alone, or that V is an element of β alone, ends up being prohibited. Without a doubt, not exclusively are α, β, and γ not *thermometric parameters* (v. Art. 133), however no thermometric boundaries exist, and no conceivable circulation can make our anchor rings or our circles a bunch of equipotential surfaces.

We can, be that as it may, streamline (2). It tends to be composed

$$D_\alpha^2(V\sqrt{r}) + D_\beta^2(V\sqrt{r}) + \frac{1}{\sinh^2 \alpha} D_\gamma^2(V\sqrt{r}) - V(D_\alpha^2 \sqrt{r} + D_\beta^2 \sqrt{r}) = 0. \quad (3)$$

$D_\alpha^2 \sqrt{r} + D_\beta^2 \sqrt{r}$ demonstrates equivalent to $-\frac{\sqrt{r}}{4 \sinh^2 \alpha}$; thus if $U = V\sqrt{r}$ (3) becomes

$$\sinh^2 \alpha (D_\alpha^2 U + D_\beta^2 U) + D_\gamma^2 U + \tfrac{1}{4} U = 0, \quad (4)$$

for which specific arrangements can promptly be tracked down by our typical interaction.

(4) can be separated into the three conditions

$$\frac{d^2N}{d\gamma^2} + (m+\tfrac{1}{2})^2 N = 0 \qquad (5)$$

$$\frac{d^2M}{d\beta^2} + n^2 M = 0 \qquad (6)$$

$$\sinh^2\alpha \frac{d^2L}{d\alpha^2} - [m(m+1) + n^2 \sinh^2\alpha] L = 0. \qquad (7)$$

$$N = A\cos(m+\tfrac{1}{2})\gamma + B\sin(m+\tfrac{1}{2})\gamma$$

$$M = A_1 \cos n\beta + B_1 \sin n\beta.$$

Assuming we bring into (7) $x = \operatorname{ctnh}\alpha$ it becomes

$$(1-x^2)\frac{d^2L}{dx^2} - 2x\frac{dL}{dx} + \left[m(m+1) - \frac{n^2}{1-x^2}\right] L = 0,$$

an answer of which is

$$L = P_m^n(x) = (1-x^2)^{\frac{n}{2}} \frac{d^n P_m(x)}{dx^n} \qquad \text{(v. Art. 102)}.$$

It is to be noticed that since $\operatorname{ctnh}\alpha$ is more noteworthy than 1

$$P_m^n(\operatorname{ctnh}\alpha) = i^{\frac{n}{2}} \operatorname{csch}^n \alpha \frac{d^n P_m(\operatorname{ctnh}\alpha)}{(d\operatorname{ctnh}\alpha)^n}.$$

The steady coefficient $i^{\frac{n}{2}}$ can be dismissed and we get

$$U - [A\cos(m+\tfrac{1}{2})\gamma + B\sin(m+\tfrac{1}{2})\gamma](A_1 \cos n\beta + B_1 \sin n\beta) \operatorname{csch}^n \alpha \frac{d^n P_m(\operatorname{ctnh}\alpha)}{(d\operatorname{ctnh}\alpha)^n}$$

as a specific arrangement of (4).

$$\frac{1}{i^{\frac{n}{2}}} P_m^n(\operatorname{ctnh}\alpha) = \operatorname{csch}^n \alpha \frac{d^n P_m(\operatorname{ctnh}\alpha)}{(d\operatorname{ctnh}\alpha)^n}$$

has been known as a Toroidal Symphonious.

EXAMPLES.

1. Given the worth of the expected capability at all focuses on the outer layer of an anchor ring; track down its worth anytime inside the ring.

Suggestion: If $V = f(\beta,\gamma)$ when $\alpha = \alpha_0$, the capability to be created is

$$\sqrt{r}.f(\beta,\gamma) \quad \text{i.e.} \quad \left[\frac{a\sinh\alpha_0}{\cosh\alpha_0 \mp \cos\beta}\right]^{\frac{1}{2}} f(\beta,\gamma)$$

what's more, the improvement will be in a twofold Fourier's Series (v. Art. 71).

2. Show that assuming we let α range from 0 to ∞, β from $-\pi$ to π, and γ from 0 to 2π, every one of the twofold signs on page 264 might be supplanted by the less sign without loss of consensus.

Chapter 9

HISTORICAL SUMMARY

The strategy for advancement in series which has empowered us in the first parts to take care of issues in different parts of numerical physical science, had its starting point, as could have been normal, in the hypothesis of the melodic vibrations of an extended string. It was in the year 1753[1] that Daniel Bernoulli articulated the guideline of the concurrence of little motions, which, in association with Taylor's and John Bernoulli's hypothesis of the vibrating string, persuaded him to think that the overall arrangement of this issue could be placed in the type of a mathematical series. This rule likewise drove him and Euler to treat likewise the issues of the vibration of a segment of air and of a versatile pole. The issue of the vibration of a weighty string suspended from one end was additionally treated in similar way by these mathematicians what's more, merits unique notice here as in it Bessel's elements of the zeroth request show up for the principal time.[2] In none of these cases, be that as it may, was any technique given for deciding the coefficients of the series.

This last comment likewise applies to the more muddled issues of the vibration of rectangular and roundabout layers, which were talked about by Euler[3] in 1764, and in the remainder of which the general Bessel's elements of indispensable orders happen.

It is in issues associated with space science that the first totally effective use of the technique here considered happens. Legendre in a paper distributed in the Mémoires des Academics Ãtrangers for 1785, first presented the zonal sounds P_m and applied them to the assurance of the fascination of solids of insurgency. He was trailed by Laplace, who in one of the most astounding diaries ever written[4] decided the potential of a strong contrasting however little from a circle through the turn of events as indicated by the circular sounds Y_m.

Firmly connected with this issue is Gauss' commended treatment of the hypothesis of earth-bound magnetism,[5] which we will hence specify here, despite the fact that it was not distributed until the greater part a century after the fact. This paper is especially vital as it contains a mathematical use of the technique for a bigger scope than ever endeavored previously or since.

After the explores of Legendre and Laplace there was a delay of a quarter of 100 years until in 1812 Fourier's broad diary: *Théorie du mouvement de la chaleur dans les corps solides* was delegated by the French Institute. Albeit not printed until the years 1824-26,[6] the composition of this work was meanwhile open to the next French mathematicians as of now to be referenced. The initial segment of this journal, which was imitated with however couple of changes in the *Théorie analytique de la chaleur* (1822), contains a treatment of the accompanying issues and

---

[1] See two articles by Bernoulli and one by Euler in the Diaries of the Foundation of Berlin for this year.

[2] See the Exchanges of the Foundation of St. Petersburg for 1732-33, 1734 and 1781.

[3] Transactions of the Foundation of St. Petersburg.

[4] "Théorie des attractions des sphéroïdes et de la figure des Planètes" Mémoires de l'académie des sciences 1782. This article, albeit bearing a prior date than that of Legendre, was truly motivated by it. It is here that "Laplace's condition" first shows up, happening, notwithstanding, just in polar coördinates.

[5] Resultate aus nook Beobachtungen des magnetischen Vereins im Jahre 1838. Leipzig, 1839. Reproduced in Gauss' gathered works, Vol. V., p. 121.

[6] Mémoires de l'académie des sciences for 1819-20 and 1821-22.

of basically all of their exceptional cases:

(a) The one layered progression of intensity. (b) The two layered progression of heat in a square shape. (c) The three layered progression of intensity in a rectangular parallelopiped. (d) The progression of intensity in a circle when the temperature relies just upon the separation from the middle. (e) The progression of intensity in a right round chamber when the temperature relies just upon the distance from the pivot. In these issues not just the less complex limit conditions are thought about yet in addition the subject of radiation into an air. In unique instances of the initial three issues recently referenced (when at least one aspects become limitless) the series decline into "Fourier's integrals."

More significant even than any of these exceptional issues is the incredible advance which Fourier made the hypothesis of mathematical series make. In a post mortem paper Euler had given the formulae for deciding the coefficients,[7] yet Fourier was quick to state and to endeavor to demonstrate that any capability, despite the fact that for various upsides of the contention it is communicated by various logical formulae, can be created in such a series. The reality that the genuine significance of geometrical series was in this way interestingly shown legitimizes us in partner Fourier's name with them, in spite of the fact that, as we have seen, they were known well before his day.

Fourier's outcomes were reached out by Laplace in 1820[8] to the general (unsymmetrical) instance of the progression of intensity in a circle, and by Poisson[9] (1821) to the unsymmetrical progression of intensity in a chamber.

In 1835 Green distributed a paper[10] in which the strategy we are thinking about is utilized to decide the capability of a heterogeneous ellipsoid. This paper, in which the examination is performed on the double for space of n aspects, expects a lot of that was in this manner done by others, however has neglected to apply an impact corresponding to its significance.

At about this time LamÃ© started a progression of distributions which have associated his name indistinguishably with the issue of the extremely durable condition of temperature of an ellipsoid. In the first of these[11] the condition $\nabla^2 V = 0$ may be changed to ellipsoidal coÃ¶rdinates and is then separated into three customary differential conditions. The remainder of the arrangement, in any case, is barely addressed. LamÃ©'s most significant work on this subject[12] was distributed in Liouville's Diary in 1839, and in it the total arrangement of the issue is given. LamÃ© obviously shows in this paper how he showed up at his answer, by taking into account first the less difficult instance of a circle where, rather than the polar coÃ¶rdinates θ and ϕ, the boundaries of two groups of confocal cones of the second degree are utilized as coÃ¶rdinates. This arrangement of curvilinear coÃ¶rdinates, which, when applied to the total circle, only gives the old aftereffects of Laplace in another structure, is scarcely referenced in LamÃ©'s later distributions. In a similar volume of Liouville's Diary LamÃ© distributed a second paper in which he applies his outcomes to the extraordinary instances of ellipsoids of unrest.

These two papers structure the beginning stage for a progression of articles on the same subject by Heine and Liouville. Heine in his PCP dissertation[13] (1842) decided the potential not just for the inside of an ellipsoid of unrest when the worth of the potential is given on a superficial level, yet in addition for the outside of such an ellipsoid and for the shell between two confocal ellipsoids of upheaval. Indeed, even in the first of these issues, which is identical to that of LamÃ©, he worked

---

[7] Lagrange had basically resolved these coefficients well before however neglected to take note what he had got.

[8] Connaissance des Temps pour l'an 1823.

[9] Journal de l'Ã©cole Polytechnique, 19^e Cahier. Albeit the last structures to which Poisson lessens his outcomes are like Fourier's, his strategies are very different.

[10] "On the assurance of the outside and inside fascination of ellipsoids of variable densities." Exchanges of the Cambridge Philosophical Society.

[11] MÃ©moires des Academics Ãtrangers, Vol. V. Albeit the volume is dated 1838 this paper (which was republished in Liouville's Diary, 1837) probably showed up to some extent however right on time as 1835.

[12] "Sur l'Ã©quilibre des TempÃ©ratures dans un ellipsoÃ¯de Ã trois tomahawks inÃ©gaux." An article by a similar creator on the two layered potential will be found in Vol. I. of this Journal.

[13] Reprinted in Crelle's Diary, Vol. 26 (1843).

In a similar Diary for 1847 F. Neumann examined the connected issue of the magnetisation of a delicate iron ellipsoid of revolution.

HISTORICAL SUMMARY.

on Lamé's answer substantially by showing that the capabilities utilized might be decreased to round music, while in the other two issues he presented circular music of the second kind, which were then new. Presently afterwards[14] Heine and Liouville distributed at the same time two papers in which they showed up autonomously of each other at about similar outcomes. In every one of these papers consideration is called to the way that the result of two Lamé's capabilities is a circular consonant and this reality is utilized to toss Lamé's answer of the issue of the long-lasting condition of temperatures of an ellipsoid into a more rudimentary structure. Other than this the second arrangement of Lamé's situation is presented for taking care of the likely issue for the *exterior* of the ellipsoid. In accordingly following up the hypothesis of intensity and the connected expected issues, we have neglected to focus on the subject of little vibrations, to which during the early piece of the century a lot of consideration had been committed by Poisson, who every now and again utilized the technique for advancement in series. In his memoirs[15] the majority of the issues left incomplete by Bernoulli and Euler are entirely treated, as well as different slight alterations of them. When, be that as it may, he tackled the issue of the vibration of a flexible plate he couldn't gain a lot of headway, owing to some degree to the incorrect type of his limit conditions. He was, in any case, ready to take care of the issue of the *symmetrical* vibration of a free round plate. The total hypothesis of the vibration of a free roundabout plate was first given by Kirchhoff.[16]

Passing now to another subject, the hypothesis of the balance of a versatile round shell, we track down an answer by Lamé in Liouville's Diary for 1854, also, by Sir William Thomson (1862) in the Philosophical Exchanges for 1863. Both of these papers comprise of a use of the round symphonious examination to this fairly confounded issue. Thomson, in any case, considers other than Lamé's concern sure related questions and the type of his examination is altogether different from Lamé's, being of the very nature as that utilized in the Reference section B of his Regular Way of thinking of which we will need to talk as of now. These examinations structure the beginning stage for various ongoing journals among which those of G. H. Darwin on cosmographical questions merit extraordinary notice.

Firmly connected with this last referenced issue is the hypothesis of the little vibrations of a flexible circle. While the least complex instance of this issue was treated by Poisson in the journal alluded to over, the overall arrangement has been as of late acquired by Jaerisch (1879)[17] and Sheep (1882).[18] The capabilities included are equivalent to those which happen in the issue of the non-fixed progression of intensity in a circle as settled by Laplace.

The Index B of Thomson and Tait's Regular Philosophy,[19] to which we have previously alluded, should be viewed as one of the most significant commitments to the overall hypothesis. The manner by which circular music are presented (as homogeneous elements of the rectangular coördinates) was then, at that point, new[20] and the arrangement of the expected issue for an assortment of new solids was demonstrated; viz., for solids whose limits comprise of concentric circles, cones of upset, and planes. We will have more to as of now say concerning the technique utilized for the arrangement of these issues.

Albeit associated just in a roundabout way with the hypothesis we are examining, it will be well to make reference to as of now the strategy for electrical pictures which is additionally because of Sir William Thomson (1845). This technique empowers us to tackle numerous expected issues for the backwards of any strong when we have settled it for the strong itself. Through this strategy the vast majority of the arrangements of potential issues got by our technique might be applied immediately with very little adjustment to frameworks of curvilinear coördinates determined by

---

[14] *Heine:* Crelle's Diary, Vol. 29, 1845. *Liouville:* Liouville's Diary, Vol. X., 1845, and Vol. XI., 1846. For a treatment of the issue of the capability of an ellipsoidal shell through an improvement of $\frac{1}{r}$ with regards to Lamé's capabilities, see a paper by Heine in Crelle's Diary, Vol. 42, 1851.

[15] See particularly the one in the Mémoires de l'académie des sciences, Vol. VIII., 1829.

[16] Crelle's Diary, Vol. 40, 1850.

[17] Crelle's Diary, Vol. 88.

[18] Proc. Lond. Math. Soc.

[19] First release, 1867. This addendum was obviously composed as soon as 1862, as Thomson alludes to it in the diary cited above.

[20] The same technique was utilized at about a similar time by Clebsch.

HISTORICAL SUMMARY.   222

reversal from those we have utilized. It won't be important to make reference to independently issues of this sort, as it is plainly unimportant whether they be addressed straightforwardly or through the strategy for inversion.[21]

Returning now to the Landmass, we find as the following significant inquiry taken up the issue of the capability of an anchor ring. The main distribution regarding this matter is a monograph by C. Neumann[22] (1864), yet in Riemann's post mortem papers which were not distributed until 1876, a decade after his demise, will be tracked down a short piece regarding this matter, which (*cf.* the last page of Hattendorf's release of Riemann's talks: "Partielle Differentialgleichungen") would seem to trace all the way back to the colder time of year 1860-61. This part is of unconventional interest, as the initial passages plainly show that Riemann had as a primary concern a lengthy article on the essential standards of our subject.

We will next make reference to two papers by Mehler in which the capabilities known as "conal sounds," which had proactively been presented by Thomson in the Supplement B previously mentioned, were applied to the arrangement of two issues in electrostatics. The first of these papers[23] (1868) manages the strong limited by two converging circles, while in the second[24] (1870) the boundless cone of unrest is dealt with. Both of these issues are basically not quite the same as those examined in the "Supplement B," while the limitless series which we ordinarily have degenerate in these cases into distinct integrals, similarly as they do in a few more straightforward cases treated by Fourier. The later of the two papers just cited additionally contains significant data concerning the idea of the arrangement of comparable issues for the hyperboloids and paraboloids of insurgency. The arrangements of these issues are not, nonetheless, given.

It stays, to close the historical backdrop of this piece of the subject, to make reference to various diaries which in spite of the fact that treating altogether new issues are of far less significance than the majority of those thought about so far, incompletely in light of the fact that the arrangement isn't brought to where it tends to be of much quick use, also, somewhat in light of the fact that the majority of the strategies utilized are, for example, couldn't fall flat to introduce themselves to any one chasing down these issues.

Of these the first is a paper by Mathieu[25] on the vibration of an elliptic layer (1868), in which the elements of the elliptic chamber happen for the first time.

This was continued around the same time by a paper on firmly unified subjects by H. Weber,[26] in which not simply the situation of the total oval is momentarily thought of, yet in addition that in which the limit comprises of two circular segments of confocal ovals and two curves of hyperbolas confocal with them. The unique case in which the ovals and hyperbolas become confocal parabolas is moreover thought of, by which the elements of the explanatory chamber are for the first time presented.

In Mathieu's "Cours de physical make-up mathématique" (1873) the issue of the non-fixed progression of intensity in an ellipsoid is addressed, and an elaborate however not extremely agreeable treatment of the unique situations where we have ellipsoids of transformation is given. New capabilities show up in all of these issues.

Of late years C. Baer has provided various missing connections in the chain of issues here considered by treating in progression the likely issue for the paraboloid of revolution,[27] the allegorical cylinder[28] and the general paraboloid.[29] In the first of these issues Bessel's capabilities happen,

---

[21] A valid example would be the likely issue for the shell between two non-converging capricious circles, since these circles can be upset into concentric circles. This issue was dealt with straight by C. Neumann in a monograph distributed in Halle in 1862.

[22] "Theorie der Elektricitäts-und Wärme-Vertheilung in einem Ringe." Halle.

[23] Crelle's Diary, Vol. 68, 1868.

[24] Jahresbericht des Exercise rooms zu Elbing.

[25] Liouville's Diary, Vol. XIII.

[26] "Ueber kick the bucket Mix der partiellen Differentialgleichung $\frac{\delta^2 u}{\delta x^2} + \frac{\delta^2 u}{\delta y^2} + k^2 u = 0$." Math. Ann., Vol. I. No *physical* issue is referenced in this paper.

[27] "Ueber das Gleichgewicht und pass on Bewegung der Wärme in einem Rotationsparaboloid." Paper, Halle, 1881.

[28] "Die Funktion des parabolischen Chambers," Gymnasialprogramm Cüstrin, 1883.

[29] "Parabolische Coordinaten," Frankfurt, 1888. See likewise a paper by Greenhill in the Proc. Lond. Math. Soc.,

as had as of now been expressed by Mehler, while in the last we track down the elements of the elliptic chamber. For every one of the three frameworks of coÃ¶rdinates utilized the same creator additionally addresses the more broad issue of the non-fixed stream of intensity, in which new capabilities happen.

But on account of the anchor ring we have found up until this point just such solids regarded by our strategy as are limited by surfaces of the first or second degree. Wangerin[30] (1875-76) considered regarding the hypothesis of the potential, more broad frameworks of curvilinear coÃ¶rdinates than had recently been utilized in actual inquiries, to be specific, *cyclidic* coÃ¶rdinates.[31] He showed, nonetheless, only how to separate Laplace's condition into three conventional differential equations.[32]

A significant part of our hypothesis which we have not yet addressed traces all the way back to the year 1836, when Sturm distributed a progression of on a very basic level significant papers in the initial two volumes of Liouville's Diary. The actual inquiry which lies at the premise of these papers is the issue of the stream of intensity in a heterogeneous bar.[33] The technique here utilized depends upon the way that the capabilities which happen are portrayed by the number of times they disappear in a specific span. This equivalent thought returns in Thomson and Tait's Addendum B previously alluded to, however first views as its full articulation in this more broad field of the three layered potential in an article by Klein: "Ueber KÃ¶rper welche von confocalen FlÃ¤chen zweiten Grades begrenzt sind"[34] (1881). Even more as of late (1889-90) Klein has in his talks stretched out this hypothesis to the treatment of solids limited by six confocal cyclids, and has demonstrated how every one of the potential issues to this point treated by our technique are unique instances of this one.

Of late years, particularly since the year 1880, the more youthful English mathematicians have done an immense measure of work in the hypothesis we are here considering. Albeit a lot of this work is of incredible worth, barely any of it can be viewed just like a genuine *development* of the technique; it is fairly an utilization of it to an extraordinary assortment of issues. We should thusly satisfied ourselves with giving a simple rundown of a couple of the more significant of these papers.

Niven: On the Conduction of Intensity in Ellipsoids of Upset. Phil. Trans., 1880.

Niven: On the Acceptance of Electric Flows in Endless Plates and Round Shells. Phil. Trans., 1881.

Hicks: On Toroidal Capabilities. Phil. Trans., 1881.

Hicks: On the Consistent Movement and Little Vibrations of an Empty Vortex. Phil. Trans., 1884, 1885.

Lamb: On Ellipsoidal Current Sheets. Phil. Trans., 1887.

Chree: The Conditions of an Isotropic Versatile Strong in Polar and Barrel shaped CoÃ¶rdinates, their Answer and Application. Camb. Phil. Soc. Trans., XIV., 1889.

Hobson: On a Class of Circular Sounds of Intricate Degree with Applications to Actual Issues. Camb. Phil. Soc. Trans., XIV., 1889.

Chree: On some Compound Vibrating Frameworks. Camb. Phil. Soc. Trans., XV., 1891.

Niven: On Ellipsoidal Music. Phil. Trans., 1892.

The verifiable sketch we have recently given would normally expect as a supplement some record of the work that has been finished on the topic of the intermingling of the different series which happen. This, nonetheless, would convey us excessively far, and we will satisfy ourselves

---

Vol. XIX., 1889 (read Dec. 8, 1887). Likewise a post mortem paper by LamÃ© in Liouville's Diary for 1874, Vol. XIX.

[30] Preisschriften der Jablanowski'schen Gesellschaft, No. XVIII., and Crelle's Diary, Vol. 82. See likewise, concerning an even more expansion, the Berliner Monatsberichten for 1878.

[31] Cyclids are a sort of surface of the fourth request (see Salmon's Geom. of three Aspects, p. 527). In his most memorable diary Wangerin considers just cyclids of revolution.

[32] See likewise a paper by this creator in GrÃ¼nert's Archiv for 1873, where the issue of the balance of versatile solids of upset is treated.

[33] The comparable issue of the vibration of a heterogeneous string under the activity of an outer power was treated by Maggi (Giornale di Matematiche, 1880). A few extraordinary cases are likewise viewed as here in detail.

[34] Math. Ann., 18.

HISTORICAL SUMMARY.

with referencing the two principal diaries by Dirichlet in Crelle's Diary, one of every 1829 on Fourier's series, and one, which has been scrutinized somewhat by ensuing mathematicians, in 1837 on Laplace's round consonant turn of events.

One more subject which normally introduces itself here is the hypothesis of the different new capabilities we have met. Those properties of these capabilities, notwithstanding, which the physicist needs have for the most part been examined by the physicists themselves in the papers referenced above; while any careful record of the advancement of the hypothesis of these capabilities would lead us into the huge locale of the cutting edge hypothesis of direct differential conditions.

We will in this way nearby simply giving a rundown of books which will be seen as valuable by those wishing to proceed with their investigation of the subject further.

We start with the books relating straightforwardly to actual inquiries:

Fourier: Théorie Analytique de la Chaleur, 1822.

Lamé: Leçons sur les Capabilities inverses des Transcendantes et les Surfaces isothermes, 1857.

Lamé: Leçons sur les Coordonnées Curvilignes et leurs diverses Applications, 1859.

Mathieu: Cours de Body Mathématique, 1873.

Riemann: Partielle Differentialgleichungen, und deren Anwendung auf physikalische Fragen (altered by Hattendorf), third release. 1882.

F. Neumann: Theorie des Possibilities und der Kugelfunktionen (altered by C. Neumann), 1887.

Thomson and Tait: Normal Way of thinking, second version, 1879.

Rayleigh: Hypothesis of Sound, 1877.

Basset: Hydrodynamics, 1888.

Love: Hypothesis of Flexibility, 1892.

Heine: Handbuch der Kugelfunktionen (second release), 1878-81.

Ferrers: Circular Music, 1881.

Haentzschel: Decrease der Potentialgleichung auf gewöhnliche Differentialgleichungen, 1893.

These last three books would likewise have a place in the accompanying rundown of books connecting with the hypothesis of the different capabilities we use:

Todhunter: The Elements of Laplace, Lamé and Bessel, 1875.

Lommel: Studien über pass on Bessel'schen Funktionen, 1868.

F. Neumann: Beiträge zur Theorie der Kugelfunktionen, 1878.

Lastly concerning the topic of union:

C. Neumann: Äber pass on nach Kreis-, Kugel-und Chamber Functionen fortschreitenden Entwickelungen, 1881.

APPENDIX.

TABLES.

Table I., a table of Surface Zonal Music (Legendrians), gives the qualities of the initial seven Music $P_1(\cos\theta)$, $P_2(\cos\theta)$, $\cdots P_7(\cos\theta)$ for the contention θ in degrees. It is taken from the Philosophical Magazine for December, 1891, and was figured by Messrs. C. E. Holland, P. R. James, and C. G. Sheep, under the heading of Teacher John Perry.

Table II., a table of Surface Zonal Music (Legendrians), gives the upsides of the initial seven Music $P_1(x)$, $P_2(x)$, $\cdots P_7(x)$ for the contention x. It is decreased from the Tables of Legendrian Capabilities figured under the course of Dr. J. W. L. Glaisher, and distributed in the Report of the English Relationship for the Progression of Science for the year 1879.

Table III., the table of Exaggerated Capabilities, gives the upsides of e^x, e^{-x}, $\sinh x$, $\cosh x$, and $\operatorname{gd} x$ (Gudermannian of x) for upsides of x from 0.00 to 1.00; furthermore, the upsides of $\log \sinh x$ and $\log \cosh x$ for upsides of x from 1.00 to 10.0. The upsides of $\operatorname{gd} x$, $\log \sinh x$, and $\log \cosh x$ are taken from the Numerical Tables ready by Teacher J. M. Peirce (Boston: Ginn & Co.).

The $\log \sinh x$ and $\log \cosh x$ for upsides of x somewhere in the range of 0.00 and 1.00 can be gotten from the qualities given for the Gudermannian of x in the table by the help of the relations

$$\log \sinh x = \log \tan(\operatorname{gd} x) \qquad \log \cosh x = \log \sec(\operatorname{gd} x).$$

Table IV. gives the initial twelve foundations of $J_0(x) = 0$ and $J_1(x) = 0$ each separated by π. The table is taken from Master Rayleigh's Sound, Vol. I., page 274, and is because of Teacher Stirs up, Camb. Phil. Trans., Vol. IX., page 186.

Table V. gives the initial nine foundations of $J_0(x) = 0$, $J_1(x) = 0$, $\cdots J_5(x) = 0$. The table is taken from Rayleigh's Sound, Vol. I., page 274, and is expected to Teacher J. Bourget, Ann. de l'Ecole Normale, T. III., 1866, page 82.

Table VI., the table of Bessel's Capabilities, gives the upsides of the Bessel's Capabilities $J_0(x)$ and $J_1(x)$ for the contention x from $x = 0$ to $x = 15$. It is taken from Rayleigh's Sound, Vol. I., page 265, and from Lommel's Bessel'sche Functionen.

APPENDIX

TABLE I. — Surface Zonal Harmonics

θ	$P_1(\cos\theta)$	$P_2(\cos\theta)$	$P_3(\cos\theta)$	$P_4(\cos\theta)$	$P_5(\cos\theta)$	$P_6(\cos\theta)$	$P_7(\cos\theta)$
0°	1.0000	1.0000	1.0000	1.0000	1.0000	1.0000	1.0000
1	.9998	.9995	.9991	.9985	.9977	.9967	.9955
2	.9994	.9982	.9963	.9939	.9909	.9872	.9829
3	.9986	.9959	.9918	.9863	.9795	.9713	.9617
4	.9976	.9927	.9854	.9758	.9638	.9495	.9329
5	.9962	.9886	.9773	.9623	.9437	.9216	.8961
6	.9945	.9836	.9674	.9459	.9194	.8881	.8522
7	.9925	.9777	.9557	.9267	.8911	.8476	.7986
8	.9903	.9709	.9423	.9048	.8589	.8053	.7448
9	.9877	.9633	.9273	.8803	.8232	.7571	.6831
10	.9848	.9548	.9106	.8532	.7840	.7045	.6164
11	.9816	.9454	.8923	.8238	.7417	.6483	.5461
12	.9781	.9352	.8724	.7920	.6966	.5892	.4732
13	.9744	.9241	.8511	.7582	.6489	.5273	.3940
14	.9703	.9122	.8283	.7224	.5990	.4635	.3219
15	.9659	.8995	.8042	.6847	.5471	.3982	.2454
16	.9613	.8860	.7787	.6454	.4937	.3322	.1699
17	.9563	.8718	.7519	.6046	.4391	.2660	.0961
18	.9511	.8568	.7240	.5624	.3836	.2002	.0289
19	.9455	.8410	.6950	.5192	.3276	.1347	−.0443
20	.9397	.8245	.6649	.4750	.2715	.0719	−.1072
21	.9336	.8074	.6338	.4300	.2156	.0107	−.1662
22	.9272	.7895	.6019	.3845	.1602	−.0481	−.2201
23	.9205	.7710	.5692	.3386	.1057	−.1038	−.2681
24	.9135	.7518	.5357	.2926	.0525	−.1559	−.3095
25	.9063	.7321	.5016	.2465	.0009	−.2053	−.3463
26	.8988	.7117	.4670	.2007	−.0489	−.2478	−.3717
27	.8910	.6908	.4319	.1553	−.0964	−.2869	−.3921
28	.8829	.6694	.3964	.1105	−.1415	−.3211	−.4052
29	.8746	.6474	.3607	.0665	−.1839	−.3503	−.4114
30	.8660	.6250	.3248	.0234	−.2233	−.3740	−.4101
31	.8572	.6021	.2887	−.0185	−.2595	−.3924	−.4022
32	.8480	.5788	.2527	−.0591	−.2923	−.4052	−.3876
33	.8387	.5551	.2167	−.0982	−.3216	−.4126	−.3670
34	.8290	.5310	.1809	−.1357	−.3473	−.4148	−.3409
35	.8192	.5065	.1454	−.1714	−.3691	−.4115	−.3096
36	.8090	.4818	.1102	−.2052	−.3871	−.4031	−.2738
37	.7986	.4567	.0755	−.2370	−.4011	−.3898	−.2343
38	.7880	.4314	.0413	−.2666	−.4112	−.3719	−.1918
39	.7771	.4059	.0077	−.2940	−.4174	−.3497	−.1469
40	.7660	.3802	−.0252	−.3190	−.4197	−.3234	−.1003
41	.7547	.3544	−.0574	−.3416	−.4181	−.2938	−.0534
42	.7431	.3284	−.0887	−.3616	−.4128	−.2611	−.0065
43	.7314	.3023	−.1191	−.3791	−.4038	−.2255	.0398
44	.7193	.2762	−.1485	−.3940	−.3914	−.1878	.0846
45°	.7071	.2500	−.1768	−.4062	−.3757	−.1485	.1270

APPENDIX

TABLE I. — Surface Zonal Harmonics

θ	$P_1(\cos\theta)$	$P_2(\cos\theta)$	$P_3(\cos\theta)$	$P_4(\cos\theta)$	$P_5(\cos\theta)$	$P_6(\cos\theta)$	$P_7(\cos\theta)$
45°	.7071	.2500	−.1768	−.4062	−.3757	−.1485	.1270
46	.6947	.2238	−.2040	−.4158	−.3568	−.1079	.1666
47	.6820	.1977	−.2300	−.4252	−.3350	−.0645	.2054
48	.6691	.1716	−.2547	−.4270	−.3105	−.0251	.2349
49	.6561	.1456	−.2781	−.4286	−.2836	.0161	.2627
50	.6428	.1198	−.3002	−.4275	−.2545	.0563	.2854
51	.6293	.0941	−.3209	−.4239	−.2235	.0954	.3031
52	.6157	.0686	−.3401	−.4178	−.1910	.1326	.3153
53	.6018	.0433	−.3578	−.4093	−.1571	.1677	.3221
54	.5878	.0182	−.3740	−.3984	−.1223	.2002	.3234
55	.5736	−.0065	−.3886	−.3852	−.0868	.2297	.3191
56	.5592	−.0310	−.4016	−.3698	−.0510	.2559	.3095
57	.5446	−.0551	−.4131	−.3524	−.0150	.2787	.2949
58	.5299	−.0788	−.4229	−.3331	.0206	.2976	.2752
59	.5150	−.1021	−.4310	−.3119	.0557	.3125	.2511
60	.5000	−.1250	−.4375	−.2891	.0898	.3232	.2231
61	.4848	−.1474	−.4423	−.2647	.1229	.3298	.1916
62	.4695	−.1694	−.4455	−.2390	.1545	.3321	.1571
63	.4540	−.1908	−.4471	−.2121	.1844	.3302	.1203
64	.4384	−.2117	−.4470	−.1841	.2123	.3240	.0818
65	.4226	−.2321	−.4452	−.1552	.2381	.3138	.0422
66	.4067	−.2518	−.4419	−.1256	.2615	.2996	.0021
67	.3907	−.2710	−.4370	−.0955	.2824	.2819	−.0375
68	.3746	−.2896	−.4305	−.0650	.3005	.2605	−.0763
69	.3584	−.3074	−.4225	−.0344	.3158	.2361	−.1135
70	.3420	−.3245	−.4130	−.0038	.3281	.2089	−.1485
71	.3256	−.3410	−.4021	.0267	.3373	.1786	−.1811
72	.3090	−.3568	−.3898	.0568	.3434	.1472	−.2099
73	.2924	−.3718	−.3761	.0864	.3463	.1144	−.2347
74	.2756	−.3860	−.3611	.1153	.3461	.0795	−.2559
75	.2588	−.3995	−.3449	.1434	.3427	.0431	−.2730
76	.2419	−.4112	−.3275	.1705	.3362	.0076	−.2848
77	.2250	−.4241	−.3090	.1964	.3267	−.0284	−.2919
78	.2079	−.4352	−.2894	.2211	.3143	−.0644	−.2943
79	.1908	−.4454	−.2688	.2443	.2990	−.0989	−.2913
80	.1736	−.4548	−.2474	.2659	.2810	−.1321	−.2835
81	.1564	−.4633	−.2251	.2859	.2606	−.1635	−.2709
82	.1392	−.4709	−.2020	.3040	.2378	−.1926	−.2536
83	.1219	−.4777	−.1783	.3203	.2129	−.2193	−.2321
84	.1045	−.4836	−.1539	.3345	.1861	−.2431	−.2067
85	.0872	−.4886	−.1291	.3468	.1577	−.2638	−.1779
86	.0698	−.4927	−.1038	.3569	.1278	−.2811	−.1460
87	.0523	−.4959	−.0781	.3648	.0969	−.2947	−.1117
88	.0349	−.4982	−.0522	.3704	.0651	−.3045	−.0735
89	.0175	−.4995	−.0262	.3739	.0327	−.3105	−.0381
90°	.0000	−.5000	.0000	.3750	.0000	−.3125	.0000

TABLE II.—SURFACE ZONAL HARMONICS.

x	$P_1(x)$	$P_2(x)$	$P_3(x)$	$P_4(x)$	$P_5(x)$	$P_6(x)$	$P_7(x)$
0.00	0.0000	−.5000	0.0000	0.3750	0.0000	−.3125	0.0000
.01	.0100	−.4998	−.0150	.3746	.0187	−.3118	−.0219
.02	.0200	−.4994	−.0300	.3735	.0374	−.3099	−.0436
.03	.0300	−.4986	−.0449	.3716	.0560	−.3066	−.0651
.04	.0400	−.4976	−.0598	.3690	.0744	−.3021	−.0862
.05	.0500	−.4962	−.0747	.3657	.0927	−.2962	−.1069
.06	.0600	−.4946	−.0895	.3616	.1106	−.2891	−.1270
.07	.0700	−.4926	−.1041	.3567	.1283	−.2808	−.1464
.08	.0800	−.4904	−.1187	.3512	.1455	−.2713	−.1651
.09	.0900	−.4878	−.1332	.3449	.1624	−.2606	−.1828
.10	.1000	−.4850	−.1475	.3379	.1788	−.2488	−.1995
.11	.1100	−.4818	−.1617	.3303	.1947	−.2360	−.2151
.12	.1200	−.4784	−.1757	.3219	.2101	−.2220	−.2295
.13	.1300	−.4746	−.1895	.3129	.2248	−.2071	−.2427
.14	.1400	−.4706	−.2031	.3032	.2389	−.1913	−.2545
.15	.1500	−.4662	−.2166	.2928	.2523	−.1746	−.2649
.16	.1600	−.4616	−.2298	.2819	.2650	−.1572	−.2738
.17	.1700	−.4566	−.2427	.2703	.2769	−.1389	−.2812
.18	.1800	−.4514	−.2554	.2581	.2880	−.1201	−.2870
.19	.1900	−.4458	−.2679	.2453	.2982	−.1006	−.2911
.20	.2000	−.4400	−.2800	.2320	.3075	−.0806	−.2935
.21	.2100	−.4338	−.2918	.2181	.3159	−.0601	−.2943
.22	.2200	−.4274	−.3034	.2037	.3234	−.0394	−.2933
.23	.2300	−.4206	−.3146	.1889	.3299	−.0183	−.2906
.24	.2400	−.4136	−.3254	.1735	.3353	.0029	−.2861
.25	.2500	−.4062	−.3359	.1577	.3397	.0243	−.2799
.26	.2600	−.3986	−.3461	.1415	.3431	.0456	−.2720
.27	.2700	−.3906	−.3558	.1249	.3453	.0669	−.2625
.28	.2800	−.3824	−.3651	.1079	.3465	.0879	−.2512
.29	.2900	−.3738	−.3740	.0906	.3465	.1087	−.2384
.30	.3000	−.3650	−.3825	.0729	.3454	.1292	−.2241
.31	.3100	−.3558	−.3905	.0550	.3431	.1492	−.2082
.32	.3200	−.3464	−.3981	.0369	.3397	.1686	−.1910
.33	.3300	−.3366	−.4052	.0185	.3351	.1873	−.1724
.34	.3400	−.3266	−.4117	−.0000	.3294	.2053	−.1527
.35	.3500	−.3162	−.4178	−.0187	.3225	.2225	−.1318
.36	.3600	−.3056	−.4234	−.0375	.3144	.2388	−.1098
.37	.3700	−.2946	−.4284	−.0564	.3051	.2540	−.0870
.38	.3800	−.2834	−.4328	−.0753	.2948	.2681	−.0635
.39	.3900	−.2718	−.4367	−.0942	.2833	.2810	−.0393
.40	.4000	−.2600	−.4400	−.1130	.2706	.2926	−.0146
.41	.4100	−.2478	−.4427	−.1317	.2569	.3029	.0104
.42	.4200	−.2354	−.4448	−.1504	.2421	.3118	.0356
.43	.4300	−.2226	−.4462	−.1688	.2263	.3191	.0608
.44	.4400	−.2096	−.4470	−.1870	.2095	.3249	.0859
.45	.4500	−.1962	−.4472	−.2050	.1917	.3290	.1106
.46	.4600	−.1826	−.4467	−.2226	.1730	.3314	.1348
.47	.4700	−.1686	−.4454	−.2399	.1534	.3321	.1584
.48	.4800	−.1544	−.4435	−.2568	.1330	.3310	.1811
.49	.4900	−.1398	−.4409	−.2732	.1118	.3280	.2027
.50	.5000	−.1250	−.4375	−.2891	.0898	.3232	.2231

TABLE II.—Surface Zonal Harmonics.

x	$P_1(x)$	$P_2(x)$	$P_3(x)$	$P_4(x)$	$P_5(x)$	$P_6(x)$	$P_7(x)$
.50	.5000	−.1250	−.4375	−.2891	.0898	.3232	.2231
.51	.5100	−.1098	−.4334	−.3044	.0673	.3166	.2422
.52	.5200	−.0944	−.4285	−.3191	.0441	.3080	.2596
.53	.5300	−.0786	−.4228	−.3332	.0204	.2975	.2753
.54	.5400	−.0626	−.4163	−.3465	−.0037	.2851	.2891
.55	.5500	−.0462	−.4091	−.3590	−.0282	.2708	.3007
.56	.5600	−.0296	−.4010	−.3707	−.0529	.2546	.3102
.57	.5700	−.0126	−.3920	−.3815	−.0779	.2366	.3172
.58	.5800	.0046	−.3822	−.3914	−.1028	.2168	.3217
.59	.5900	.0222	−.3716	−.4002	−.1278	.1953	.3235
.60	.6000	.0400	−.3600	−.4080	−.1526	.1721	.3226
.61	.6100	.0582	−.3475	−.4146	−.1772	.1473	.3188
.62	.6200	.0766	−.3342	−.4200	−.2014	.1211	.3121
.63	.6300	.0954	−.3199	−.4242	−.2251	.0935	.3023
.64	.6400	.1144	−.3046	−.4270	−.2482	.0646	.2895
.65	.6500	.1338	−.2884	−.4284	−.2705	.0347	.2737
.66	.6600	.1534	−.2713	−.4284	−.2919	.0038	.2548
.67	.6700	.1734	−.2531	−.4268	−.3122	−.0278	.2329
.68	.6800	.1936	−.2339	−.4236	−.3313	−.0601	.2081
.69	.6900	.2142	−.2137	−.4187	−.3490	−.0926	.1805
.70	.7000	.2350	−.1925	−.4121	−.3652	−.1253	.1502
.71	.7100	.2562	−.1702	−.4036	−.3796	−.1578	.1173
.72	.7200	.2776	−.1469	−.3933	−.3922	−.1899	.0822
.73	.7300	.2994	−.1225	−.3810	−.4026	−.2214	.0450
.74	.7400	.3214	−.0969	−.3666	−.4107	−.2518	.0061
.75	.7500	.3438	−.0703	−.3501	−.4164	−.2808	−.0342
.76	.7600	.3664	−.0426	−.3314	−.4193	−.3081	−.0754
.77	.7700	.3894	−.0137	−.3104	−.4193	−.3333	−.1171
.78	.7800	.4126	.0164	−.2871	−.4162	−.3559	−.1588
.79	.7900	.4362	.0476	−.2613	−.4097	−.3756	−.1999
.80	.8000	.4600	.0800	−.2330	−.3995	−.3918	−.2397
.81	.8100	.4842	.1136	−.2021	−.3855	−.4041	−.2774
.82	.8200	.5086	.1484	−.1685	−.3674	−.4119	−.3124
.83	.8300	.5334	.1845	−.1321	−.3449	−.4147	−.3437
.84	.8400	.5584	.2218	−.0928	−.3177	−.4120	−.3703
.85	.8500	.5838	.2603	−.0506	−.2857	−.4030	−.3913
.86	.8600	.6094	.3001	−.0053	−.2484	−.3872	−.4055
.87	.8700	.6354	.3413	.0431	−.2056	−.3638	−.4116
.88	.8800	.6616	.3837	.0947	−.1570	−.3322	−.4083
.89	.8900	.6882	.4274	.1496	−.1023	−.2916	−.3942
.90	.9000	.7150	.4725	.2079	−.0411	−.2412	−.3678
.91	.9100	.7422	.5189	.2698	.0268	−.1802	−.3274
.92	.9200	.7696	.5667	.3352	.1017	−.1077	−.2713
.93	.9300	.7974	.6159	.4044	.1842	−.0229	−.1975
.94	.9400	.8254	.6665	.4773	.2744	.0751	−.1040
.95	.9500	.8538	.7184	.5541	.3727	.1875	.0112
.96	.9600	.8824	.7718	.6349	.4796	.3151	.1506
.97	.9700	.9114	.8267	.7198	.5954	.4590	.3165
.98	.9800	.9406	.8830	.8089	.7204	.6204	.5115
.99	.9900	.9702	.9407	.9022	.8552	.8003	.7384
1.00	1.0000	1.0000	1.0000	1.0000	1.0000	1.0000	1.0000

TABLE III.—Hyperbolic Functions.

x	e^x	e^{-x}	sinh x	cosh x	gd x
0.00	1.0000	1.0000	0.0000	1.0000	0°.0000
.01	1.0100	0.9900	.0100	1.0000	0.5729
.02	1.0202	.9802	.0200	1.0002	1.1458
.03	1.0305	.9704	.0300	1.0004	1.7186
.04	1.0408	.9608	.0400	1.0008	2.2912
.05	1.0513	.9512	.0500	1.0012	2.8636
.06	1.0618	.9418	.0600	1.0018	3.4357
.07	1.0725	.9324	.0701	1.0025	4.0074
.08	1.0833	.9231	.0801	1.0032	4.5788
.09	1.0942	.9139	.0901	1.0040	5.1497
.10	1.1052	.9048	.1002	1.0050	5.720
.11	1.1163	.8958	.1102	1.0061	6.290
.12	1.1275	.8869	.1203	1.0072	6.859
.13	1.1388	.8781	.1304	1.0085	7.428
.14	1.1503	.8694	.1405	1.0098	7.995
.15	1.1618	.8607	.1506	1.0113	8.562
.16	1.1735	.8521	.1607	1.0128	9.128
.17	1.1853	.8437	.1708	1.0145	9.694
.18	1.1972	.8353	.1810	1.0162	10.258
.19	1.2092	.8270	.1911	1.0181	10.821
.20	1.2214	.8187	.2013	1.0201	11.384
.21	1.2337	.8106	.2115	1.0221	11.945
.22	1.2461	.8025	.2218	1.0243	12.505
.23	1.2586	.7945	.2320	1.0266	13.063
.24	1.2712	.7866	.2423	1.0289	13.621
.25	1.2840	.7788	.2526	1.0314	14.177
.26	1.2969	.7711	.2629	1.0340	14.732
.27	1.3100	.7634	.2733	1.0367	15.285
.28	1.3231	.7558	.2837	1.0395	15.837
.29	1.3364	.7483	.2941	1.0423	16.388
.30	1.3499	.7408	.3045	1.0453	16.937
.31	1.3634	.7334	.3150	1.0484	17.484
.32	1.3771	.7261	.3255	1.0516	18.030
.33	1.3910	.7189	.3360	1.0549	18.573
.34	1.4049	.7118	.3466	1.0584	19.116
.35	1.4191	.7047	.3572	1.0619	19.656
.36	1.4333	.6977	.3678	1.0655	20.195
.37	1.4477	.6907	.3785	1.0692	20.732
.38	1.4623	.6839	.3892	1.0731	21.267
.39	1.4770	.6771	.4000	1.0770	21.800
.40	1.4918	.6703	.4108	1.0811	22.331
.41	1.5068	.6636	.4216	1.0852	22.859
.42	1.5220	.6570	.4325	1.0895	23.386
.43	1.5373	.6505	.4434	1.0939	23.911
.44	1.5527	.6440	.4543	1.0984	24.434
.45	1.5683	.6376	.4653	1.1030	24.955
.46	1.5841	.6313	.4764	1.1077	25.473
.47	1.6000	.6250	.4875	1.1125	25.989
.48	1.6161	.6188	.4986	1.1174	26.503
.49	1.6323	.6126	.5098	1.1225	27.015
0.50	1.6487	0.6065	0.5211	1.1276	27°.524

TABLE III.—Hyperbolic Functions.

x	e^x	e^{-x}	sinh x	cosh x	gd x
0.50	1.6487	0.6065	0.5211	1.1276	27°.524
.51	1.6653	.6005	.5324	1.1329	28.031
.52	1.6820	.5945	.5438	1.1383	28.535
.53	1.6989	.5886	.5552	1.1438	29.037
.54	1.7160	.5827	.5666	1.1494	29.537
.55	1.7333	.5770	.5782	1.1551	30.034
.56	1.7507	.5712	.5897	1.1609	30.529
.57	1.7683	.5655	.6014	1.1669	31.021
.58	1.7860	.5599	.6131	1.1730	31.511
.59	1.8040	.5543	.6248	1.1792	31.998
.60	1.8221	.5488	.6367	1.1855	32.483
.61	1.8404	.5433	.6485	1.1919	32.965
.62	1.8589	.5379	.6605	1.1984	33.444
.63	1.8776	.5326	.6725	1.2051	33.921
.64	1.8965	.5273	.6846	1.2119	34.395
.65	1.9155	.5220	.6967	1.2188	34.867
.66	1.9348	.5169	.7090	1.2258	35.336
.67	1.9542	.5117	.7213	1.2330	35.802
.68	1.9739	.5066	.7336	1.2402	36.265
.69	1.9937	.5016	.7461	1.2476	36.726
.70	2.0138	.4966	.7586	1.2552	37.183
.71	2.0340	.4916	.7712	1.2628	37.638
.72	2.0544	.4867	.7838	1.2706	38.091
.73	2.0751	.4819	.7966	1.2785	38.540
.74	2.0959	.4771	.8094	1.2865	38.987
.75	2.1170	.4724	.8223	1.2947	39.431
.76	2.1383	.4677	.8353	1.3030	39.872
.77	2.1598	.4630	.8484	1.3114	40.310
.78	2.1815	.4584	.8615	1.3199	40.746
.79	2.2034	.4538	.8748	1.3286	41.179
.80	2.2255	.4493	.8881	1.3374	41.608
.81	2.2479	.4449	.9015	1.3464	42.035
.82	2.2705	.4404	.9150	1.3555	42.460
.83	2.2933	.4360	.9286	1.3647	42.881
.84	2.3164	.4317	.9423	1.3740	43.299
.85	2.3396	.4274	.9561	1.3835	43.715
.86	2.3632	.4232	.9700	1.3932	44.128
.87	2.3869	.4190	.9840	1.4029	44.537
.88	2.4109	.4148	.9981	1.4128	44.944
.89	2.4351	.4107	1.0122	1.4229	45.348
.90	2.4596	.4066	1.0265	1.4331	45.750
.91	2.4843	.4025	1.0409	1.4434	46.148
.92	2.5093	.3985	1.0554	1.4539	46.544
.93	2.5345	.3946	1.0700	1.4645	46.936
.94	2.5600	.3906	1.0847	1.4753	47.326
.95	2.5857	.3867	1.0995	1.4862	47.713
.96	2.6117	.3829	1.1144	1.4973	48.097
.97	2.6379	.3791	1.1294	1.5085	48.478
.98	2.6645	.3753	1.1446	1.5199	48.857
.99	2.6912	.3716	1.1598	1.5314	49.232
1.00	2.7183	0.3679	1.1752	1.5431	49°.605

TABLE III.—Hyperbolic Functions.

x	$l\sinh x$	$l\cosh x$	x	$l\sinh x$	$l\cosh x$	x	$l\sinh x$	$l\cosh x$
1.00	0.0701	0.1884	1.50	0.3282	0.3715	2.00	0.5595	0.5754
1.01	.0758	.1917	1.51	.3330	.3754	2.01	.5640	.5796
1.02	.0815	.1950	1.52	.3378	.3794	2.02	.5685	.5838
1.03	.0871	.1984	1.53	.3426	.3833	2.03	.5730	.5880
1.04	.0927	.2018	1.54	.3474	.3873	2.04	.5775	.5922
1.05	.0982	.2051	1.55	.3521	.3913	2.05	.5820	.5964
1.06	.1038	.2086	1.56	.3569	.3952	2.06	.5865	.6006
1.07	.1093	.2120	1.57	.3616	.3992	2.07	.5910	.6048
1.08	.1148	.2154	1.58	.3663	.4032	2.08	.5955	.6090
1.09	.1203	.2189	1.59	.3711	.4072	2.09	.6000	.6132
1.10	.1257	.2223	1.60	.3758	.4112	2.10	.6044	.6175
1.11	.1311	.2258	1.61	.3805	.4152	2.11	.6089	.6217
1.12	.1365	.2293	1.62	.3852	.4192	2.12	.6134	.6259
1.13	.1419	.2328	1.63	.3899	.4232	2.13	.6178	.6301
1.14	.1472	.2364	1.64	.3946	.4273	2.14	.6223	.6343
1.15	.1525	.2399	1.65	.3992	.4313	2.15	.6268	.6386
1.16	.1578	.2435	1.66	.4039	.4353	2.16	.6312	.6428
1.17	.1631	.2470	1.67	.4086	.4394	2.17	.6357	.6470
1.18	.1684	.2506	1.68	.4132	.4434	2.18	.6401	.6512
1.19	.1736	.2542	1.69	.4179	.4475	2.19	.6446	.6555
1.20	.1788	.2578	1.70	.4225	.4515	2.20	.6491	.6597
1.21	.1840	.2615	1.71	.4272	.4556	2.21	.6535	.6640
1.22	.1892	.2651	1.72	.4318	.4597	2.22	.6580	.6682
1.23	.1944	.2688	1.73	.4364	.4637	2.23	.6624	.6724
1.24	.1995	.2724	1.74	.4411	.4678	2.24	.6668	.6767
1.25	.2046	.2761	1.75	.4457	.4719	2.25	.6713	.6809
1.26	.2098	.2798	1.76	.4503	.4760	2.26	.6757	.6852
1.27	.2148	.2835	1.77	.4549	.4801	2.27	.6802	.6894
1.28	.2199	.2872	1.78	.4595	.4842	2.28	.6846	.6937
1.29	.2250	.2909	1.79	.4641	.4883	2.29	.6890	.6979
1.30	.2300	.2947	1.80	.4687	.4924	2.30	.6935	.7022
1.31	.2351	.2984	1.81	.4733	.4965	2.31	.6979	.7064
1.32	.2401	.3022	1.82	.4778	.5006	2.32	.7023	.7107
1.33	.2451	.3059	1.83	.4824	.5048	2.33	.7067	.7150
1.34	.2501	.3097	1.84	.4870	.5089	2.34	.7112	.7192
1.35	.2551	.3135	1.85	.4915	.5130	2.35	.7156	.7235
1.36	.2600	.3173	1.86	.4961	.5172	2.36	.7200	.7278
1.37	.2650	.3211	1.87	.5007	.5213	2.37	.7244	.7320
1.38	.2699	.3249	1.88	.5052	.5254	2.38	.7289	.7363
1.39	.2748	.3288	1.89	.5098	.5296	2.39	.7333	.7406
1.40	.2797	.3326	1.90	.5143	.5337	2.40	.7377	.7448
1.41	.2846	.3365	1.91	.5188	.5379	2.41	.7421	.7491
1.42	.2895	.3403	1.92	.5234	.5421	2.42	.7465	.7534
1.43	.2944	.3442	1.93	.5279	.5462	2.43	.7509	.7577
1.44	.2993	.3481	1.94	.5324	.5504	2.44	.7553	.7619
1.45	.3041	.3520	1.95	.5370	.5545	2.45	.7597	.7662
1.46	.3090	.3559	1.96	.5415	.5587	2.46	.7642	.7705
1.47	.3138	.3598	1.97	.5460	.5629	2.47	.7686	.7748
1.48	.3186	.3637	1.98	.5505	.5671	2.48	.7730	.7791
1.49	.3234	.3676	1.99	.5550	.5713	2.49	.7774	.7833
1.50	0.3282	0.3715	2.00	0.5595	0.5754	2.50	0.7818	0.7876

TABLE III.—Hyperbolic Functions.

x	$l \sinh x$	$l \cosh x$	x	$l \sinh x$	$l \cosh x$	x	$l \sinh x$	$l \cosh x$
2.50	0.7818	0.7876	2.75	0.8915	0.8951	3.0	1.0008	1.0029
2.51	.7862	.7919	2.76	.8959	.8994	3.1	1.0444	1.0462
2.52	.7906	.7962	2.77	.9003	.9037	3.2	1.0880	1.0894
2.53	.7950	.8005	2.78	.9046	.9080	3.3	1.1316	1.1327
2.54	.7994	.8048	2.79	.9090	.9123	3.4	1.1751	1.1761
2.55	.8038	.8091	2.80	.9134	.9166	3.5	1.2186	1.2194
2.56	.8082	.8134	2.81	.9178	.9209	3.6	1.2621	1.2628
2.57	.8126	.8176	2.82	.9221	.9252	3.7	1.3056	1.3061
2.58	.8169	.8219	2.83	.9265	.9295	3.8	1.3491	1.3495
2.59	.8213	.8262	2.84	.9309	.9338	3.9	1.3925	1.3929
2.60	.8257	.8305	2.85	.9353	.9382	4.0	1.4360	1.4363
2.61	.8301	.8348	2.86	.9396	.9425	4.1	1.4795	1.4797
2.62	.8345	.8391	2.87	.9440	.9468	4.2	1.5229	1.5231
2.63	.8389	.8434	2.88	.9484	.9511	4.3	1.5664	1.5665
2.64	.8433	.8477	2.89	.9527	.9554	4.4	1.6098	1.6099
2.65	.8477	.8520	2.90	.9571	.9597	4.5	1.6532	1.6533
2.66	.8521	.8563	2.91	.9615	.9641	4.6	1.6967	1.6968
2.67	.8564	.8606	2.92	.9658	.9684	4.7	1.7401	1.7402
2.68	.8608	.8649	2.93	.9702	.9727	4.8	1.7836	1.7836
2.69	.8652	.8692	2.94	.9746	.9770	4.9	1.8270	1.8270
2.70	.8696	.8735	2.95	.9789	.9813	5.0	1.8704	1.8705
2.71	.8740	.8778	2.96	.9833	.9856	6.0	2.3047	2.3047
2.72	.8784	.8821	2.97	.9877	.9900	7.0	2.7390	2.7390
2.73	.8827	.8864	2.98	.9920	.9943	8.0	3.1733	3.1733
2.74	.8871	.8907	2.99	.9964	.9986	9.0	3.6076	3.6076
2.75	0.8915	0.8951	3.00	1.0008	1.0029	10.0	4.0419	4.0419

APPENDIX

TABLE IV.—Roots of Bessel's Functions.

	$\frac{x}{\pi}$ for $J_0(x)=0$	$\frac{x}{\pi}$ for $J_1(x)=0$		$\frac{x}{\pi}$ for $J_0(x)=0$	$\frac{x}{\pi}$ for $J_1(x)=0$
1	0.7655	1.2197	7	6.7519	7.2448
2	1.7571	2.2330	8	7.7516	8.2454
3	2.7546	3.2383	9	8.7514	9.2459
4	3.7534	4.2411	10	9.7513	10.2463
5	4.7527	5.2428	11	10.7512	11.2466
6	5.7522	6.2439	12	11.7511	12.2469

TABLE V.—Roots of $J_n(x)=0$.

	$n=0$	$n=1$	$n=2$	$n=3$	$n=4$	$n=5$
1	2.405	3.832	5.135	6.379	7.586	8.780
2	5.520	7.016	8.417	9.760	11.064	12.339
3	8.654	10.173	11.620	13.017	14.373	15.700
4	11.792	13.323	14.796	16.224	17.616	18.982
5	14.931	16.470	17.960	19.410	20.827	22.220
6	18.071	19.616	21.117	22.583	24.018	25.431
7	21.212	22.760	24.270	25.749	27.200	28.628
8	24.353	25.903	27.421	28.909	30.371	31.813
9	27.494	29.047	30.571	32.050	33.512	34.983

TABLE VI.— Bessel's Functions.

x	$J_0(x)$	$J_1(x)$	x	$J_0(x)$	$J_1(x)$	x	$J_0(x)$	$J_1(x)$
0.0	1.0000	0.0000	5.0	−.1776	−.3276	10.0	−.2459	.0435
0.1	.9975	.0499	5.1	−.1443	−.3371	10.1	−.2490	.0184
0.2	.9900	.0995	5.2	−.1103	−.3432	10.2	−.2496	−.0066
0.3	.9776	.1483	5.3	−.0758	−.3460	10.3	−.2477	−.0313
0.4	.9604	.1960	5.4	−.0412	−.3453	10.4	−.2434	−.0555
0.5	.9385	.2423	5.5	−.0068	−.3414	10.5	−.2366	−.0789
0.6	.9120	.2867	5.6	.0270	−.3343	10.6	−.2276	−.1012
0.7	.8812	.3290	5.7	.0599	−.3241	10.7	−.2164	−.1224
0.8	.8463	.3688	5.8	.0917	−.3110	10.8	−.2032	−.1422
0.9	.8075	.4060	5.9	.1220	−.2951	10.9	−.1881	−.1604
1.0	.7652	.4401	6.0	.1506	−.2767	11.0	−.1712	−.1768
1.1	.7196	.4709	6.1	.1773	−.2559	11.1	−.1528	−.1913
1.2	.6711	.4983	6.2	.2017	−.2329	11.2	−.1330	−.2039
1.3	.6201	.5220	6.3	.2238	−.2081	11.3	−.1121	−.2143
1.4	.5669	.5419	6.4	.2433	−.1816	11.4	−.0902	−.2225
1.5	.5118	.5579	6.5	.2601	−.1538	11.5	−.0677	−.2284
1.6	.4554	.5699	6.6	.2740	−.1250	11.6	−.0446	−.2320
1.7	.3980	.5778	6.7	.2851	−.0953	11.7	−.0213	−.2333
1.8	.3400	.5815	6.8	.2931	−.0652	11.8	.0020	−.2323
1.9	.2818	.5812	6.9	.2981	−.0349	11.9	.0250	−.2290
2.0	.2239	.5767	7.0	.3001	−.0047	12.0	.0477	−.2234
2.1	.1666	.5683	7.1	.2991	.0252	12.1	.0697	−.2157
2.2	.1104	.5560	7.2	.2951	.0543	12.2	.0908	−.2060
2.3	.0555	.5399	7.3	.2882	.0826	12.3	.1108	−.1943
2.4	.0025	.5202	7.4	.2786	.1096	12.4	.1296	−.1807
2.5	−.0484	.4971	7.5	.2663	.1352	12.5	.1469	−.1655
2.6	−.0968	.4708	7.6	.2516	.1592	12.6	.1626	−.1487
2.7	−.1424	.4416	7.7	.2346	.1813	12.7	.1766	−.1307
2.8	−.1850	.4097	7.8	.2154	.2014	12.8	.1887	−.1114
2.9	−.2243	.3754	7.9	.1944	.2192	12.9	.1988	−.0912
3.0	−.2601	.3391	8.0	.1717	.2346	13.0	.2069	−.0703
3.1	−.2921	.3009	8.1	.1475	.2476	13.1	.2129	−.0489
3.2	−.3202	.2613	8.2	.1222	.2580	13.2	.2167	−.0271
3.3	−.3443	.2207	8.3	.0960	.2657	13.3	.2183	−.0052
3.4	−.3643	.1792	8.4	.0692	.2708	13.4	.2177	.0166
3.5	−.3801	.1374	8.5	.0419	.2731	13.5	.2150	.0380
3.6	−.3918	.0955	8.6	.0146	.2728	13.6	.2101	.0590
3.7	−.3992	.0538	8.7	−.0125	.2697	13.7	.2032	.0791
3.8	−.4026	.0128	8.8	−.0392	.2641	13.8	.1943	.0984
3.9	−.4018	−.0272	8.9	−.0653	.2559	13.9	.1836	.1166
4.0	−.3972	−.0660	9.0	−.0903	.2453	14.0	.1711	.1334
4.1	−.3887	−.1033	9.1	−.1142	.2324	14.1	.1570	.1488
4.2	−.3766	−.1386	9.2	−.1367	.2174	14.2	.1414	.1626
4.3	−.3610	−.1719	9.3	−.1577	.2004	14.3	.1245	.1747
4.4	−.3423	−.2028	9.4	−.1768	.1816	14.4	.1065	.1850
4.5	−.3205	−.2311	9.5	−.1939	.1613	14.5	.0875	.1934
4.6	−.2961	−.2566	9.6	−.2090	.1395	14.6	.0679	.1999
4.7	−.2693	−.2791	9.7	−.2218	.1166	14.7	.0476	.2043
4.8	−.2404	−.2985	9.8	−.2323	.0928	14.8	.0271	.2066
4.9	−.2097	−.3147	9.9	−.2403	.0684	14.9	.0064	.2069
5.0	−.1776	−.3276	10.0	−.2459	.0435	15.0	−.0142	.2051

www.ingramcontent.com/pod-product-compliance
Lightning Source LLC
Chambersburg PA
CBHW060411220526
45465CB00008B/2841